Adaptationist Evo-Devo

Adaptationist Evo-Devo

MARK E. OLSON

Instituto de Biología,
Universidad Nacional Autónoma de México, Mexico City, Mexico

OXFORD
UNIVERSITY PRESS

OXFORD
UNIVERSITY PRESS

Oxford University Press is a department of the University of Oxford. It furthers the University's objective of excellence in research, scholarship, and education by publishing worldwide. Oxford is a registered trade mark of Oxford University Press in the UK and in certain other countries.

Published in the United States of America by Oxford University Press
198 Madison Avenue, New York, NY 10016, United States of America.

CIP data is on file at the Library of Congress.

ISBN 9780197831502

ISBN 9780197831496 (hbk.)

DOI: 10.1093/9780197831533.001.0001

Paperback Printed by Marquis Book Printing, Canada

The manufacturer's authorized representative in the EU for product safety is Oxford University Press España S.A. of Parque Empresarial San Fernando de Henares, Avenida de Castilla, 2 – 28830 Madrid (www.oup.es/en or product.safety@oup.com). OUP España S.A. also acts as importer into Spain of products made by the manufacturer.

MIX
Paper | Supporting responsible forestry
FSC® C103567
FSC www.fsc.org

To my dear Julieta

To my dear teachers, Allan Larson, Jonathan Losos, and Peter Raven

Acknowledgments

The cover illustration is by Mexican artist Eduardo Martínez Díaz, from the Náhuatl-speaking town of Xalitla, in Guerrero State, Mexico. Xalitla has a long tradition of exuberant, animated imagery, often painted on bark paper handmade from trees or on ceramics. The town lies within a tropical dry forest extraordinarily rich in species and in morphological disparity among them, a perfect illustration of nature's developmental diversity. Eduardo depicts hallmark species of this landscape: towering *Pachycereus* cacti; aromatic *Bursera* ("elephant trees") used for copal incense; wild *Ceiba* (kapok trees); leguminous trees like *Acacia*; wild "hog plum" (*Spondias purpurea*); azúchil trees (*Astianthus*) that embellish arroyos with their yellow flowers; and more. His animals include iguanas (*Ctenosaura*), horned lizards (*Phrynosoma*), rattlesnakes (*Crotalus*), foxes, birds, butterflies, rabbits, deer, and others. Eduardo offers, as he puts it, "*a mi modo*," his own take, on some of the case studies discussed in the book, including the diversity of domestic goldfish, selection's effects on dog skull shape, fore- and hind-wing scaling in butterflies, and a few scientists. For many hours of truck conversations on misty roads, I am especially grateful to Erika Edwards and Michael Donoghue. I'm deeply grateful to many colleagues for conversations and insights on the topics explored here and on their research, or for reviews of parts of the book, including Gilberto Alemán-Sancheschúlz, Tommaso Anfodillo, Alfonso Arroyo-Santos, Alex Badyaev, David Baum, Darren Burton, Taggert Butterfield, Felipe Campos-Cerda, Sherwin Carlquist, Matiss Castorena, Kuo-Fang Chung, Jeff Conner, René Duckworth, Alberto Echeverría, Brian Enquist, Jed Fahey, Alex Fajardo, Frietson Galis, R. Geeta, Deborah Gordon, Giuseppe Fusco, Arnon Levy, Sandro Minelli, Pablo Padilla, A. Isabel Pérez-Maussán, Emilio Petrone-Mendoza, Massimo Pigliucci, Katherine Renton, Barbara Schaal, Sahotra Sarkar, Alfredo Saynes Vásquez, Alí Segovia-Rivas, Enrico Scarpella, Emmanuele Serrelli, Dani Sol, Alan Templeton, Francisco Vergara-Silva, Susanne Vetter, and Stephen Yablo. I am thankful to Paul Griffiths for hosting me during sabbatical, and to the much-missed Karola Stotz. Their representinggenes.org project (now closed) and associated publications inspired the discussion in Chapter 3, Section 3.6 on "how many genes" may be present at a locus (Stotz and Griffiths, 2004; Griffiths and Stotz, 2006, 2008; Stotz, 2009). I also thank UNAM and the Sydney Centre for the Foundations of Science for sabbatical support; I am grateful for conversations with Pierrick Bourrat, David Braddon-Mitchell, John Dupré, Kristie Miller, Maureen O'Malley, Steve Orzack, and Kim Sterelny during that time and beyond. Frietson Galis, Michal Polak, and Enrico Scarpella warmly provided images for me to use, and Atsushi Yabe (National Museum of Nature and Science, Japan) provided the *Nyssa* leaf clearing. Norberto Martínez Méndez provided guidance on pentadactyly, A. Isabel Pérez-Maussán prepared the lovely *Cistanthe* leaf clearing, and Jorge Nieto Sotelo helped with Chapter 3. I am grateful for the suggestions of anonymous reviewers whose comments substantially restructured the book, I hope making it more useful.

After years of conversation with them, I must report that evo-devo adaptationist practitioners such as those whose work is profiled here are bright, generous, inspired and inspiring, friendly, cheerful, humorous, and talking to them reveals shimmering windows onto both the human world of science as well as the natural one.

Contents

Introduction

Why you want to be an evo-devo adaptationist

This book is intended for practicing evolutionary biologists, evo-devo practitioners, ecologists, physiologists, and anyone curious about evolution. For standard-issue evolutionary biologists, it is intended to help provide inroads into the contributions that the study of ontogeny can make to studies of adaptive evolution. For evo-devoers and developmental biologists, it provides specific guidance and rationale for beefing up their reasoning regarding adaptation. It provides a welcoming framework for ecologists and physiologists who increasingly encounter "why?" questions in their work. It also provides a roadmap for understanding persistent scientific controversies for those outside of evolutionary biology. There are useful insights to be gained by crossing over between perspectives within biology, and I provide a rationale for and examples of how to do so.

For students having a hard time navigating the forest of terminology in evolutionary biology: it's not you, it's us. Many of our terms are vague and have multiple definitions, so it's completely natural to find them confusing or uncomfortable. A lot of these terms (adaptive radiation... niche conservatism... constraint, etc.) sound like they capture something definitive but often blur over key missing details about what we think is actually going on in nature. This book offers a friendly guide to exploring biological form by starting from what is possible and impossible in development, and understanding the causes of those possibilities and limits. It links these to comparisons of the relative fitness of different viable variants. Framing things this way helps make many familiar but murky biological terms more transparent and easier to evaluate rigorously.

The main point of this book is that it is essential to combine evidence from different perspectives within evolutionary biology. The guidance provided is based on what seems most helpful to working biologists, and it doesn't require asserting things like the Modern Synthesis, which unified Darwinian theory with Mendelian genetics in the mid-20th century (Depew and Weber, 1997; Bowler, 2020; Berry and Browne, 2022; Stenseth *et al.*, 2022), is way off-base, nor that evo-devo or the biologists calling for "expansion" of evolutionary biology don't have anything novel to add (see Baedke and Fábregas-Tejeda, 2023; Brown, 2023). Standard Modern Synthesis biology, evo-devo and developmental biology, as well as many flavors of ecology and physiology, all really do stand to inform one another in day-to-day practice. Real biological practice in all of these areas is necessarily bolstered by incorporating some of the other approaches. Some biologists are already doing so, and a taste of their brilliant work is explored below.

It's their work that inspires the term "adaptationist evo-devo," not as a grand claim for theoretical overhaul, but because it seems like a helpful label. While it is true that

Adaptationist Evo-Devo. Mark E. Olson, Oxford University Press. © Oxford University Press (2026).
DOI: 10.1093/9780197831533.003.0001

evolutionary biologists can point to cogent theory and practice regarding the study of pleiotropy, epistasis, and "constraints" throughout the history of the field, at the same time, it is also true that in practice biologists often don't think about what is and is not possible in development as much as would be useful (Baedke and Fábregas-Tejeda, 2023). Evo-devo biologists can point to talk of adaptation in their literature, but just as often, they show thinking that is unclear regarding the action of selection. Ecologists and physiologists who find themselves drawn into evolutionary questions often arrive without much orientation to the long-standing debates surrounding adaptation, selection, and developmental possibility, and they've rarely been offered a clear foothold. The intent here is to provide such a foothold with a useful series of pointers and case studies.

With this guide, I want to make sure that biologists feel authorized, and in some situations required, to be able to tell standard adaptationist stories. If for no other reason, it is important for biologists to be able to tell these stories because otherwise it is hard to see where exactly the criticisms of adaptationism, such as the famous but hard-to-understand *Spandrels of San Marco* essay (Gould and Lewontin, 1979; see Chapter 5.10), find their footing. Without recognizing where exactly the *Spandrels* and related criticisms land in standard accounts makes it difficult to take their spirit into account to improve empirical studies. In addition, as we will see, being able to deploy standard adaptationist scenarios provides many other benefits too.

After making sure readers are expert traditional adaptationists in Chapter 1, I then outline key modifications to standard accounts in Chapter 2. For anyone who might be puzzled, I try to explain in Chapter 3 some lines of reasoning, for the reader to take or leave, that lead to persistent calls to "expand" evolutionary biology. I conclude that chapter with the cheerful suspicion that calls for expansion or resistance to it might be, for the most part, moot, because the natural world itself is leading all of us biologists in the same general direction. Chapter 4 examines the important but often confusing issues of phenotypic plasticity and phenotypic accommodation, and their key role in adaptationist evo-devo explanations. We are then ready to turn in Chapter 5 to the ways that attempts to explain organismal form implement all of these considerations in the field and lab. Based on frequently encountered difficulties in current practice, I include an important series of observations for building explanations, and outline some of the research directions that adaptationist evo-devo points to. Ultimately—this is the central conclusion—the goal is an evolutionary biology in which there is better collaboration across its subdisciplines and traditions. Such integration is essential, because as we will see, it produces better explanations of organismal form than any one subdiscipline working separately.

As part of this unification, labels describing points of view such as externalist, internalist, structuralist, or even adaptationist really don't apply. "Externalist" views are those that see variation within populations as wide and unencumbered by gaps, limits, or preferred directions, so that the main factor explaining organismal form is natural selection acting via the external environment, hence "externalist" (Sterelny and Griffiths, 1999; Sansom, 2009; Brown, 2023). This view can also be called "adaptationist" because adaptation is its main recourse in explaining organismal form. "Internalist," "developmentalist," or "structuralist" views tend to see the evidence for features of development that don't seem to have anything to do with natural selection as being the main factors shaping organismal form (Dwyer, 1984; Goodwin, 1990; Griffiths, 1996a; Amundson, 2001; Goodwin, 2001; Steinberg, 2003; West-Eberhard, 2003; Newman, 2014, 2015; see also Held, Jr, 2021). These views focus less on the molding of the external environment and more on features of development, "internal" to the organism, hence "internalist."

None of these views really make any sense as stand-alone frameworks, because development and selection are inextricable (e.g. Schwenk and Wagner, 2004; Breuker *et al.*, 2006; Shanahan, 2008; Lala *et al.*, 2024). As we will see, it doesn't make any sense to ask if a given pattern is the result of selection or developmental factors. All patterns in organisms involve both, in ways that can't be clearly parsed into one category or another (Fisher, 1985; Olson, 2019a). So, in truth, there is no adaptationism, no evo-devo, just the effort to build explanations of why organisms have the forms that they do and not others. "Form" here is meant in the broadest sense—morphology, physiology, biochemistry, and so on. A shorter name than "the effort to build explanations of why organisms have the forms that they do and not others," and one that captures the necessary complementarity between camps, is adaptationist evo-devo, and its practitioners evo-devo adaptationists. What follows is a guide showing the basic outlines of how to become an evo-devo adaptationist, and the very good reasons for wanting to become one.

1
Standard adaptation

To become an evo-devo adaptationist, the best place to start is to make sure that you are an expert at formulating standard adaptationist accounts. The literature in certain academic circles makes it sound like evolutionary biologists are constantly and expertly deploying adaptationist reasoning at every turn. I am not sure what literature they are reading: there is a surprising unevenness in evolutionary biology and ecology in expertise in articulating such standard accounts, to the detriment of not just traditional evolutionary biology and ecology, but evo-devo too. In this chapter, I will give examples to illustrate this unevenness. These examples include the proliferation of the term "constraint" as a many-headed, confused reference to natural selection (Section 1.4). If biologists were being clear about the processes that they are invoking, they would provide their reasoning in terms of natural selection when they mean natural selection. The frequency with which natural selection reasoning is shrouded in vague "constraint" terminology tells us of a generalized need in biology to sharpen standard-issue adaptationist reasoning.

Another example covered below is the mistaken use of the term "macroevolutionary" to refer to studies of convergent evolution (Section 1.12.1). Convergent evolution is the action of natural selection leading to similar organismal outcomes under similar selective contexts, starting from different ancestral states. Natural selection is a standard population-level process. Adding the term "macroevolutionary" makes it sound like an additional process is at work. This frequent larding of comparative studies with the fancy-sounding but, in this context, empty term "macroevolution" is more evidence that many biologists need to articulate their thinking more clearly regarding adaptation. I also explore (Section 1.12.2) the way biologists treat metaphors such as interspecific "competition," "stress," and environmental "filtering" as though they were real phenomena and not, at least some of the time, also a sign that adaptationist thinking could be clearer (see Section 5.7.10 for more on metaphors). Teleological language is another way covered below in which scientists commonly cloud thinking about adaptation (Section 5.7.6). Likewise, Chapter 4 illustrates that notions of "maladaptive plasticity" are also symptoms of faulty population-level reasoning. None of these ideas could proliferate so freely in a biology in which thinking in the most standard adaptationist terms were clearer. I hope these examples will help convince you that it is very much worthwhile pausing for a moment to recognize what a standard adaptation scenario implies.

In addition, becoming an expert adaptationist is the best way to appreciate the importance of the adaptationist evo-devo perspective. If you are starting from traditional evolutionary biology or ecology, then a refresher on the key points of a standard adaptationist account will best highlight the aspects that the adaptationist evo-devo approach adds. These additions are not so much to biological theory as in addressing the tacit assumptions and practices that we biologists often employ.

Adaptationist Evo-Devo. Mark E. Olson, Oxford University Press. © Oxford University Press (2026).
DOI: 10.1093/9780197831533.003.0002

Evolutionary biologists have acknowledged since the inception of our field that development is important, that phenomena such as pleiotropy and epistasis are of crucial importance, and that organisms are inextricably tied to their environments. However, when we don't examine directly what is possible and what is not possible in development ("developmental potential") as part of an explanation of organismal form, then we are invoking, tacitly, the assumption that developmental potential isn't an important consideration. Becoming an expert at deploying standard adaptationist accounts sharpens awareness of these tacit assumptions and practices in a helpful way.

A standard account is structured around the three conditions necessary for observing natural selection, and breaks fitness down into three components. Presenting a list of three conditions necessary for observing natural selection is sometimes called the "recipe approach" to natural selection (Godfrey-Smith, 2009; Bourrat, 2015). It is like a recipe because it gives a list of essential ingredients. If all three conditions are met, then we will observe selection. There are other ways of depicting selection, but because it provides a list of aspects that need empirical investigation, the recipe approach is a very helpful way for working biologists to frame their hypotheses, so it is used here throughout. There are many excellent treatments of selection from the "recipe" and other perspectives (for example, the series of papers including Frank, 2012; also Templeton, 2021), so we can focus here on the essentials, with special emphasis on certain pitfalls of each.

The intent here is to build an essential foundation for an adaptationist evo-devo view. Even if you are one of those biologists who think that adaptation hypotheses are naive, you need to be able to generate a rigorous standard adaptationist hypothesis in order to show where you think adaptationism goes wrong. You need to be able to do this *without* getting distracted by your convictions regarding the other processes you think might be generating the pattern of interest. Fusco (2001) offers a menu of possible processes. If you *do* think that selection is involved in generating your pattern of interest, then it goes without saying that you need to be able to generate your hypotheses with predictions that are as explicit as possible. As part of this effort, in this section, we start with the first condition necessary for observing natural selection, variation.

1.1 The three conditions necessary for observing natural selection

1.1.1 Variation

The first condition necessary for observing natural selection is variation within populations (Hallgrímsson and Hall, 2005). Expectations about variation are also a main target of criticisms of standard accounts, and so understanding the "variation" criterion is absolutely central to evo-devo adaptationism. Expectations about variation can be surprisingly difficult for even experienced scientists to formulate, so let's take a moment to look at what you need to do to generate a solid standard adaptationist hypothesis.

A standard scenario of natural selection is "given heritable variation associated with fitness differences in a trait, some variants of the trait will come to increase in proportion over others in a population over generations." In generating a specific hypothesis that applies to your specific case, it is important to specify the "variation" part clearly. Natural selection is a population-level phenomenon, so "variation" is manifest in differences in the trait of interest between members of the same species. Take a moment to practice picturing the appropriate variation being referred to.

Statements about variation imply at a minimum two aspects, the traits and the individuals involved, and to start, it is helpful to picture familiar situations. You can picture the trait "height" that varies between siblings in a family. You can picture variation in the color patterns between puppies in the same litter. You can picture variation in leaf shape between the seedlings that sprout from the same seed pod. All of these situations include both of the aspects mentioned above. In the example of siblings, the trait is height, and the individuals in question are the siblings. In the puppies example, the trait is coat color and the individuals are the members of a litter. In the example of the seed pod, leaf shape is the trait and the individuals are the seedlings from the same pod.

Do not let the obviousness of these examples make you lower your guard and risk glossing over this step. Remember to specify clearly both the trait and the individuals involved. A common error many biologists fall into is to picture variation across species when thinking about natural selection. However, the raw material of selection is variation within species, across the individuals that can potentially coexist and share the same conditions of natural selection. Picturing variation across species when thinking about natural selection is an error with regard to the relevant population of individuals. Another common error is to be vague about the trait. For example, it is often easy just to invoke "variation" without specifying what trait might be varying and how. Instead, we want to make a guess that is as educated as possible and as specific as possible, given the information at hand, regarding what we think might be happening in nature. If picturing variation is tricky, the second condition turns out to be, in some ways, even more so.

1.1.2 Heritability

The second condition necessary for observing natural selection is heritability. Like variation, heritability seems straightforward, but what needs to be pictured is important to examine carefully. The key point is that heritability is a pattern, not the process of inheritance (see Section 5.7.11 on the pattern-process distinction). It is the tendency for progeny to resemble their parents more closely than they do other members of the species. This resemblance is with regard to the trait of interest. So, if we are interested in height, we notice that tall parents tend to have tall children. If we are interested in oil content in corn, we notice that parent plants from lines of corn that have been bred for high oil content tend to have progeny with seeds with high oil content. If we are interested in dog leg length, we notice that short-legged dog parents tend to have short-legged puppies.

Though the notion of heritability seems obvious and clear-cut, ask a class of biology graduate students what "heritability" is and most will talk about inheritance, the mechanisms by which traits are transmitted from generation to generation. Rather than the pattern of heritability, they talk about the process that leads to heritability.

It is, of course, important to understand how parents pass traits along to their offspring, but for selection to be observed, it doesn't actually matter how it occurs. As long as there is a mechanism—any mechanism—that causes offspring to resemble their parents reliably in the trait of interest, generation after generation, natural selection can occur. That Darwin came up with natural selection with incorrect ideas about the mechanics of inheritance illustrates that the details of the specific mechanism of inheritance are not necessary to study selection (Berry and Browne, 2022; Stenseth *et al.*, 2022).

Heritability is important because, without it, the variants with higher performance will not be differentially represented generation after generation. A classic illustration of this point is August Weismann's experiment of removing the tails from mice. In his book *Essays upon Heredity and Kindred Biological Problems*, Weismann (1891) describes an experiment that he conducted. Over five generations, Weismann cut the tails off of mice and bred the tailless mice. Over the course of the experiment, he bred 901 mice, all with tails, from tailless parents. In the vocabulary of a standard natural selection account, the fitness of the tailless variants was very high because no variants with long tails were allowed to breed. So, we have variation in the form of long and short tails. We also have marked fitness differences because the mating success of tailless individuals was maximal, and that of long-tailed variants was zero, because they didn't participate in any mating. Weismann's experiment thus meets two of the three conditions necessary for observing natural selection. What is missing is the second criterion, heritability. The tailless condition in Weismann's experiment was not heritable. Tailless parents always had offspring with tails. This example illustrates that even when there is intrapopulational variation and fitness differences between variants, if variation is not heritable, then there will be no change between generations.

Heritability comes in degrees. For some traits, the values observed in parents predict those in offspring very well. For others, the resemblance is loose, with offspring only vaguely resembling their parents (Falconer and Mackay, 2009). One of the central dances of selection is the interaction between trait heritability (how reliably offspring resemble their parents) and the "strength" of selection (how strongly survival and reproduction differ among trait variants). When both heritability and selection are strong, change can be rapid: variants that confer a marked survival or reproductive advantage will quickly increase in frequency. But when heritability is low and selection is weak, little to no change in trait frequency occurs between generations.

1.1.3 Fitness

For selection to occur, the heritable variation between individuals within a population needs to be associated with fitness differences. Fitness is reflected in the fact that some variants survive to adulthood or reproduce better than others. It is helpful to think of fitness as being made up of three components that follow the life of an individual. These are survivorship, mating success, and fecundity. Let's look at these in order before touching briefly on how they interact with one another.

1.1.3.1 Survivorship

Even if all of the variants in a population are identical in their likelihood of mating and in the numbers of offspring they produce, they can still differ markedly in their representation across generations if they differ in survivorship. One useful way to think about survivorship is as survival to the reproductive stage. This is the first thing that a growing individual must do before reproducing, and in this sense survivorship is the first component of fitness. Imagine two heritable variants in a population. One variant almost always fails to reach the reproductive stage, whereas the other almost always does. Even though reproductive individuals of this variant reproduce at the same rate, the one with higher survivorship will nevertheless come to predominate because there are so many more individuals of that variant reproducing. Therefore, survivorship is an important potential contributor to which variants are differentially represented between generations.

1.1.3.2 Mating success

Once an individual has survived to the reproductive stage, to contribute to the next generation, it then needs to convince another member of the species to mate with it. The most obvious expression of mating success is found in sexual animals such as mammals. In these cases, mating success is reflected in variables such as the number of times an individual copulates with another individual with at least the potential for a zygote to be formed. In plants, mating success can be reflected in the number of pollen grains exported or the number of pollen grains deposited on the female flower parts. Asexual species don't have mating success. As is always the case, which variables are relevant to measure will depend on the biology of a specific study system (see Section 1.7).

1.1.3.3 Fecundity

Now that an individual has survived to the reproductive stage and has successfully mated, we want to know how many offspring are produced from that mating event. This aspect is fecundity, as reflected by variables such as the number of eggs laid, seeds set, or live offspring born (or in asexual species, how many daughter individuals are produced).

1.1.3.4 Interaction among the components of fitness

In generating predictions regarding fitness, remember to separate the components. It is helpful to invoke an "all else being equal" assumption when thinking about each component (see Section 5.7.5 on this assumption). To do this, if you are thinking about survivorship, assume that all of the individuals in your population are the same with regard to mating success and fecundity; that is, the individuals that survive to reproductive age are otherwise the same. If you are thinking about mating success, assume that all of the individuals in your population are the same with regard to survivorship and fecundity. Sometimes you will find that you only have reasons to predict differences in just one or two components of fitness between members of a population. In making your predictions, it is also helpful to think about the ways that mating success and fecundity can trade off within populations in ways that lead to equivalent fitness between variants. Two variants could have equivalent fitness with very different combinations of mating success and fecundity. You could have 100 progeny if you mate 100 times and have one offspring per event. You could also have 100 progeny if you mate one time but have a single litter of 100. In the end, the specifics of predictions regarding fitness depend on the biology of the species in question (see Section 1.7 for encouragement on studying fitness). Fortunately, there is a great deal of helpful thinking on fitness and its conceptual and empirical challenges that you can draw on as a working biologist (Rosenberg and Williams, 1986; Ariew and Lewontin, 2004; Abrams, 2009; MacColl, 2011, and many others).

1.2 The "force" of natural selection is metaphorical

Natural selection is often regarded as a "force," and while this can be a useful shorthand, it's important to remember that the "force" of selection is metaphorical. Wind really does push a sailboat through the water, but there is nothing, really, pushing populations through the space of possible trait combinations. When viewing the results of selection, though, it can look as though something is pushing populations. The differences between generations are often directional, either pushing trait means to different values over the generations, or maintaining trait values around some mean over the generations. This makes it look as though some invisible wind is blowing the population through trait space, or coming at it from both sides.

Instead of there being an invisible wind pushing populations, it just happens to be the case that some variants survive and reproduce better than others (Cofnas, 2024). When these variations are heritable, then there tends to be a change in trait frequencies between generations. This tendency is forceless. See Section 5.7.10 for more on how to recognize and use the metaphors of evolutionary biology, and Pigliucci and Kaplan (2006) for an excellent discussion of the "force" metaphor (also Matthen and Ariew, 2002).

Underscoring the forcelessness of natural selection is that it transcends biology. Any system meeting the three conditions will show intergenerational change in trait frequencies. Jokes are a good example. They are repeated over and over, from forwarded emails to tellings around the water cooler. They vary with different tellings. These different tellings are analogous to reproduction, and the different versions are analogous to intrapopulational variation. So, jokes meet the conditions of variation and heritability. Some versions are funnier than others, and let's assume that versions that get a lot of laughs get repeated more than those that bomb. Over time, the funniest versions should come to be the most-forwarded, and thus have higher joke-fecundity than less-funny ones. Science fiction robots that can build new versions of themselves invoke the transcendence of selection. If they produce variation between iterations, and preferentially build versions that meet some performance standard, then they too will show intergenerational change as a result of heritable fitness differences. Recognizing that some variants become overrepresented under the right conditions even in nonliving systems helps avoid invoking a literal "force" of selection when generating expectations about natural phenomena.

Recognizing that no force corresponds to natural selection is a helpful antidote to teleological thinking, which is the very seductive tendency to think of organisms as exerting purposeful volition in adapting themselves to challenges (Section 5.7.6 provides more detail on teleology and 5.7.10 on metaphors). Avoiding teleology is crucial for building evolutionary studies around postulated biological processes that are potentially real (in this case, selection), rather than ones that for sure aren't (consciously striving organisms). Here is another helpful distinction.

1.3 Sorting versus selection

A common way to get tripped up when coming up with hypotheses regarding natural selection is to call any intergenerational change in frequencies "selection." A helpful remedy is the distinction between sorting and selection (Vrba and Gould, 1986). Sorting is the differential representation of some variants over others within populations across generations. Sorting can be caused by many things, chance, for example. Imagine a population in which the individuals are of two colors, red and white, and mill around randomly. As the individuals bumble around, at some point, all of the red ones find themselves together and all of the white ones together. At one of those moments, all the white ones all happen to be in the same spot, and that spot happens to be under a cliff. At the same time, all the red ones were out on the plain at the moment that the cliff collapsed and squashed the white individuals. Given the heritability of the variants, there would be a marked change in the representation of variants from one generation to the next as the result of the chance event of the white individuals all getting squashed. Assume that this was the one and only time that a cliff collapse ever affected this species. Because cliff collapse is not a recurrent factor in the life of the species, it does not constitute an agent of selection that can shape the species (Section 1.6 has more on what constitutes a selective agent).

Natural selection occurs when some aspect of the biophysics of one variant causes it to have higher performance, and thus fitness, than some other variant. This connection between the biophysics of the different variants and their fitness is thus not random. Our example of the white variants getting squashed by the cliff, in contrast, was not determined by the biophysics of the white variants. They simply all happened to be in the same place at the time that the cliff collapsed. So, chance events provide good examples of sorting without selection.

In the same way that, in principle, selection can be observed in non-biological systems, sorting can as well. For example, beach sands tend to be sorted by size, thanks to the steady movement of waves and currents that naturally separate out grains by size. Grains found higher up on the beach, like in the foreshore and backshore, are usually more uniform than those found in the shallow water, where the changing energy of waves mixes together different grain sizes (Carranza Edwards, 2001). Beaches with large waves, which have more kinetic energy that can move larger particles, tend to have higher particle size variances. The result is that sand on beaches with high-energy waves is coarser. Gentler waves tend to lead to the accumulation of finer particles.

Different conditions throughout the year, say from the hurricane season to the dry season, lead to different patterns of accumulation of particle sizes even on the same beach. So, here we have a situation of sorting—differential representation of some variants (particle size)—owing to differing conditions. In contrast to selection, in sand grains, there is no reproduction or heritability, no differential representation over time, and no generations for that matter, just the same grains getting moved around in different ways throughout the year.

Selection is one cause of sorting. In biological contexts, sorting is an intergenerational change in the frequencies of intrapopulational variants. As alluded to above, under selection, which variant or variants prevail over others is not random but instead is the result of the biophysics of the system. This non-random association between trait performance and biophysical properties is what motivates evolutionary optimality models (discussed in Section 5.1.3), which assume that certain trait values lead to higher fitness because of the biomechanical and energetic interactions between different variants and their selective contexts. Selection is the differential representation between generations of some variants over others because of some characteristics that lead to some variants surviving and reproducing better than others. Thus, all selection involves sorting but not all sorting involves selection.

Sorting can result from selection at the level of organization under study, or from selection at higher levels. "At the level of organization under study" is a reference to the hierarchical organization of biological entities: species are made up of individuals, individuals of parts, parts of organs, organs of tissues, tissues of cells, cells contain many different components, and so on. Say you are studying flower pigments, and you notice that one allele involved in anthocyanin production is becoming more common over generations in your focal species. This is a case of sorting—one allelic variant is coming to predominate over another across generations. As you investigate, it turns out that a change in pollinators in the area has led to purple flowers having higher fruit set and seed number than red flowers. So, purple flowers come to predominate in the population given pollinator preference. This situation describes selection involving traits observed at the level of individual plants as they interact with their selective context. This selection involving individual-level traits drags all of the lower-level factors that are correlated with the higher-level interaction along with it. In this case, sorting at the level of alleles is not caused by selection at the allele level. Instead, sorting at the allele level is caused by selection occurring because of differences observed between individual organisms. As pollinators preferentially pollinate some colors, the alleles involved in the production of the favored colors sort differentially because of selection that is the result of phenotypic differences among individual organisms. In this example, sorting among genetic variants is caused by selection occurring between individuals.

So, a helpful distinction is that between sorting and selection. Selection is a major cause of sorting, but it is not the only cause. And even when selection is the cause of sorting, the biophysical interactions responsible for differential representation may occur at a different level of biological organization than the one where sorting is observed.

1.4 Watch out for "selection in constraint's clothing"

Perhaps the most treacherous single term in evolutionary biology is "constraint." It is treacherous because it is used very frequently but at the same time it is very vague (Pigliucci and Kaplan, 2000). This vagueness means that different biologists often have different ideas about what the term refers to. You can make one statement about "constraint" and the five biologists listening will come away with five different ideas about what you meant. A useful experiment is to ask a gathering of biologists to define the term (the chat window in an online meeting is ideal for this), and you will see how diverse the notions are. The crucial point to watch out for, and which we will explore briefly here, is that notions of "constraint" span the range from factors that have nothing to do with natural selection to natural selection itself.

There is nothing inherently wrong with vague terms in science (see Section 5.7.10 on metaphors). The vagueness of terms such as "adaptive radiation" or "niche" actually helps inspire diverse research and bring researchers together (Olson *et al.*, 2019; Neto, 2020). But unless biologists define what they mean when they use a term, vagueness does invite misunderstanding. Moreover, the use of "constraint" usually covers over unclear thinking, and this unclear thinking translates into hypotheses that are more confused than they should be, and thus hard to test.

In the case of "constraint," a key lack of clarity can be called "selection in constraint's clothing." If there is a useful definition of the term "constraint," it is to designate phenomena that limit the production of variants that would be favored by natural selection if only they could be produced. This definition includes everything from tradeoffs to contingent (i.e. evolution just happened to occur in a certain way) characteristics of development. But there are many other uses of the term that usually just refer to selection. This use of "constraint" language when really referring to selection is what is meant by "selection in constraint's clothing." Here is an example.

In discussing the factors that lead snakes to have the tail lengths they do, King (1989) entertains what he calls the "morphological constraint hypothesis" (p. 145). The notion is that "male tail length may be constrained by the presence of the hemipenes and retractor muscles" (p. 145). It's impossible to tell exactly what this means based on the language used, but given that the study primarily examined issues of adaptation, a plausible interpretation of this "constraint" is simply as follows. Tail length varies heritably within a population, relative to body size. Males with tails that are too short may lack the space to house the hemipenes and retractor muscles fully, leading to partial eversion, hemipenes dragging on the ground, and reduced survivorship and mating success. Males with very long tails, on the other hand, incur energetic costs in growing, maintaining, and moving around extra tissue, costs that could otherwise be spent seeking mates or producing sperm. Longer tails may also increase predation risk by offering a convenient handle. The "just right" tail length balances these competing pressures, an evolutionary compromise between genital function, energy allocation, and predator avoidance.

Lest the reader excuse "morphological constraint" as harmless shorthand, it's worth noting that nothing like the scenario just sketched is offered in King's paper, only vague gestures toward "the role of morphological constraint in limiting adaptive evolution" (p. 146). This illustrates the dull power of constraint language:

it never clarifies evolutionary logic, only impersonates it. In the scenario just described, there is nothing standing outside adaptation. The observed tail lengths reflect a balance among multiple selective vectors: selection against genital malfunction favors longer tails; selection against unnecessary energetic expenditure and increased predation risk favors shorter ones. Organisms are always the result of such negotiations. King's "morphological constraint" simply *is* adaptive evolution. The "constraint" label looks explanatory, but in fact conceals a selection hypothesis behind a façade of causal clarity, a textbook example of selection in constraint's clothing.

To make this kind of confusion easier to spot, it helps to have a list of phrases that usually, though not always, signal selection in constraint's clothing. In these cases, replacing "constraint" with an explicit hypothesis about what kinds of variants are developmentally possible, and which of those would plausibly have higher or lower fitness, makes the evolutionary logic more transparent. Below are examples of such terms, along with brief examples of ways that they often function as covert references to selection:

- "Allometric constraints": two variables covary, with those in the empty quadrants of morphospace above and below the scaling line having lower fitness than those along the line. Allometry shaped by natural selection (it practically always is shaped by selection) is discussed in Sections 1.11.1, 2.3.6, 2.3.8, and 2.3.11.
- "Biophysical constraints": some variants are biomechanically or otherwise unviable and so not favored by selection.
- "Competitive constraints": for example, in interspecific "competition," species B is one of the factors in the selective context of species A (see Section 1.12.2).
- "Design limits": beyond some value of size, metabolism, or trait combination, variants have low fitness.
- "Developmental bias": at least some notions of this multi-meaning term imply that natural selection favors developmental systems that produce functional variants over those with low fitness.
- "Energetic constraints": Variants with higher energy expenditure for a given performance have lower fitness.
- "Environmental constraints": the conditions of selection are such that some (or even all) variants are not favored.
- "Functional constraint": language that appeals to function implies selection. So, "functional constraint" simply means that some variants have higher performance, and thus fitness, than others. This includes situations in which a single structure has multiple roles, with selection favoring the "least bad" combination.
- "Genetic constraints": some uses of this term refer to selection. For example, when selection favors some trait combinations over others, from a quantitative genetic point of view, this "genetic covariation" is often regarded as a "constraint." Likewise, when selection reduces or eliminates variation, the resulting lack is sometimes framed as a constraint. Some other senses of "genetic constraint" that don't so straightforwardly reflect selection are mentioned below.
- "Geometric constraints": some morphological combinations, such as length-width relations, are favored over others.

- "Large genome constraint": selection eliminates variants with genomes that are so large that their metabolic costs outweigh their benefits.
- "Mechanical constraint": see biophysical constraint.
- "Metabolic constraint": some metabolic rates are favored over others in the context of other features, such as body size.
- "Model constraints": In an optimality model, a constraint is a condition that a scientist specifies that limits the range of possible values a trait can take. Though the term sounds like a valid technical definition, often model constraints smuggle in selection. For example, in a model of optimal leaf size for a plant, the model constraint that leaf size can't be larger than what the stem can support mechanically implies that plants with leaves that would buckle the stem would have low fitness, not that the combination is ontogenetically impossible. Again, just selection.
- "Morphological constraint": some morphological variants have lower fitness than others.
- "Physiological constraints": given variation in physiology within a population, some variants are favored over others.
- "Selective constraint": natural selection favors some variants over others (this term can always be replaced by just reference to "selection"; adding "constraint" doesn't make prose clearer or fancier).
- "Universal constraint": Conditions of natural selection that apply to all organisms. Maynard Smith *et al.* (1985) cite the physics of levers. For the same force applied (e.g. the same muscle area), a longer lever (e.g. longer limb bones) generates more moment but slower running speed; this does not mean that organisms cannot produce a very wide range of muscle-limb bone proportions. Lever physics are simply a factor of selection that applies in similar ways across all of them.

If you find yourself disagreeing with any of these interpretations, note that this just proves the point that "constraint" is so vague that any number of interpretations are possible for any given use. Other confused uses of constraint usually don't refer to selection quite so directly. Examples include:

- "Architectural constraint" and "spandrel": often confused terms to refer to incorrectly delimited traits (Olson, 2019b), see Section 5.10.
- "Phylogenetic constraint": usually means "a very long way in morphospace between species A and B" or "closely related species resemble one another" (see Section 5.7.11).

Notions of "constraint" that might be useful if the term didn't have so many meanings include:

- "Developmental constraint": One meaning is "contingent features of development (i.e. ones that just happen to have evolved that way but that in principle could have evolved a different way) that impede the production of certain variants, often variants that would be favored by selection if they could be produced"

(Sections 2.1, 2.3.13–15). In practice, this term often just refers to selection in constraint's clothing.
- "Variational constraint": Sometimes variation that could plausibly be expected is not present in a population, as when developmental variation is either absent or is less in one direction than another. A lack of variation is just a pattern, and understanding its cause requires investigation, rather than a vague name that implies that something is known about its cause, as terms like "variational" or "genetic" constraint sometimes do.
- "Genetic constraint," pleiotropy, and epistasis: When a gene affects multiple traits (pleiotropy), selection on one trait can cause changes in others, often reducing the performance of both as evolution settles on a compromise. These trait correlations may be "contingent," not favored from a wide range of alternatives, but from how the organism happened to evolve. Epistasis, where a gene's effect depends on interactions with other genes, can make mutations advantageous only in specific combinations, excluding others even if they would be favored by selection. In both cases, interactions can limit which trait changes are possible, regardless of what selection would favor.
- Resource allocation tradeoffs: Finite resources can be invested in many small eggs or few large ones, but not many large eggs. Garland *et al.* (2022) provide a helpful classification of different types of tradeoffs, with the exhortation to uncover their mechanistic causes.

The best policy with regard to "constraint" is to avoid its use entirely. Here's a slogan to live by:

> *there is nothing to be gained by replacing the complex but precise vocabulary of selection with the vague language of "constraint"*

Instead, it is much more helpful to specify exactly what is meant by referring to what is possible or not in development, and, of the possible, what is favored. This makes thinking clearer, hypotheses much more accessible to testing, and instantly improves communication with colleagues. As part of this exercise, notice how often "constraint" is really just a vague and confusing reference to natural selection. To demonstrate that it is possible to write an instant classic in evolutionary biology without ever needing the term, in what follows, I avoid the use of "constraint" except in scare quotes, to denote a vague term with any number of meanings.

1.5 Beware the ad hoc hypothesis

A genuine criticism of adaptationist explanations is that with a little effort, it is possible to save a favored hypothesis in the face of results that would seem to reject it. Evidence arises that rejects a favored hypothesis. We then posit additional considerations as necessary to account for each objection and thus be able to maintain the favored hypothesis. These are what critics of adaptationism call *ad hoc* hypotheses, and they have a point.

For example, most ecologists think that thick bark on woody stems is always an adaptation that protects plants from fire (Pausas, 2015). One way to test this hypothesis is to go to habitats that practically never burn naturally. If the plants there have always lived in fire-free habitats and nevertheless have very thick bark, then selection favoring thick bark by virtue of its fire-resistant properties cannot be the only explanation for thick bark in woody plants. Indeed, many desert plants have very thick bark, but deserts practically never burn naturally. This rarity of burning means that fires cannot possibly be selective agents shaping desert plant morphology (see the next section on generation time and recurrence of selection pressures). These observations alone reject the hypothesis that thick bark is always the product of natural selection protecting stems from fire (Rosell, 2016, 2019).

To explain thick bark in desert plants or other fire-free habitats, fire ecologists resort to *ad hoc* gymnastics. For example, they can propose that deserts do indeed experience fires, just on intervals that are so long that these fires have never been observed by ecologists. Because factors such as fire need to occur repeatedly within the lifetime of an individual to exert a potent selective pressure (see Section 1.6), fire ecologists would then be obligated to make another *ad hoc* somersault. If fire return intervals in deserts are so long that biologists have never observed them, and if fires must be experienced repeatedly in the life of an individual plant, then desert plants with thick bark—a humble cholla cactus, for example—must be extremely long-lived and much older than anyone currently imagines. Fire ecologists can then call it a day, without finding ways to test the notion that deserts do indeed burn but no one has happened to notice, or that fat-barked desert plants live for thousands of years.

Instead, they embody the *ad hoc* protection of their favored hypothesis perfectly. Confronted by evidence that clearly rejects their hypothesis (thick bark in non-burned habitats), they then propose as-necessary (*ad hoc*) hypotheses that mean that they do not have to reject their hypothesis (deserts burn on very long intervals; fat-barked desert plants live many thousands of years). And keep on proposing as many *ad hoc* hypotheses as necessary. Even the most frequently burned habitats on Earth, such as the Brazilian savannah, include species with thin bark. Fire ecologists can invent *ad hoc* mini fire-free pockets that somehow spare the thin-barked species as these habitats burn year after year, or that just before you went to measure bark thickness, there was an inexplicable fire-free period that allowed uncharacteristically thin-barked species to spring up. By proposing as many *ad hoc* hypotheses as are necessary to confront each new piece of contradictory evidence, and by not subjecting these *ad hoc* hypotheses to testing, fire ecologists can protect their hypothesis that thick bark is always an adaptation to fire.

The alternative to this hypothesis-protecting is to allow a hypothesis to be rejected, to propose alternatives or refinements, and to test them again. In the case of bark thickness, there is plenty of evidence that thicker bark stores more water and starch, provides more mechanical support, often carries out photosynthesis, and protects against herbivores and not just fire. In addition, bark contains the phloem, the tissue that exports sugars from the leaves, nourishing the rest of the plant. Species in fire-free dry areas often have very thick bark, filled with living cells storing water and nutrients. Water storage in fire-free dryland plants is very plausibly an adaptation, much more plausible than fire resistance. Climbing plants have much thicker bark for a given stem diameter than self-supporting plants do. These reasons alone are enough to think that the thickness of bark can be shaped by other factors in addition to just fire (Rosell, 2016, 2019). Allowing the hypothesis that thick bark is always an adaptation to fire to be rejected reveals the multiple causes that bark thickness might have.

In the same way, you must be vigilant of the distinction between proposing new hypotheses for testing in the face of new evidence and proposing hypotheses to brush away inconsistencies. The example of bark and fire resistance illustrates that it is important to keep an eye on the fine line that, in practice, separates the natural scientific process of iteratively testing and refining hypotheses and proposing *ad hoc* ones to protect a particular one from rejection.

1.6 What constitutes a selective pressure? Generation time and recurrence

The example of bark and fire resistance illustrates that biologists sometimes generate ideas about function in biological features that cannot possibly be adaptations. For example, most tropical forests, whether tropical rainforest, cloud forest, or tropical dry forest, only rarely experience forest fires. When they do burn, some species tend to survive and others tend to die. Which species survive and which succumb generally has something to do with the particular traits of a given species. For example, some species make abundant sprouts from the base, whereas others never seem to do so. After a fire, "resprouting" species can quickly start to replace lost stems, whereas non-resprouters often die (Clarke *et al.*, 2013). Biologists sometimes see evidence for adaptation in such differences and interpret resprouting in rainforest species as an adaptation to fire in rainforest environments. But resprouting cannot possibly be an adaptation to fire if the species evolved in anything like the current, fire-free regime. The key in determining why has to do with two considerations.

The considerations for identifying potential adaptations are the frequency of a supposed selection "pressure" (see Section 1.2 on the force metaphor) versus the generation time of a species (Rollo, 1995). Selection tends to shape organisms in response to factors that come up more frequently. If a factor comes up frequently within each generation, then there are more opportunities for differences in survivorship and reproduction to occur between heritable variants in a population. If a factor comes up just once each generation, then there are fewer opportunities. If a factor comes up only once over many generations, then it is extremely unlikely that populations can experience change in their trait distributions in a way that makes them more resilient to the factor.

In other words, if a factor is very infrequent with respect to generation time, then it is unlikely that a species will be adapted to this factor. Take a situation in which natural fires occur in a tropical forest only every few thousand years, and a given tree species in which the lifespan tends to be 200 years. Let's assume that the species and its ancestors arose and have always lived in moist tropical forests, which is the case for most plant species on Earth (Slik *et al.*, 2015). Given that practically all generations of the species are born, live their lives, and die without ever experiencing fire, it is extremely unlikely that the traits that the species has represent adaptations to fire. Therefore, the explanation for resprouting almost certainly needs to be sought elsewhere than as an adaptation to fire.

An even more extreme example comes from mass extinctions. Mass extinctions—global catastrophes that cause many species to go extinct at about the same time—occur about every 30 million years (Gould, 1985). No species has generation times that are so long that they experience multiple mass extinction events. Most species do not even exist on Earth for this long. And yet some species survive mass extinctions thanks to the traits that they happen to have, which allow them to persist despite the radically changed post-catastrophe conditions. Like resprouting in trees of fire-free forests, having a trait that allows a species to survive a mass extinction is just luck—the explanation for the presence of the trait must be sought elsewhere than as the result of selection in the context of asteroid strikes. These traits could be adaptations, but if they are, they are adaptations to factors that occur more frequently with respect to their generation times instead of adaptations to asteroid impacts.

The relevant relationship between frequency and generation time necessarily varies with the biology of a given species. The factors that constitute important selective agents might differ between a tree species that routinely lives thousands of years versus a fly with a generation time of two weeks. Remember that your task is not to be rigid about these principles, but instead to discover what might really be happening in nature. An insect species with multiple generations per year might still have adaptations to winter conditions even though some generations never experience winter. The exact balances of causes between generation time and the frequency of factors that make up the selective environment of a species is an object of investigation. Similar considerations apply to "fitness."

1.7 Measuring fitness empirically

Above, we saw the outlines of thinking about fitness in a way that is a very useful starting point for studying evolution in general in any given case. This outline almost never captures all of the important considerations for understanding why some variants come to be better represented across generations than others in any particular case, though. Do not get hung up on the limitations of the general outline of fitness. Some biologists conclude that fitness is at best a slippery concept limited in usefulness.

Here is a more helpful view. Selection involves the greater survivorship and reproduction of some variants over others within populations. In other words, when natural selection occurs, there must be *some* factors that lead to some variants being differentially represented between generations. As a working biologist, what you need to do for any given case is to figure out what the important factors are that cause differential representation across generations, even if they are more subtle or complicated than the textbook definition of fitness (Matthen and Ariew, 2002; MacColl, 2011; Bourrat, 2015; Evans and Postma, 2025). Some examples of these considerations include the following.

In some cases, longevity can be as important as survivorship. We defined survivorship as survival to the reproductive stage. But in many species, individuals live for a long time after reaching the reproductive stage. Trees are a good example. They often reach reproductive size quite quickly, but then can live for hundreds and even thousands of years, reproducing each year (Piovesan and Biondi, 2021). Individuals within populations can vary in their longevity, and this variation can interact with other components of fitness, such as lifetime fecundity. All else being equal, such as annual reproductive output, the variants that live longer will have more reproductive events and therefore greater lifetime fecundity. In a few animal species, some individuals can live well beyond the reproductive stage, and there are plausible reasons why this might be favored by selection, perhaps because post-reproductive individuals aid in the survivorship and reproductive success of other individuals (Monaghan and Ivimey-Cook, 2023). This interaction among the components of fitness illustrates not that the textbook definition is incorrect but instead that the cause of differential representation between generations is something to be discovered through close study of the species in question—if selection is occurring, your task is to discover the details.

Likewise, questions arise regarding fecundity. For one, it doesn't make reference entirely to the focal individual but to the next generation as well. Survivorship is straightforwardly an attribute of a single individual. Whether it reached reproductive age or not is an attribute of that individual. But fecundity involves knowing how many offspring a given individual bearing a certain variant gave rise to. These individuals are a product of the individual in question, but they are also their own individuals, with their own survivorship, mating success, and fecundity to be measured. Should all progeny be counted, or just the ones that survive? Of those that survive, should we only count those that reproduce? In the face of this continuity between parents and progeny, where and why to draw a meaningful line in measuring variables that reflect the reasons why some variants are differentially represented between generations requires research, not just the application of definitions.

A seductive pitfall is to find that a given notion of fitness does not apply perfectly to a given biological case and to conclude that something is wrong with the concept of fitness. Asexual species lack mating success, though there can be variation in survivorship and fecundity. In superorganisms such as ant colonies, the relevant variation in fitness is between colonies, not individual bodies (Pigliucci and Kaplan, 2006; Okasha, 2008; Gordon, 2010, 2023; Simpson *et al.*, 2024, on the "levels of selection" debate). In other cases, there are favored variants, not a single variant. In some species, sex is decoupled from reproduction, raising the question of how mating success should be measured. What an appropriate variable for comparison between individuals is—per day, per season, per lifetime, per body mass—will differ between species, and finding these variables is central in discovering the details of the biology of a species. In all of these situations, in any given case of natural selection, there are indeed variants that have been favored, and are being favored, over others. In the face of traits that manifestly have been and are being favored, what your job is as a biologist is to discover what the causes of those differences are, no matter how much richer the natural world is as compared to any given textbook definition of fitness.

1.8 The importance of energetic efficiency in postulating fitness differences

In coming up with biologically plausible standard adaptationist hypotheses that are as well-justified as possible, it is always worth considering the potential ways that energetic efficiency might be involved in generating differences between variants in the three components of fitness. This is because the three components of fitness all involve the ways that metabolic resources are allocated. If you are having a hard time coming up with plausible predictions for the ways that variation in your focal trait might be involved in fitness differences, one good strategy is to think about energetic efficiency (a good place to start, and not just for the surname, is Watt, 1986). Let's take the three components of fitness in order.

With regard to survivorship, it is easy to come up with plausible and testable predictions. Hydrodynamic shape in fast-moving aquatic animals provides a good example (again, not just because of the surname, Fish and Lauder, 2006). We can postulate that, given variation in shape, some individuals move far and fast, whereas other conspecifics move only slowly for the same energy expenditure (see Webb, 1984; Eloy, 2013; Nesteruk *et al.*, 2014; Saadat *et al.*, 2017; Li *et al.*, 2021; Jahn and Seebacher, 2022). For the blocky variants, swimming at a sufficient speed so as to escape a pursuing predator will require much more energy than for the streamlined variants. This difference might mean that some variants cannot muster the energy expenditure to escape predators at all, and don't contribute to the next generation because they are eaten. Also, more energy expenditure in predator avoidance means less energy directed to cellular maintenance and immune responses. In this way, variants that invest a great deal of energy in movement could have lower survivorship because they are afflicted with diseases. More energy efficiency could also mean that less drag-inducing variants can grow to predator-defying sizes more quickly. Differences in energy efficiency can almost always be postulated to be associated with differences in survivorship, with less-efficient variants having lower survivorship than more-efficient ones (see the example in Section 2.3.5 on Daniel Sol's work on pigeon leg-body mass covariation).

In contrast, it can often seem tricky or even impossible to generate expectations regarding differences in mating success between variants in a population. If there really does seem to be no reason to expect differences, so be it. Not all components of fitness necessarily contribute to differences in the representation of variants across generations in every species. As with survivorship, though, a good place to start in thinking about the ways that mating success could differ between variants is energetic efficiency. In species that have mating displays or mate choice, energetic efficiency can easily be expected to intervene. Individuals that made it to the reproductive stage can differ in their size and appearance. If some individuals have large energy surpluses and can grow to preferred sizes, engage in more frequent or elaborate displays, have glossier coats or better preened feathers, and if these traits are involved in mate choice, then individuals with greater energetic surpluses will be associated with greater mating success. Greater energy surpluses can even be associated with greater mating success in organisms such as plants. Imagine a population of a tree species, in which some individuals have large amounts of living wood that the leaves need to supply with photosynthetically captured energy, and other individuals in which there is less energy-demanding wood, but still enough to meet the needs of the tree, such as mechanical support. The individuals with just the right amount of wood will support themselves, transport water to the leaves, resist wind, and do all the things the trees with more wood do. But because they have fewer mitochondrial mouths to feed in the wood per unit leaf area (= available energy), then the less-wood individuals will have greater energy surpluses. In the case of mating success, we would predict that greater energy surpluses should be associated with more flowers, larger flowers, more and higher quality floral pigments, floral rewards like nectar, floral attractants like scents, and more pollen grains that can be exported. Differences in energy efficiency can thus often be linked to easily tested expectations regarding mating success, with more-efficient variants likely to be more represented in subsequent generations because they mate more often than less-efficient ones.

Finally, expectations regarding differences in fecundity associated with differences in energetic efficiency are easy to generate. Variants with more surplus energy, say per body mass, can produce more ovules, more eggs, more babies, more fruits, or more seeds. In consequence, more-efficient variants are likely to be more represented in subsequent generations because they produce more propagules than less-efficient ones.

If any of these predictions are biologically implausible, it has yet to be shown where and why. They are also entirely testable, and readily so, all helping to build standard adaptationist accounts. Here are a couple more considerations that offer encouragement to explore where textbook definitions might chafe against biological realities.

1.9 There can be multiple favored configurations and multiple "functions"

We saw in the previous sections that if selection is leading to differences between generations, then there must be some causal interaction between organismal traits and their selective environments such that some variants survive and reproduce over others. It is our job as biologists to discover what this interaction might consist of, however complicated it might be and however it might depart from standard textbook definitions of fitness. Considerations similar to those offered for fitness in the previous section also apply to notions of "the" configuration favored by selection, as well as "the" function of a biological feature. We examine these in turn.

1.9.1 "The" favored configuration or configurations

In discussing the action of selection, biologists often refer to "the" trait or condition favored by selection, as though there were just one. The definition of selection as "the differential survivorship and/or reproduction of one variant out of the wide range of heritable, developmentally possible alternatives" implies just one favored variant. When biologists talk about selection pushing toward a mean value, it can be construed as meaning that the mean value is the single favored value. These and similar expressions make perfect sense in discussing the generalities of selection.

As with fitness, though, it's necessary for each biologist to understand the nature of the set of favored variants in their particular cases under examination. In most cases, there is a range of quantitative variation about a mean value. Talk of a "single" favored variant corresponding to the mean is therefore just an abstraction, a way of thinking about the metaphorical vectors that selection is "pushing" along. Why ranges are wide or narrow about the mean is a key aspect of evolutionary biology, as in the focus on trait variances in quantitative genetics (Pigliucci, 2006; Falconer and Mackay, 2009), and as we will explore below, it's also a focus of adaptationist evo-devo. In some cases, there are reasons to think that selection is quite strong and favors wide trait ranges, or that selection is weak and simply allows wide trait ranges. In other cases, selection may plausibly favor multiple mean values for the same trait (Rueffler *et al.*, 2006). As with fitness, it's important to recognize discussions in the form of generalities for what they are, and to make the range of the favored as well as the possible variants (we explore many techniques for doing so below) a focus of empirical research in any given specific case.

1.9.2 "The" function of a biological feature

Just as biologists often refer to "the" variant favored by selection, we also often speak of "the" function of a trait (Pigliucci and Kaplan, 2006). The "function" of a trait is the role that it performs by virtue of which that variant became differentially represented in populations, and so which helps explain its presence in populations (Garson, 2016). Just as with "fitness" or "the" favored configuration, the reasons can be very involved. An example is the phosphoenolpyruvate carboxylase (*PEPC*) locus in plants, which is involved in the production of the enzyme phosphoenolpyruvate carboxylase (PEPC). First, in respiration, PEPC helps maintain the supply of oxaloacetate, a key intermediate of the Krebs cycle. As the cycle churns away, a critical early step consumes oxaloacetate, and by catalyzing its formation from phosphoenolpyruvate (PEP) and CO_2, mitochondrial PEPC prevents oxaloacetate from becoming limiting. Second, oxaloacetate is a precursor for aspartate and other molecules involved in nitrogen assimilation, storage, and transport, and PEPC contributes here as well. Third, PEPC plays a key role in photosynthetic carbon fixation in plants with CAM or C_4 photosynthesis, by helping form oxaloacetate as an early product of CO_2 capture in chloroplasts (see more in Section 2.3.9). These multiple roles illustrate an important point: with critical functions across respiration, photosynthesis, and nitrogen metabolism, it clearly doesn't make sense to speak of "the" function of PEPC. And yet, there is (or must be) an evolutionary explanation for why PEPC is present in each of these roles.

Like PEPC, practically all biological features are involved in multiple roles. As a result, discussing "the" function of a feature is often a simplification for the sake of generalization. Discovering the multiple roles a feature has and how one or more contributed to the molding of organismal form is precisely the purview of evolutionary biologists. Adaptationist evo-devo approaches are a key part of the necessary toolbox.

1.10 "Continuous" variation: the main reaction point for adaptationist evo-devo

Biologists have long recognized that variation isn't always continuous; thanks to pleiotropy, epistasis, and developmental interactions, this point was already clear to early architects of the Modern Synthesis and before (Geoffroy de Saint-Hilaire, 1832; Bateson, 1894). Yet despite this, we often tacitly assume that variation is continuous. Sometimes this assumption is unconscious, other times it's a convenient starting point. This section introduces that idea and explains why noticing the assumption, when we make it and when we don't, can be useful. In this context, "continuous" means four things: (1) evolution proceeds via small steps; (2) phenotypic space is smooth, with no major holes, and any point B can be reached from point A by any route; (3) variation is unbiased—equally likely in any direction; and (4) variation is essentially unbounded. Each of these ideas is taken up in turn below.

1. The fine grain of interindividual variation. "Continuous" variation implies that variation is "fine-grained," meaning that differences between individuals are small and continuous. This assumption holds that adaptive evolution proceeds mainly through selection on small, incremental differences. While few biologists deny the existence of pleiotropy, epistasis, or tradeoffs—indeed, many explore these complexities directly—in many situations, there is tacit reliance on the idea that variation within populations is smooth and evenly distributed. This fine-grained phenotypic space is often modeled or implicitly treated as densely filled: individuals are assumed to differ only slightly from their neighbors, and intermediate forms are expected between any two phenotypic extremes. Selection, in this view, gradually shifts the population mean through normally distributed, incremental variation.
2. No holes in phenotypic space. "Continuous" variation also implies that variation describes a continuous field, without holes or gaps in it. It is possible to step from individual to individual, from one extreme to another, through a perfectly graded series of intermediates that only differ very slightly from one another. There aren't any "jumps" or "holes" in any part of this field of variation. From the continuous variation perspective, the existence of any two variants, no matter how extreme, means that many individuals, each of which differed only slightly from the next, must exist, or have existed, as intermediates directly between them, not going around any gaps in the space of possibilities.
3. Variation is unbiased. No tendency exists for there to be more abundant variation in one direction than another. The notion of unbiased variation tacitly assumes that phenotypic variation is as likely in one direction as it is in another. More significantly, it assumes that the effects on performance and fitness are also just as likely in one direction as in any other. Biologists widely recognize this is not the case, but again, what is of interest here is to detect when we assume this tacitly, as well as the usefulness of invoking it explicitly.
4. Variation is, for practical purposes, unbounded. This doesn't mean that there are no biophysical limits to biological form. What it means, instead, is that the

> biophysical limits lie so far outside the range of functionally relevant variation that they can be safely ignored.

Referring to variation that is "continuous" in the sense explained here, Svensson (2023, p. 11) notes, correctly, that "no leading evolutionary biologist has such a naive view of unconstrained variation." But any time a biologist invokes natural selection without studying the range of variants that can be produced in development, they implicitly invoke continuous variation in the sense explained here. The traditional account of why giraffes have long necks makes tacit assumptions about what is and what is not possible in development. The story implies that shorter necks are developmentally possible in giraffes, and that these short-necked variants would have lower fitness compared to longer-necked conspecifics. This assumption certainly is plausible but what we want to notice here is that the assumption is tacit, that it is perfectly testable, and that it provides a useful starting point when made explicit, as treated in the next chapter.

Assuming continuous interindividual variation does make for a tractable first approximation to many biological situations. It allows researchers to explore the possible role of natural selection in shaping the distribution of traits within a population. In these scenarios, the space of possible variation is very wide, such that very different phenotypic combinations can be reached. Nothing in natural populations of wolves would make anyone suspect the disappointing fact that Chihuahua dogs are possible under selection. Looking at the tiny, inedible cob of the wild ancestor of maize does nothing to suggest giant, succulent corn on the cob that barely fits on a plate. Domestic dogs and corn certainly appear consistent with the idea of wide, essentially unbounded variation. Assuming continuous variation provides a reasonably plausible and certainly pragmatic starting point for empirical studies of these and many other cases.

While keeping the pragmatic value of the assumption in mind, this notion of continuous, unbounded variation is the main reaction point for adaptationist evo-devo. Being able to deploy paradigm unbounded variation scenarios is the essential foundation on which an adaptationist evo-devo perspective is built. So, never stop working on your standard adaptationist and continuous variation scenarios, in which essentially any point in morphological space can be reached, and in which morphological space is bound only by selection, even if you recognize the importance of pleiotropy, epistasis, or tradeoffs—even if you study them. If you are proficient in generating "continuous" scenarios, then this will situate you in the best position possible to see what the adaptationist evo-devo perspective brings to the table. The next section provides a proving ground for standard adaptationist scenarios.

1.11 Standard adaptation case studies

The point of the examples that follow is to sharpen your skills at generating standard adaptationist accounts. No matter what your actual beliefs are about the cause of a given pattern, you must be able to invoke the reasoning that *if a given pattern is the result of selection in the context of continuous variation*, then certain states of affairs involving variation, heritability, and fitness must be true. You might really think that, rather than being an adaptation, some other cause is responsible for the pattern in one or more of the examples that follow. Again, Fusco (2001) provides a menu of possible processes. You might have seen studies, or even followed long debates, that show that it is unlikely that adaptation is the main explanation for a particular observed pattern or phenomenon. However, that is not your assignment. Right now, you are tasked with being an expert in generating standard adaptationist scenarios. Your task, as a critical thinker and as an aspiring evo-devo adaptationist, is to answer the following question: if a given pattern is the result of natural selection, then what sorts of things must be true with regard to variation, heritability, and fitness?

1.11.1 Standard adaptation case studies: mammalian long bone-body mass relations

Variation in the proportions of certain bones across animal species has been noted for centuries. Galileo, in his *Dialogues Concerning Two New Sciences* even provided a drawing (Figure 1.1) (Galilei, 1638). He noted that (in the translation of Galileo, 1954, p. 131) "I have sketched a bone whose natural length has been increased three times and whose thickness has been multiplied until, for a correspondingly large animal, it would perform the same function that the small bone performs for its small animal."

By "perform the same function," Galileo is referring, over a hundred years before the term was coined, to what we now call mechanical stress: the ratio of body weight to the cross-sectional area of bone must remain constant as size increases. Because body mass scales with the cube of linear dimensions, an increase in length implies a much greater increase in mass. If bone proportions remain the same, meaning diameter increases linearly with mass, then bone area increases only with the square of length, while mass increases with the cube. This mismatch means that slightly larger animals would experience much higher mechanical stress on their bones. For stress to remain constant, bone area must also increase with the cube of length. This would require disproportionately thicker bones in larger animals, a point Galileo illustrated in Figure 1.1 by drawing an enlarged bone with dramatically increased girth.

Now, contrasting the real world with his imagined situation of constant stress, Galileo went on to say that "From the figures here shown you can see how out of proportion the enlarged bone appears. Clearly then if one wishes to maintain in a great giant the same proportion of limb as that found in an ordinary man he must either find a harder and stronger material for making the bones, or he must admit a diminution of strength in comparison with men of medium stature; for if his height be increased inordinately he will fall and be crushed under his own weight." In other words, in contrast to Galileo's postulated situation, as animals get larger, bones are increasingly thinner than expected given constant mechanical stress. Galileo made these breathtakingly perspicacious observations long before the key concepts of mechanics—stress, Young's modulus, strain (Gordon, 2003)—were formalized, to say nothing of evolutionary optimality models. His insights laid the foundation for centuries of thought about bones.

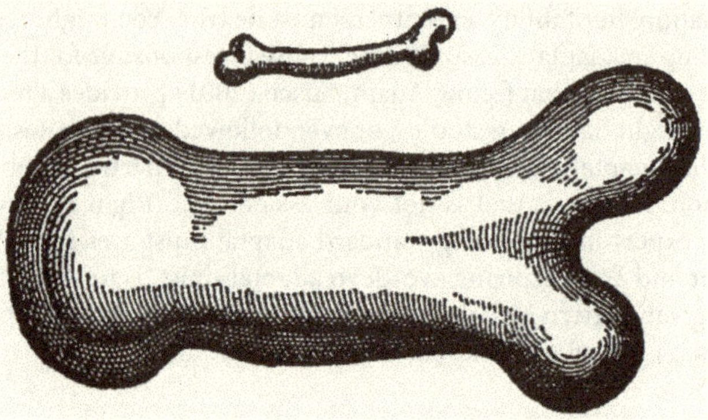

Figure 1.1 Galileo's counterfactual. In *Dialogues Concerning Two New Sciences*, Galileo drew two bones, the top one with natural proportions. In the second, he increased the length three times. If an animal's linear dimensions (length, height) are scaled up by a factor of three, its volume and mass increase by a factor of $3^3 = 27$. To maintain the same amount of body mass per unit cross-sectional area, the cross-sectional area of the bone must increase by a factor of $3^2 = 9$. The empirical reality that bone length-diameter proportions do not scale in this way indicates that selection maintaining "mechanical similarity" (constant structural load per unit area) is not the only factor of natural selection shaping bone length-diameter proportions.

Reproduced from Galilei, G. 1638. *Discorsi e Dimostrazioni Matematiche Intorno a Due Nuove Scienze.* Appresso gli Elsevirii, Leiden, archive.org/details/bub_gb_E9BhikF658wC/page/n141/mode/2up, public domain.

Much of this subsequent thought about bones is in the form of quantitative measurements within and across species. Christiansen (1999) provides a beautiful example from terrestrial mammals. Mostly based on bones of dead zoo animals in the Zoological Museum in Copenhagen, Christiansen's methods are arresting in their simplicity: "The long bones... from 98 specimens of extant terrestrial mammals, representing 79 species from seven orders, were measured using calipers and a measuring tape" (p. 169), yielding, among many other similarly trim graphs, Figure 1.2A, in which femur length scales beautifully—shockingly, perhaps, to biologists accustomed to noisier relationships—with body mass, surely one of the best bangs in statistical fit and biological import as you are going to find for a 98-specimen-measuring tape-and-caliper buck.

The fit of Figure 1.2A is even more remarkable when we contemplate both sample sizes and the phylogenetic breadth of his measurements. First of all, the sample sizes are, frankly, minuscule. The largest replicate Christiansen had was his sample of okapi, for which he had a whopping four specimens. He measured three jaguars, three Siberian tigers, and two samples of eight other species. The 68 remaining species, which included animals as different in humor as a koala, a thylacine, a springbok, a giraffe, a warthog, a cheetah, an otter, a coatimundi, a hyrax, a beaver, and some elephants and rhinos, were single-species measurements. Think of that—a smattering of measurements from boxes of dead zoo animals produced the fit in Figure 1.2A. Surely this is a biological monument, as striking a pattern as can be found in the natural

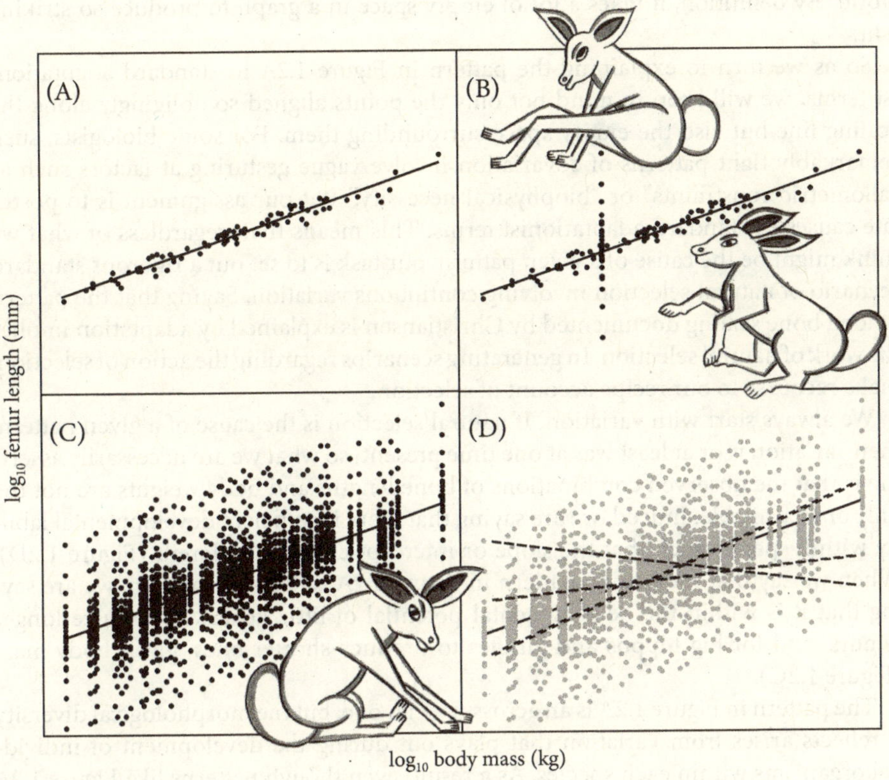

Figure 1.2 Long bone length-body mass scaling across terrestrial mammals. (A) Imagine that this graph is of femur length against body mass across terrestrial mammals. Each point in A is a mean value of a species. It shows a remarkable fit, and such a conspicuous pattern demands explanation. (B) If this pattern is the result of selection, then intraspecific developmental alternatives above and below the line must be, or have been, possible. Using the example of Eduardo's rabbits (Christiansen didn't measure rabbits, though), variants above the line would have longer-than-average femurs, and those below, shorter ones. The vertical spread of points denotes a range of heritable variants in femur length within a species—Eduardo's long-, normal-, and short-femured rabbits—approximating a normal distribution about each species mean. (C) If the y-intercept and slope of the line in A reflect selection acting across species, then such heritable variation in limb length given body size must be present in all the species. (D) If the pattern in A is due to selection, then this implies that developmental variation is sufficiently wide that functionally significantly different interspecific patterns are in principle possible—higher or lower slopes or even negative slopes (black dashed lines). All of these "must be" inferences are all entailed if the pattern in A is explained by a standard adaptation scenario.

Drawings courtesy of Eduardo Martínez Díaz.

world. By definition, it takes a lot of empty space in a graph to produce so striking a fit.

So as we turn to explaining the pattern in Figure 1.2A in standard adaptationist terms, we will keep in mind not only the points aligned so obligingly along the scaling line but also the empty space surrounding them. For some biologists, such remarkably tight patterns of covariation involve vague gesturing at factors such as "allometric constraints" or "biophysical necessity." But our assignment is to postulate causes in standard adaptationist terms. This means that, regardless of what we think might be the cause of a given pattern, our task is to set out a rigorous standard scenario of natural selection involving continuous variation. Saying that the pattern of long bone scaling documented by Christiansen is explained by adaptation implies the work of natural selection. In generating scenarios regarding the action of selection, make recourse to our recipe account of selection.

We always start with variation. If natural selection is the cause of a given pattern, then variation is or at least was at one time present, so what we are necessarily asserting is that the observed combinations of bone lengths and body weights are not the only ones available. Indeed, we are saying that there is sufficient developmental lability within species such that the slope or intercept could be different (Figure 1.2D). When we say that the scaling pattern in Figure 1.2A reflects adaptation, we are saying that it is within the developmental potential of little koalas to produce longer femurs, and for big hippos and giraffes to produce shorter ones, given body mass (Figure 1.2C).

The pattern in Figure 1.2A is an across-species one, but the morphological diversity it reflects arises from variation that plays out during the development of individual organisms within each species. As a result, even though patterns like Figure 1.2A are *across*-species, adaptationist interpretations of them are necessarily imbued with assumptions regarding what happens *within* species. Let's zoom in on the fennec fox, the smallest point in Figure 1.2A. If the observed trait values are the products of selection, then other, presumably less fit, variants are developmentally possible (Figure 1.2B). Now do this for each species (Figure 1.2C). The space potentially filled is impressively ample. If we pick values, one for each species, from our large cloud and then fit a line to them, we can generate slopes very different from the one Christiansen found, even, perhaps, converting a positive slope to a negative one (Figure 1.2D). By reaching the interspecific level via our assumptions regarding developmental potential in individual species or populations, we can see how our adaptive interpretations are ultimately based on assumptions regarding what happens at the intimate level of the seedpod or the litter, the potentiality in development inherent in the development of single individuals, and that variation is the ideal starting point for explanations of organismal form.

As if it weren't audacious enough to posit that, in every single one of the 79 species in Christiansen's graph, there is interindividual variation in bone length, our standard adaptationist account also makes assumptions about heritability. It postulates that within these populations in which bone length-body mass proportion varies among individuals, this variation is also heritable. In other words, fennecs that have relatively long femurs for a given body mass have litters of fennec puppies that tend to have relatively long femurs for a given body mass, and the same for those with relatively short femurs. It postulates the same thing for warthogs and viscachas and hippos and all the other species in the pattern. There is no escaping this assumption. If the pattern is explained by selection, and selection requires heritable intrapopulational variation, then we are forced to postulate heritable intrapopulational variation in all of the species making up the pattern. If all this seems entirely natural, then you are very well along in your development as a teller of standard adaptationist stories. If, on the other hand, this feels like an oddly specific and perhaps unwarranted scenario to infer from a simple pattern, then you're beginning to see the value in unpacking the assumptions baked into a standard adaptationist account.

Now that we have articulated the role of variation and heritability, we can turn to fitness. Saying that the interspecific scaling pattern is "due to selection" means not only that there is heritable variation in femur length-body mass relationships within all of the species studied (as in Figure 1.2C), but also that this variation is associated with differences in fitness. Let's take the components of fitness one by one. The bending moments experienced by bones are calculated as the fourth power of bone length times mass. Therefore, variation in femur length-body mass relationships could plausibly be associated with variation in survivorship. For a given body mass, even a slightly longer bone would experience much more bending. While a lengthy bone might hold up an animal's mass under static loading, as when it is just standing around, it would very likely fracture, in a small or a catastrophic way, under the dynamic loads of walking, running, or jumping that terrestrial mammals routinely subject their bones to.

As to survivorship, we can postulate that individuals with highly breakable bones would have a lower propensity to reach reproductive size as compared to those with less breakable ones. Perhaps individuals with broken bones are more likely to be eaten by predators or starve before they heal. So, we are postulating that there is some bone length given body mass that is "too long," that is, breaks or otherwise requires excessive metabolic attention as compared to shorter ones. If the pattern in Figure 1.2A is the product of selection, then we are also saying that intrapopulational variants with bones that are "too short" for a given body mass have lower survivorship (below the line in Figure 1.2C). These short-limbed variants would be resistant to bone breakage, but they would require large energetic investments for a given speed of movement. If a herd of a species were escaping from a predator or pursuing prey, and all were running at the same speed (let's make all the individuals of the same mass), we postulate that the long-boned individuals would suffer the most injuries involving leg bones, and the short-boned individuals would expend the most energy and have the lowest energy surpluses. Short-limbed variants might have low survivorship due to exhaustion of energy reserves, low investment in immune defenses, or even being eaten by predators if they cannot run fast enough. In between would be the favored long bone-body mass combination, just long enough that locomotion is relatively energy-efficient but short enough such that fractures are uncommon over the range of activities a given species engages in routinely. Standard adaptationist explanations involve considering similar examinations of mating success and fecundity, plausibly invoking the metabolic surpluses that characterize the favored variant (see Section 1.8).

When we make adaptationist interpretations, we are saying that the space surrounding the line is developmentally accessible and that the observed pattern is just one of many possibilities, culled from an expansive field of contenders. The example in Figure 1.2 is a pattern across many species, but our adaptationist interpretation of the interspecific pattern actually carries with it assumptions about what can happen in the development of individuals of each species. Perhaps you have other intuitions about what might cause the pattern in Figure 1.2, or perhaps you are convinced that the scenario just postulated is the correct one. Either way, the reasoning laid out is the structure of a standard adaptationist explanation for a pattern in nature, in this case, a striking one across species.

1.11.2 Standard adaptation case studies: the human female orgasm

Why women have orgasms is an excellent example of a trait that isn't explained very well by standard adaptationist accounts (Gould, 1993; Lloyd, 2005; Pavličev and Wagner, 2016; Wagner and Pavličev, 2017). Adaptationist accounts attempt to explain patterns in nature by appeal to the process of natural selection. The going explanation does involve natural selection, but the selection is not on orgasm in women. It seems that selection potently favors dependable ejaculation during sex in men because of its role in reproduction. This is the aspect of the explanation of the human female orgasm that involves standard natural selection. Given heritable variation among men, the men who most dependably ejaculate during sex would have the highest probability of fathering children and thus fecundity.

The aspect that is nonstandard is that men and women share so much of their development. Embryos begin relatively undifferentiated with regard to sex, and it isn't until the fetus is three months or so old that the penis and clitoris begin to show size differences. This shared developmental pathway apparently involves clitorises inheriting a lot of the characteristics of penises, among them, excitability, and under the right stimulation, producing an orgasmic response. This is nonstandard because it suggests that variation is in some way limited, that selection on male characteristics apparently drags female characteristics regardless of whether they are favored by selection or not in women. So, the human female orgasm appears to be a happy accident of the shared developmental pathway between males and females, and why men and women should share so much of our development remains at least in part fairly inscrutable and not entirely explained by standard adaptationist accounts (Lloyd, 2005; Pavličev and Wagner, 2016; Wagner and Pavličev, 2017; Barresi and Gilbert, 2023).

But such nonstandard explanations are not our mission now. Our mission is to produce a completely standard adaptationist account for why women have orgasms. The reason we do this is to sharpen our reflexes for generating hypotheses for empirical testing. In general, when confronted with a pattern in nature, it's almost always unclear whether the non-observed phenotypes are so because they are developmentally impossible for some reason, or because they are developmentally possible but not favored by selection. Inquiring about these two alternatives is a very useful place to start in empirical investigation of the cause of a given pattern (we end up with this conclusion in Chapter 6). In doing so, notice that, for any given pattern, it's possible to come up with "developmentally inaccessible" as well as "accessible but low fitness" scenarios.

Setting out these scenarios is essential to understand the contribution of adaptationist evo-devo approaches. The current example illustrates that it is possible to come up with a plausible adaptation hypothesis even in cases that we know perfectly well don't fit standard adaptation scenarios. Such exercises not only hone our ability to come up with directions for empirical work, but they also remind us that it's very easy to come up with plausible scenarios. Noticing how easy coming up with these scenarios is helps guard against accepting a given scenario just on the basis of its plausibility (see Section 1.5).

Coming up with adaptationist hypotheses for the human female orgasm is well-trodden territory, with biologists over the decades having been pretty insistent that such an arresting phenomenon should have *some* functional explanation. These range from "pair-bonding" accounts, in which orgasm induces languorous attachment in a couple, which then translates into better care for children. Other explanations include the entertainingly named "uterine upsuck" account, which postulates that orgasm is more likely to occur when mating with a male with good genes. Orgasm, under this account, induces negative pressure that draws sperm toward the ovum, making fertilization more likely. Another posits that pleasurable sensations in women mean that they are more likely to have sex more often, meaning they have higher mating success. Higher mating success means that the attribute of women having orgasms would be favored in humans. Another one is that muscular contractions during the female orgasm induce orgasm in men, increasing the probability of fertilization by preferred males. There is even an account that involves orgasm inducing abortion. Under this abortion-inducing orgasm hypothesis—take a breath before reading this convoluted scenario—a woman who was fertilized by one man but then had sex with a more charming lover and had an orgasm with this second man would spontaneously abort the previous man's fetus because the second man would be likely to kill it. These apparently are all real adaptationist hypotheses that have been proposed in the literature (Lloyd, 2005). So, our task is to develop one of these as a standard adaptationist explanation.

Let's use the uterine upsuck hypothesis, because of its appealingly biomechanical tone. First, we begin with variation. In this case, we postulate that orgasm is associated with negative pressure in the uterus. So, the variation in question is the presence or absence across women of an orgasm response, or some gradient of response. We then need to postulate that women with a certain response have daughters who tend to have that certain response too. Finally, we need to postulate fitness differences between variants. In this case, it is the negative pressure caused by the uterus during orgasm that draws sperm toward the ovum, promoting fertilization. In this way, orgasm-prone women should have higher fecundity than non-orgasm-prone women. Available data reject this hypothesis pretty conclusively, with not only there being little evidence of uptake of fluid during orgasm, but even it being possible that the uterus expels fluids during orgasm rather than producing negative pressure. Never mind—we have fulfilled our task, which is to generate a possible explanation for a trait in nature that would be the case *if the trait were the product of adaptation*.

1.11.3 Standard adaptation case studies: developmental bias

In ontogeny, some variants arise more readily than others, a tendency often referred to as "developmental bias" (Arthur, 2002, 2004; Salazar-Ciudad, 2006; Brigandt, 2020; Moczek, 2020; Lala *et al.*, 2024); while, like many biologists, your first inclination in explaining this tendency might be to appeal to some sort of non-adaptive "constraint" (see Section 1.4), our mission here is to provide a standard adaptationist hypothesis.

First, here is an example of the meaning of the term as intended for the purposes of the present exercise. In humans, there is a galaxy of developmental variants that can arise, but most are very rare (Wilder, 1908; Alberch, 1989; Blumberg, 2009). Humans can be born with three eyes, two noses, or two mouths (Figure 1.3A, B). They can be born with just one eye, and with a tubular proboscis above the eye rather than a conventional nose below the eyes. They can be born with two heads (Figure 1.3C, D). All of these possibilities are just as natural and just as much a part of the possibilities of our development as is the normal configuration of one head with two eyes, one nose, situated in the middle of the face, below the eyes, and one mouth.

Notice that I used the term "normal." The "normal" configuration is the common variant, the one usually observed in populations. "Normal" carries no normative

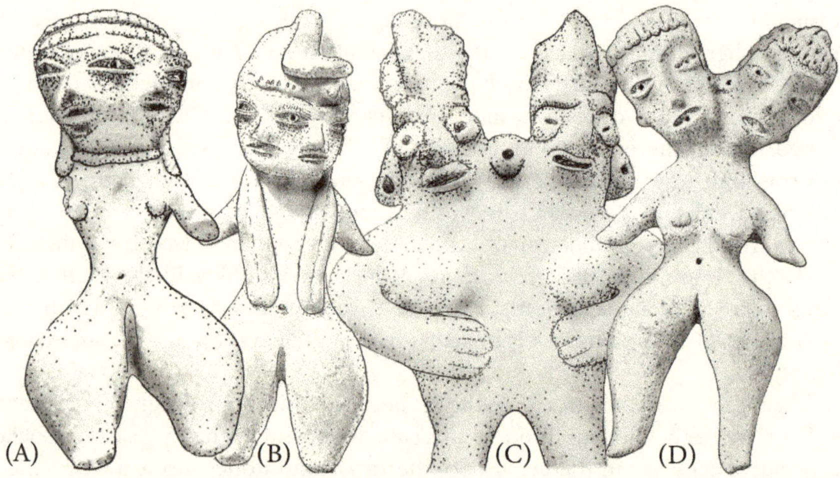

Figure 1.3 Diprosopia and dicephalus as natural facets of human developmental possibility. (A, B) The "three eyes-two noses" condition is known as diprosopia, a rare outcome involving duplication or abnormal expansion of craniofacial organizer regions during early development. It results in partial or complete duplication of facial features and has been observed in many vertebrates (Blumberg, 2009). (C, D) In contrast to diprosopia, which arises from a single embryo, dicephalus involves the incomplete separation of two embryos (a form of conjoined twinning). Like diprosopia, dicephalus also occurs across a wide range of vertebrates. According to Bendersky (2000), the earliest known depictions of both conditions come from the Tlatilco culture of the Valley of Mexico, dated to approximately 1200–700 BCE, as seen in these terracotta figurines, A and C from the Sala Preclásico of the Museo Nacional de Antropología, Mexico City, and B and D from the Museo Amparo, Puebla, Mexico.

weight; it doesn't mean good in some subjective sense, natural, or correct. It simply refers to the most frequent configuration that tends to arise under typical developmental conditions. This term highlights that, while development in our species and in all species can produce a multiplicity of configurations, the majority of these configurations are practically never seen. Instead, the most common outcome of development is a small subset of the possible, which we designate by calling it the "normal" range of variation. If development is perturbed by some challenge, and it always is, say by a temperature spike, an exposure to a toxin, or a small mutation in some genetic locus, the most usual outcome is that these challenges are buffered in such a way that more or less the normal configuration results (see Section 4.9). So, this is what is meant by developmental bias—a sweep of developmental potential in all species, but that only some variants tend to be produced, and when development is perturbed in some way, the normal variant or something close to it usually results. This is the phenomenon we want to try to explain with a standard adaptationist hypothesis.

We always start with a variation that is heritable. So, we must envision a population in which there is variation in the tendency to produce some developmental variants over others. To return to our example of developmental potential in humans, we posit the existence of some variants that produce three-eyed configurations as their normal outcome. Others under normal circumstances would produce the usual two-eyed configuration, but in these variants, development is easily pushed by temperature variation, exposure to toxins, or small mutations into three-eyed territory. In others, the normal two-eyed variant is reliably produced across a very wide range of developmental conditions. We then predict that these tendencies can be heritable. For instance, parents with the tendency for challenges in development to be buffered in such a way as to produce something like the normal configuration have offspring who also tend to do so.

With our picture of heritable variation among individuals within a population in hand, we can now turn to the possible differences in fitness among these individuals. In the matter of survivorship, the differences between typical and highly unusual phenotypes can be stark. For instance, in dicephalic animals, the two heads can disagree over who eats what or over which direction to move, an especially important disagreement when crossing a busy road (Blumberg, 2009). Fecundity may also be reduced, perhaps due to the metabolic costs of constant internal conflict. Individuals with three eyes might suffer from visual confusion if the brain lacks the circuitry to integrate the extra input, or from respiratory difficulties if that eye compresses two airways instead of one.

Our standard adaptationist account thus envisions the following. There is a sprawling topography of developmental possibilities within all species, including our own. Within this ocean of variation, some variants perform better in the contexts in which we usually find ourselves. The adaptationist view of developmental bias is that natural selection favors variants that tend to produce maximally functional configurations across as wide an array of developmental conditions as possible. Variants with the heritable tendency to produce high-performance configurations across a range of developmental conditions—configurations with the highest survivorship, mating success, and fecundity—will come to predominate over those that reliably result in lower-fitness outcomes, or whose development is more easily displaced into low-fitness regions of morphological space. So, mission accomplished—we have provided a plausible adaptationist hypothesis accounting for why development tends to result in the normal configuration and does not routinely produce the huge universe of outcomes that are just as possible and just as fundamental as the normal one.

1.11.4 Congratulations on expertly adaptationizing

All of these examples necessarily posited variation about the commonly observed variants. The long bone scaling example required positing long- and short-legged rabbits, and in all the other terrestrial mammal species, too. The example of the human female orgasm required positing variation in orgasm propensity connected to mating success or fecundity. The adaptationist explanation for developmental bias also required positing variation in developmental tendencies. Section 5.9 gets into more detail about what biologists mean when they say that a pattern is "explained by adaptation." For now, it's sufficient to note that an assumption of functionally significant variation around the commonly observed configuration is essential for a standard adaptationist account.

1.12 The benefits of being an expert adaptationist (and dangers of not being one)

You are now an expert in generating standard adaptationist explanations in terms of the fitness consequences of heritable intrapopulational variation. The most important aspect of being such an expert is that you can recognize tacit assumptions about variation where they arise, and deploy them on purpose when you want to. But there are many other benefits too, such as being able to recognize confusing language regarding adaptation and read between the lines to testable formulations in a multitude of situations. Here are some examples.

1.12.1 Macroevolution that isn't

The term "macroevolution" has been used in many ways over its long history, some uses of which helpfully highlight key biological processes, and others that masterfully obscure them. One such enduringly misapplied use of "macroevolutionary" is to describe standard comparative studies of convergent evolution (Olson, 2021). Convergent evolution denotes the situation in which natural selection favors similar forms under similar conditions, starting from significantly different ancestral states. Wings of flying animals are a classic example. In birds, bats, insects, and pterosaurs, they all permit (or permitted) sustained heavier-than-air powered flight, and they all were derived from very different ancestral states. As we will see in Chapter 5, convergent evolution across many species is one source of evidence regarding adaptation.

One expression of the study of convergence is to examine quantitative relationships between variables across many species. For example, Mincer and Russo (2020) examined the relationship between tail length and body length across over 1000 species spanning the mammal phylogeny. They found that tail length scales predictably with body length, with larger mammals having longer tails. Within this general pattern, they found that the intercept of the tail length-body length scaling line was higher in arboreal mammals as compared to terrestrial ones. In other words, taking body length into account, arboreal mammals tend to have longer tails than terrestrial ones. Mincer and Russo present a compelling argument, consistent with experimental evidence, to explain this pattern. Arboreal animals use their tails in balancing as they make their way through trees, and in some cases, have prehensile tails. In general, longer tails are more effective than shorter ones in these roles.

So, a standard natural selection scenario operating independently in each species is consistent with the striking pattern that Mincer and Russo describe. There is variation in tail lengths within a population of an arboreal mammal, with some being born with longer tails given body length than others. This variation is heritable, with relatively long-tailed individuals tending to have babies with longer tails. This heritable variation is associated with fitness differences. For example, survivorship is higher in variants with longer tails that balance better and fall from branches less. But even though this explanation invokes only standard population-level processes, the authors advertised their findings with the title of their paper, "Substrate use drives the macroevolution of mammalian tail length diversity." They say that the "goal of this study is to systematically evaluate the macroevolution of mammalian tail length diversity" (p. 2).

"Macroevolution" implies that some additional process is going on rather than just the usual population-level evolutionary ones producing heritable variation with fitness consequences. This is not to pick on Mincer and Russo—their usage is totally acceptable given the current parlance in which "macroevolution" makes fuzzy allusions to anything "large-scale." But if the biologists who use "macroevolution" in this way were clear on the process that they were invoking, they would simply couch their discussion in the vocabulary of the population-level process they postulate as producing the observed pattern. Calling a study of convergence "macroevolutionary" actually adds nothing. Cross out "macro," and Mincer and Russo's prose makes even more sense than it did before, for example, "[their results provide] compelling evidence that substrate use drives the evolution of mammalian tail length diversity" (p. 6). Because you are an expert at telling standard adaptationist stories, you can detect such scenarios even when authors write about them with unnecessary or confusing language, such as the trappings of "macroevolution" to refer to a study of standard population-level processes.

1.12.2 Biological metaphors: interspecific "competition," "stress," and environmental "filtering"

Now that you are an expert in standard adaptationist reasoning, you can recognize when authors are unclear regarding the causes that they are invoking, as reflected in their use of terminology that hides rather than makes clearer the causal content of their postulates. Use of "constrain" or "constraint" almost always signals the need for translating vague reasoning to clearer terms, and very often this clarification ends up taking the form of a standard adaptation scenario (Section 1.4). Another example is provided by some common metaphors. Key conceptual tools for many biologists are notions of interspecific "competition" or "facilitation," environmental "filtering," and "stress." These all often come down to standard adaptation scenarios. Three examples follow.

The notion of "competition" between species plays a very important role in ecology. For example, Keddy (2001), in his book *Competition*, states that "competition is one of the three fundamental forces (competition, predation, mutualism) that connects organisms in living systems" (p. 3). Twenty years after Keddy, the notion that species or even phylogenetic lineages "compete" with one another remains a central motivator for many studies. Condamine *et al.* (2020) speculate that not just species but "entire branches on the Tree of Life may actively compete for ecological dominance" (p. 28867). "Competition" vividly evokes the willful effort of vying parties to obtain some mutually exclusive objective. But species, to say nothing of "entire branches of the Tree of Life," certainly do not know that they are vying for the same resource. Therefore, it is important to ask what real biological process could underwrite reference to interspecific "competition." Interspecific "competition" can be reduced to a standard adaptation scenario, as follows. "The selective environment of species A consists of multiple factors. Which variants of species A come to predominate depends on these factors. One of these factors is species B." The metaphor might be useful, but being an expert adaptationist helps keep the bedrock processes it implies in view.

Another metaphor is "stress"; many biologists use the term, including some evolutionary biologists (Wagner *et al.*, 2019), and it is the centerpiece of the field known as "stress ecology" (Barrett *et al.*, 1976; Steinberg, 2011). "Stress," as in expressions such as "drought stress," "salinity stress," or "temperature stress," all evoke conditions that lower performance relative to other conditions, which are perceived as "optimal." From the outlook of agriculture or husbandry, "stress" makes perfect sense. You are a farmer who has purchased a variety of beans, which is advertised as yielding 6 tons per hectare. Any factor—lack of nutrients, drought, or herbivores—that lowers your yield can be called a "stress." But there isn't any optimum in nature, no baseline expected performance, just the conditions a population happens to experience at any given moment throughout its history (Anfodillo *et al.*, 2016). Like interspecific "competition," the term "stress" usually just rephrases a standard selection scenario: the selective environment varies across space and time; within any population, some variants survive or reproduce more under certain conditions; in some environments, very few, or none, do. Metaphors certainly have roles in painting broad-brush pictures of large-scale processes, but they also often hide the bedrock processes involved; an expert adaptationist lens helps clarify some of what the vague metaphor might imply.

A final vague metaphor, very important in some areas of biology, is the notion of environmental "filtering." Environmental filtering evokes a series of factors that prevent or allow some species but not others to arrive and persist in a given area. Environmental filters include factors such as temperature or salinity. The species that can reach a given spot but can't tolerate the conditions there have thus been filtered out of the local species pool. On the other hand, those that can tolerate the conditions at a given spot have made it through the environmental filter. The notion of an environmental "filter" evokes factors that keep some species out while letting others in from a regional species pool. There obviously isn't a literal filter somewhere keeping some species out and letting others through. If we want to be maximally clear regarding the processes that we think might be shaping the species pool at a given site, then we need to ask what real biological process might underlie environmental "filtering." Here again, in many cases, environmental "filtering" can be rephrased as a standard natural selection scenario: "the conditions of natural selection acting on species A are such that no variants are able to survive and reproduce at a given site. Some individuals in species B survive and reproduce there." Being able to deploy standard selection scenarios helps clarify specific predictions that many terms might otherwise obscure, surely an ability that is helpful.

1.13 Helpfulness is a helpful criterion

The standard adaptationist scenario is a very useful place to start even if no one at all claims to be a standard adaptationist. What matters is that often in practice it is as though we were, when we fail to test the implicit assumption that developmental alternatives to the commonly observed variants are possible. It is even a useful starting point in many cases. We are now going to look at the reasons that so many biologists criticize the standard adaptationist view or find it incomplete, and after that we will look at the empirical response that biologists have developed in the form of adaptationist evo-devo. We can arrive at adaptationist evo-devo from many starting points. Standard adaptationism is not the only or necessary starting point. But because it starts with reasoning and practice that is familiar to most biologists, it is a particularly pragmatic one to start with. Also, because standard adaptationism was so demonized in the 1970s and 1980s, thinking in explicitly adaptationist terms went underground to an extent. Today, in many circles, the ability to apply adaptationist reasoning without embarrassment, as we did in exercises on long bone scaling, the human female orgasm, and developmental bias, has become something of a lost art. If those exercises gave you pause, if you felt any resistance to explaining the pattern strictly with reference to selection, or found it difficult to do so, then you've already encountered what I mean.

And if you didn't have any trouble at all, and you're skeptical that adaptationist reasoning is in decline among biologists, consider the confusing and inconsistent ways in which terms like "constraint" (Section 1.4), "macroevolution" (Section 1.12.1), and "stress" and other metaphors are deployed (Section 1.12.2), often as veiled references to selection itself. That confusion is hard to explain if clear adaptationist thinking were widespread and fluent. To build robust explanations of organismal form, you need to be able to deploy such explanations knowingly and easily. Being able to navigate tricky concepts such as those above, and to make our causal claims as clear and explicit as possible, is one of the great strengths of being a skilled adaptationist. So, with standard adaptationism as our starting point, let's now see where the empirical realities are that motivate adaptationist evo-devo.

2
Developmental potential and adaptation

Whether or not, in one's calmest moments, anyone adheres fully to the standard adaptationist view, it is the perfect foil for motivating adaptationist evo-devo. This is because the heart of any adaptationist explanation is variation, and it is precisely variation that is the concern of adaptationist evo-devo. Maybe no one, when asked to reflect carefully about intrapopulational variation, will cleave with conviction to the notion of continuous variation described in Section 1.10. But every time we build an explanation of organismal form *without* contemplating what is possible and what is not in development, we implicitly assume that variation is continuous.

In motivating studies of development, or collaboration with those who do, it is helpful to have examples of exceptions to the standard account to point to. So, we will examine some case studies here, starting with now classic cases that have served as foundational inspiration for adaptationist evo-devo practitioners. If the state of affairs in one species does not fit the standard adaptationist account, then it's hard to justify assuming that the standard account necessarily obtains in any other species. Then, we'll see that such examples abound in the natural world, in every clade and species. First, let's turn to the notion of variation that is limited, as opposed to continuous in the sense discussed earlier.

2.1 Motivating adaptationist evo-devo: limited variation

Biologists are concerned to a surprising degree with limited variation. We often make a big deal about studying the "grand variety" of living things, the "stunning diversity" of form, color, sound, and function from the deep sea to coral reefs, to deserts and rainforests and in the sky. And yet, biologists trying to explain why organisms take the forms that they do actually spend most of their time studying the factors that *restrict* variation. This is obvious when we look at a study of natural selection. Of the wide array of variants that arise in populations, only some are favored and thus commonly observed. In the example of the remarkable positive scaling across terrestrial mammal species in long bone length-body mass proportions (Section 1.11.1), appeal to selection implies that the variants above the commonly observed scaling line can be, or at least were at one time, produced in populations. These variants have bones that are "too long" given body mass (i.e. very thin), breaking more readily than the commonly observed variants. Those below the line, with bones that are "too short" given body mass (i.e. very thick), are breakage-resistant but require more energy to produce and move around than the common variants. So, a standard adaptationist explanation requires that there be a sufficient number of alternative variants such that the

Adaptationist Evo-Devo. Mark E. Olson, Oxford University Press. © Oxford University Press (2026).
DOI: 10.1093/9780197831533.003.0003

question of "why is this variant commonly observed rather than the large universe of possible alternative variants?" is reasonably satisfyingly answered by "the action of selection" (more on this in Section 5.9).

However, if one or both of the spaces above or below the scaling line are impossible to produce, then the explanation is modified. In that case, the answer to "why is this variant commonly observed rather than the large universe of possible alternative variants?" would clearly need to include not just selection but also the factors that cause the impossibility of producing variants that would fall in the empty space close to the scaling line.

This way of looking at explanations of form is a great place to start because it throws into sharp relief the need to study what is possible and what is not in development. Putting things this way might make it seem as though there are two distinct situations—ones in which selection is an explanation and ones in which it isn't—but we'll put this apparent dichotomy into perspective as we go along. To get started, let's now turn to studies that showcase ways that researchers investigate the accessibility of "empty space." Notice that in all of these examples, functional considerations provide the relevant reference point, illustrating why the standard adaptationist explanation is such a valuable point of departure.

2.1.1 Examples of limited variation: Always odd segment number in geophilomorph centipedes

Segment number in geophilomorph centipedes is a celebrated example motivating biologists to take developmental possibilities into account in their studies of the causes of form. Geophilomorpha are an order of centipedes which Bonato *et al.* (2014, p. 485) call "the major predators in the litter and soil interstitial environment throughout the world," a wonderfully grand role in a habitat I hadn't realized had majors. Across the more than 1,000 or so species of geophilomorphs, segment number varies across species from 27 to 191 (Leśniewska *et al.*, 2009). There are some species in which segment number even varies among individuals within a species. What is conspicuous about this pattern is that segment number is invariably odd. The exception is a single specimen of *Stigmatogaster subterranea*, an apparently unique developmental outcome whose cause is unknown. In all other cases, even across the members of the same population, no individuals with even segment numbers are ever produced. The starting point of a standard adaptationist explanation shows why this example is so conspicuous.

This example is conspicuous not in isolation but with reference to the contrast it makes with a standard adaptationist explanation. As in a standard adaptationist explanation, the geophilomorph example involves a wide space of developmental possibilities that plausibly should be producible. In the long bones example in Section 1.11.1, we observe certain proportionalities in extant species, and plausibly expect that bones that are a bit too long or too short for a given body mass can be produced, and that if and when they are produced, their bearers have lower fitness than the common variants. In the geophilomorph example, if individuals within a species can be observed with 97, 99, and 101 segments, there is nothing to make us think that individuals with 98 or 100 segments shouldn't be just as frequent. The long-bone scaling example is one of differential survivorship and reproduction as a result of the biophysical consequences of variation on organismal performance, that is, different variants having different propensities to breakage or different metabolic costs for the same speed of locomotion. An analogous situation for the geophilomorph centipedes would be one in which even or odd segment numbers have strong effects on locomotion and thus survivorship or reproduction.

The striking contrast with the standard adaptationist explanation is that no biologist has been able to propose a plausible hypothesis that would account for why selection would favor, say, 99 or 101 segments but not 100, in fossorial, long, slender, nearly wormlike animals that move via rhythmic waves of limb motion along the body axis (notwithstanding the joke of the centipede with the loose shoelace tripping, and tripping, and tripping...) (Arthur and Farrow, 1999; Leśniewska *et al.*, 2009; Minelli, 2009). Odd segment numbers in geophilomorph centipedes is a celebrated case study not because it obviously supports adaptationist reasoning, but because it firmly resists it. It's hard to imagine how the absence of even-numbered individuals could reflect selection on organismal performance (though see Arthur, 2002). There's no compelling reason to expect that locomotion would differ so dramatically between, say, 98 and 99 segments that selection alone would explain the consistent absence of even-numbered individuals. Segment number in these centipedes offers a clear case of conspicuously limited variation for which no selective explanation has been offered.

Instead, the particulars of how segmentation is achieved in development in these animals make an explanation that appeals to evolutionary contingency seem more satisfactory (Minelli, 2009). "Evolutionary contingency" in this sense highlights traits that become fixed not because the fittest variants were favored from among a wide field of developmental contenders, but because of some historical accident that left no alternatives. It might be imagined that a growing baby centipede would add segments one after the other as it grows in length. Along the way, it would traverse odd and even numbers on the way to its mature segment number. If this were the case, then it would be possible for segment production to cease at either odd or even numbers. But this isn't the way they form. Instead, segmentation in geophilomorphs appears to arise from standing waves of signaling molecules that form stable, periodic patterns in the embryo. The peaks and troughs of these waves determine where segments will form, setting up the repeated units along the body axis (Chipman *et al.*, 2004). Computer simulations of reaction-diffusion systems show that in such systems, even-numbered segment counts are entirely possible given the right conditions, so nature isn't lacking the means. The conclusion scientists tend to come to is that in this clade, only odd segment numbers are observed not because of selection weeding out even-numbered segment variants, but almost certainly because these variants never arose. If in cases like this the "empty spaces" in phenotype space so clearly betoken structured developmental variation outside the standard adaptationist framing, then it is entirely reasonable to ask, in any given case, whether unobserved variants are absent because selection eliminated them or because they are developmentally impossible.

2.1.2 Examples of limited variation: The pentadactyl hand

Another conspicuous example of limited variation is the pentadactyl or five-fingered hand, which characterizes the tetrapods (Held, Jr, 2009). Across groups as diverse as amphibians, "reptiles," and mammals, the production of five digits is remarkably constant across extant tetrapods, even as the pathways for digit development vary (there are at least four) (Wagner and Chiu, 2003). Except for teratologies, apparent deviations usually represent predictable loss of digits from a pentadactyl condition rather than true departures (Alberch and Gale, 1983; Alberch, 1989). The remarkable constancy in digit number across vertebrates seems unaccountable, given the diversity of functional contexts across which pentadactyly persists.

The remarkably low lability in digit number across the tetrapods is striking because of the variation in functional contexts across which it occurs. If all tetrapods had exactly the same lifestyle, then variation in digit number would be less conspicuous. Uniformity of lifestyle suggests uniformity of selective conditions, and thus little reason to expect divergence in morphology and performance. But tetrapods are famously varied in the uses they put their extremities to, from the flippers of whales or sea turtles to the hooves of horses, the wings of birds and bats, to clever-handed raccoons and primates.

Amid this variety, it is very hard to come up with a plausible standard adaptationist explanation that involves selection favoring pentadactyly consistently across the exigencies of the different lifestyles exhibited by tetrapods. Such an explanation would imply that variants with basic plans built on three or four or seven, or eight digits within a population would have lower fitness than those with a 5-digit plan (the "plan" qualification is added here because some tetrapods produce fewer than five digits but only as a modification of the 5-digit plan). At its most basic, such a scenario would show that horses, say, that develop their giant hoof starting from a 1-digit plan should have lower fitness than actual horses, which produce the single hoof as a modification of a 5-digit plan. These differences would be the result of, for example, the locomotory or other aspects of the performance of horse feet. Likewise, such a scenario would show that whales with four or six digits suffer lower survivorship or reproduction as compared to those with five.

Pentadactyly is resistant to a standard adaptationist explanation because it would seem inevitable that over the enormous evolutionary sweep of the tetrapods, if selectable variation in the basic plan were possible, then some species would, in some context, deviate significantly from it. This lack across the immense stretch of time over which tetrapod lineages have been diverging from one another leads biologists to look beyond standard adaptationist explanations. Even more than the case of the geophilomorph centipedes, the five-fingered hand provides a compelling reason to look beyond standard adaptationist explanations. Whereas no one has been able to think of a reason why an odd segment number should be better for locomotion than an even number, it stands to reason that under at least some conditions a higher number of digits would be favored over just five if the variation were present. Instead, some other process would seem to be at hand, some contingent, difficult-to-escape aspect of development not forged by selection independently favoring exactly the same digit number in all tetrapods.

One possible explanation for the absence of five digits in extant tetrapods is that, for some reason, biological systems are incapable of producing more. Cetaceans might seem to support this view: their flippers enclose concealed digits, so why not six or seven? It's hard to see why selection would specifically favor five in such a context. Ichthyosaurs, with their similarly flipper-like limbs, often had more digits. Their example shows clearly that additional digits are not only developmentally possible but can function effectively in comparable aquatic lifestyles. Early tetrapods such as *Acanthostega, Ichthyostega*, and *Tulerpeton* had more than five digits (Coates and Clack, 1990; Lebedev and Coates, 1995), showing that the potential for producing extra digits existed in the early history of the group. That no living tetrapods show this pattern suggests not that tetrapod tissues are incapable of producing more digits, but a contingent fixation of the pentadactyl plan. Again, "contingent" is the scientific term to express that evolution sometimes proceeds in a way that reflects historical accident rather than selection freely favoring from among a very wide array of possibilities. So, the pentadactyl hand is thus a striking example of a morphological feature that has changed little over hundreds of millions of years, even though selection would certainly favor, in at least some situations, a deviation from it if only it were possible—another example of a case in which a standard adaptationist explanation seems unfit.

2.1.3 Examples of limited variation: The mystery of synapomorphy

Stipules are modest, often tiny, paired projections at the bases of petioles, the stalks of leaves in flowering plants, and the coffee family (Rubiaceae) is characterized by a special stipular configuration. In Rubiaceae, one stipule that corresponds to one leaf is fused with the stipule of the other leaf at the same node, leaves being borne opposite to one another in Rubiaceae (rather than spirally along the stem, the most common arrangement). At each leaf pair in Rubiaceae, there are thus what appear to be two stipules instead of the usual four. They usually appear as small triangles that stretch from petiole base to petiole base, and so are called interpetiolar stipules (Figure 2.1). They are characteristic of the family, which is useful given how many Rubiaceae species there are, with the interpetiolar stipule greatly aiding in recognizing whether a plant belongs to the family. Ye and Ronse de Craene (2024) go so far as to call the interpetiolar stipule in Rubiaceae "iconic," a superlative probably never before lavished on such modest botanical parts.

The constancy of presence of the interpetiolar stipule commands attention, again because of reference to adaptationist considerations. There are over 13,000 species of Rubiaceae (Wikström *et al.*, 2015) and most of them have the interpetiolar stipule. Moreover, Rubiaceae span most of the world's terrestrial habitats (Figure 2.1A), on all continents but Antarctica. Probably the most common habit is a rainforest understory shrub or small tree, but some are towering rainforest trees. Other Rubiaceae are climbing lianas; thick-trunked palm-like species; tropical rheophytes (plants that live in streambeds, Figure 2.1B); dryland shrubs (Figure 2.1C); bizarre epiphytes growing high in the branches of mangroves, housing symbiotic ants in their swollen bases (Figure 2.1D); and tiny creeping herbs in cold cloud forests (Figure 2.1E). The presence of the interpetiolar stipule is a remarkable constancy given the extraordinary ecological and morphological diversity encompassed by the family. Rubiaceae species started diverging from one another perhaps 90 million years ago (Wikström *et al.*, 2015), so there has been plenty of time for loss or transition into unrecognizability of the stipule.

Stipules are often assumed to have a role in protecting developing buds (Lubbock, 1891). In some Rubiaceae, the stipules do envelop the twig tip and clearly seem to protect the young leaves (Figure 2.1F), or resemble leaves and contribute to photosynthesis (Figure 2.1G). But in practically all species (as in Figure 2.1B–E), the stipules are small and often just fall off shortly after they are produced. In many of these species, the inner face of the stipule is lined with glands that produce a gummy, polysaccharide- and protein-bearing secretion that also can contain antimicrobial or antifungal substances (Thomas and Dave, 1990; Klein, 2004; Muravnik *et al.*, 2014). Whether they are persistent or quickly fall off, they are reliably present near the stem tips. An emerging leaf is a very tender structure, not yet equipped with toughened cell walls, pigments that protect from the sun, or defensive prickles, hairs, or secretions. So, the reliable presence of stipules at the stem tip, and that they might secrete substances that plausibly protect vulnerable young tissues, is consistent with a protective function. But pause for a moment to reflect on what it truly means to explain the presence of stipules by attributing them to their protective function.

Explaining the presence of the interpetiolar stipule in Rubiaceae by virtue of their protective function would require the following. It would require that, in each current generation of each Rubiaceae species, there is relevant heritable variation, such as interpetiolar stipule presence or absence. It would require that this variation be heritable. This heritable variation would need to have fitness consequences such that the individuals within a Rubiaceae species lacking interpetiolar stipules would have lower fitness than conspecifics that have them. We would need to tell such stories for every species of Rubiaceae of every habit and habitat—that somehow these tiny structures are favored over their absence in mangrove-dwelling ant plants, climbing lianas, desert shrubs, tiny moisture-loving herbs, and rainforest trees, in all of the over 13,000 species of Rubiaceae.

If the interpetiolar stipule is indeed favored across Rubiaceae by virtue of its protective role, then the consistency of selection across such an ecologically and geographically diverse family is at the very least striking and in need of serious documentation. A common hypothesis is that stipules are favored because they secrete a protective gum that shields emerging leaves from desiccation, herbivores, pathogens, ultraviolet radiation, and other environmental threats. But surely these threats vary widely, from the cold, damp hollows inhabited by *Nertera* (Figure 2.1E), to the parched lava flows of *Bouvardia* (Figure 2.1C), or the salt-saturated mangrove bark that *Myrmecodia* clings to (Figure 2.1D). Such contrasts in humidity, temperature, light, herbivores, and pathogens raise doubts about whether qualitatively similar selective advantages could obtain with sufficiently similar intensity across all these species to yield the same evolutionary outcome of interpetiolar stipule presence. If stipule traits vary heritably, whether in size, presence, or secretory function, then it would seem plausible that many populations should face conditions in which the trait is selectively neutral or even disadvantageous, or under which different configurations would be favored. Species of other families that co-occur with Rubiaceae often lack

Figure 2.1 The mystery of synapomorphy: the interpetiolar stipule. The interpetiolar stipule of the coffee family Rubiaceae is usually a small triangular structure between neighboring leaf bases. (A) Across Rubiaceae's globe-spanning range (gray areas on the map), the stipule is found in thousands of species, in tiny herbs, desert shrubs, rainforest trees, long lianas, and in epiphytic ant-plants. (B) *Augusta rivalis* is a neotropical rheophyte, a plant that grows in riverbeds and withstands flowing water. (C) *Bouvardia ternata* is a straggly dryland shrublet, seen here growing on a recent lava flow in central Mexico. (D) *Myrmecodia beccari* is an epiphyte, here clinging to a mangrove in Queensland, Australia. Its prickly, swollen base is filled with tunnels inhabited by ants. (E) *Nertera* is a tiny, ground-hugging creeping herb that grows in cool, moist places around the Pacific, as here on the South Island of New Zealand. In some species, the stipule appears to have a conspicuous function, as in (F) *Nauclea orientalis* in a tropical savannah in the Northern Territory of Australia, where stipules envelop and protect the developing leaves. (G) In *Galium*, a common herb in the north temperate zone, the stipules are leaf-like and participate in photosynthesis. No biologist has ever proposed a scenario in which the interpetiolar stipule is favored by selection in all of these situations.

stipules altogether, and of those that do have stipules, none possess the distinctive interpetiolar stipules characteristic of the family, underscoring that such structures are not essential for survival in Rubiaceae habitats.

Selection-based scenarios that demand convergence so astonishing as to border on the implausible inspire biologists to wonder whether other processes could be involved in the patterns at hand. When confronted with such remarkable constancy, many biologists entertain what Arthur and Farrow (1999) describe as "the non-production of variant phenotypes caused by the nature of the developmental system." Often, these strangely invariant traits are synapomorphies.

The interpetiolar stipule is a synapomorphy of the coffee family; because biologists use synapomorphies in routine daily work, we often forget the profound mysteries that they usually represent. A synapomorphy is a feature that arose once in the history of life and uniquely characterizes a clade (Patterson, 1982). Many synapomorphies are shared by all extant members of a clade. Mammalian-type hair uniquely characterizes the mammals, and all mammals have hair. Flowers uniquely characterize the angiosperms, all of which produce flowers. Feathers uniquely characterize the birds, and all birds have feathers. The list, of course, goes on. It goes on because every single clade of any phylogeny at any scale has features that uniquely characterize it.

If we ask, "why is a feature produced recognizably in all the members of this clade?" we usually have no idea. In some cases, standard selection accounts undoubtedly apply. For example, both feathers in birds and leaves in Euphyllophytes (ferns and seed plants) are synapomorphies likely maintained by selection rather than by any sort of developmental factor impeding their loss. Feathers, essential for thermoregulation, flight, and display, are universal in birds, yet their absence on certain body parts, such as most bird legs or on the necks of vultures, shows that featherlessness is not only developmentally possible but can be favored by selection under some conditions. Similarly, leaves, critical for photosynthesis, are almost universally present in euphyllophytes. However, leafless plants, like cacti or many parasitic plants, illustrate that leaves can be lost under certain conditions of selection. In both cases, the widespread presence of these structures is attributable to selection favoring their presence, not to any developmental impossibility of producing organismal bodies without them.

But in the case of many synapomorphies, it is hard to imagine that standard selectionist scenarios are at work. Very few such synapomorphies have been subject to much study, so it's often anyone's guess whether or not they really do vary heritably within populations and whether the commonly observed variants are common because they have higher fitness than the other possibilities. Because of their constancy, as in the stipules of Rubiaceae, they offer fertile ground for exploring how biological form is shaped by the limits and possibilities of development (Wagner, 1989).

2.1.4 Examples of limited variation: Tradeoffs

Resource allocation tradeoffs are a major factor that limits the variation that can be produced in development. A tradeoff is said to occur when a limited resource is allocated to two or more functions (Garland *et al.*, 2022). Allocating to one necessarily reduces the allocation to the other. As a result, simultaneous maximization of both is impossible, and these zones of simultaneous maximization correspond to empty developmental space.

An example is the apparent tradeoff between storage and mechanical support in woody plants (Figure 2.2) (Pratt and Jacobsen, 2017). In the wood of trees, storage of water and metabolites occurs in cell lumina whereas support is chiefly provided by cell walls. Thicker walls leave less lumen volume but provide greater support. Therefore, a plant stem that maximizes storage will be made up of very flexible tissue. One that has very stiff tissue will have very low storage capacity (Rosell and Olson, 2014; Pratt and Jacobsen, 2017; Chen *et al.*, 2020). Because a given volume of tissue can either be allocated to storage or support, simultaneous maximization of both functions is impossible. Therefore, there is a conspicuous negative relationship between storage and support and a patent empty area of developmental space corresponding to combinations that would simultaneously maximize support and storage. The opposite combination, of low mechanical support and low storage, would seem to be entirely possible developmentally but plants don't often fall in this area of morphospace. Instead, plants tend to fall along a marked negative relationship, presumably delimited above by ontogenetic inaccessibility (high storage and high mechanical stiffness), and below by selection eliminating lower values (low storage and low mechanical stiffness) (Rosell *et al.*, 2012). If, for the same metabolic cost, plant tissue could simultaneously exhibit maximal storage and mechanical support, it is difficult to conceive of a situation in which this tradeoff-escaping variant would not be favored by selection. Because it so restricts the possibilities of selection, the tradeoff must be considered in any explanation of why a plant exhibits a given combination of storage and mechanical support.

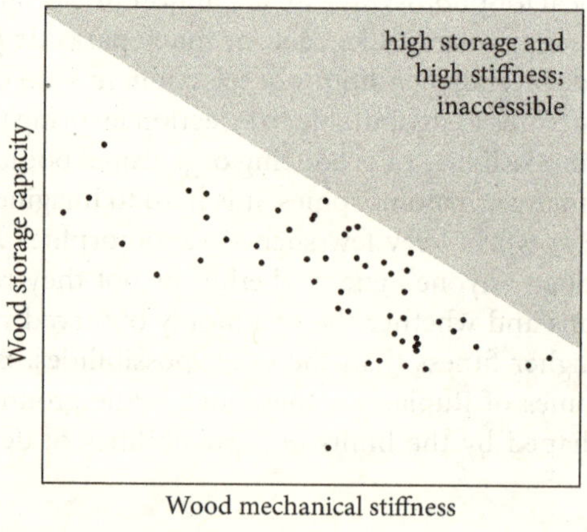

Figure 2.2 The support-storage tradeoff in the wood of trees. Mechanical support in wood is provided by stiff cell walls. Storage of water and nutrients occurs in cell lumina. A given volume can be filled by lumen or cell wall, but not both. As a result, the combination of maximal storage and maximal mechanical support is developmentally impossible, but would certainly be favored by selection if it were possible. The points are species means. "Mechanical support" was measured as Young's modulus, and "Storage" as volume per gram of wood, with data from Olson *et al.* (2018).

The same is true for all tradeoffs. Consider the tradeoff between many small propagules with low survivorship and few large propagules with high survivorship. Imagine a population of a small-propagule species, say a sea turtle. Now say that some variant arises that, for the same metabolic cost, somehow produces the same number of propagules as its conspecifics, but instead of small ones of low survivorship, produces the same number but of large ones with high survivorship. This variant would inevitably come to prevail within its population. Tradeoffs are therefore ready examples of limited variation that don't entirely lend themselves to standard adaptationist explanations.

2.1.5 Examples of limited variation: Shared developmental cascades

Another kind of tradeoff or pleiotropy in a broad sense can be called "shared developmental cascades." Shared developmental cascades are situations in which two or more features share so much of their development that the space of options for their mutual divergence is limited. The wood of trees again provides an example, in cells called fibers and vessel elements (Figure 2.3). Angiosperms (flowering plants, as distinct

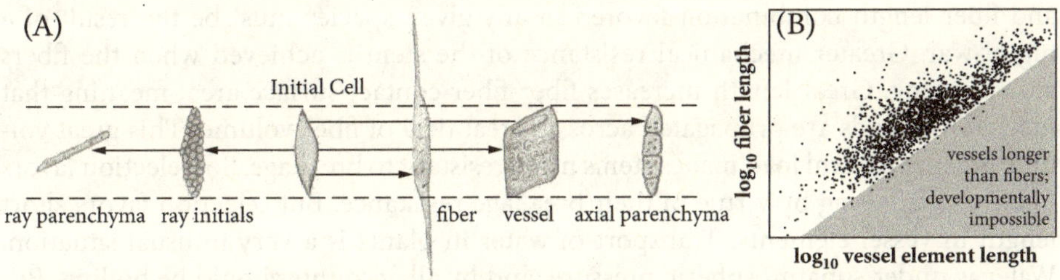

Figure 2.3 Shared developmental cascades in the cells of wood. Parts that diverge from one another later in ontogeny often share more characteristics, whereas those that diverge earlier share fewer. Such "shared developmental cascades" are a major source of similarity between traits in the natural world. (A) The wood of trees provides a good example. The wood of trees is made up of a few different cell types, all of which are derived from the same embryonic cell type, the fusiform cambial initial ("Initial Cell" in the figure). The dashed lines show the size of the initial cell relative to the mature sizes of the other cell types. While cells can grow taller than the initial cells, as fibers do, they do not grow shorter. (B) There is therefore an inevitable positive relationship between the sizes of wood cells, such as the relationship between fiber length and vessel element length. Because all of the cell types start life at the same length, the evolutionary possibilities for divergence in cell length are limited. For example, even if selection were to favor fibers that are shorter than the vessel elements, such a condition cannot be produced in development because their minimum sizes are set when they are cut off from the vascular cambium initial cells. As a result, the space below the line is inaccessible developmentally. Across living things, developmental pathways that are shared between parts to one degree or another limit the possibilities of natural selection because some developmental outcomes are impossible. Data from Echeverría *et al.* (2023).

from conifers) conduct water in long tubes called vessels. Vessels are made up of many can-shaped, hollow, dead cells stacked end-to-end, forming tubes. These individual cells that are the basic elements of vessels are called, appropriately, vessel elements. In addition to these wide, hollow, conductive cells, trees have long, slender support cells with narrow lumina and thick walls. For our discussion, we can refer to these as fibers. Vessel elements and fibers share practically their entire developmental pathways. The cells of the wood are produced from the vascular cambium, an embryonic layer of cells just below the bark. One type of embryonic cell in the vascular cambium produces both vessel elements and fibers (Figure 2.3A). Because they are produced from the same vascular cambium cells, both vessel elements and fibers start their lives at the same length (Montes-Cartas *et al.*, 2017), but a key difference separates them. Vessel elements do not elongate in ontogeny beyond the heights of the initial cells that give rise to them. Fibers, however, elongate extensively, their long tails growing interdigitated among one another. Because they start at the same length, and because vessel elements do not elongate in ontogeny, the combination "vessel elements longer than the fibers" is developmentally inaccessible (Figure 2.3B). Therefore, even if selection were to favor fibers shorter than the vessel elements, such a condition is developmentally impossible given their shared developmental cascades.

Moreover, their shared ontogenetic origin limits the mutual developmental independence between fibers and vessel elements, such that the particular vessel element and fiber length combination favored in any given species must be the result of a tug-of-war. Greater mechanical resistance of the stem is achieved when the fibers are very long. Great length increases fiber-fiber contact surface area, meaning that mechanical loads are propagated across a great deal of fiber volume. This great volume per mechanical load makes stems more resistant to breakage. So, selection favors fibers that are long by virtue of their breakage resistance. But selection favors short length in vessel elements. Transport of water in plants is a very unusual situation. Water is under subatmospheric pressure, and by all accounts should be boiling. But it is confined in the tiny, membrane-bound vessels, keeping it liquid in a precarious metastable state. A key selection pressure in woody plants is avoidance of vessel collapse under negative pressures. Vessel elements have annuli at each end, which help in resisting deformation (Echeverría *et al.*, 2023). The shorter the vessel elements, the more deformation-resisting annuli there are per unit vessel length. In this tug-of-war, selection favors shorter vessel elements in conferring resistance to implosion and longer fibers in resisting bending of the stem. In any given species, selection favoring longer fibers will at some point be opposed by selection favoring shorter vessel elements, and vice versa. This is an excellent example of a situation in which only the study of development reveals the space of possible configurations that are available for selection to favor.

2.2 Developmental potential (evolvability)

The foregoing examples of limited variation can be seen as examples of a category we can call "plausible but unobserved configurations." In the case of centipedes, if there are odd numbers of segments, and because there is variation in segment number between individuals within a population, it would seem reasonable to expect individuals with even numbers of segments. By the same token, given that extant tetrapods can have one to five digits, and some extinct ones had more than five, it seems reasonable to expect that in at least some extant species there would be more than five digits, but this doesn't seem to be the case either. How do scientists identify these holes in the space of imaginable configurations?

They do so by mapping developmental potential. Developmental potential is the sum of all of the configurations that a given individual, species, or clade can produce, from the commonly observed to the rare but still a part of everyday populations, to the teratological, as well as variation induced experimentally. Studies of developmental potential always reveal unevenness, with variation occurring more often in some directions than others, as well as limits to the variation that appears across individuals. These studies often encounter mysterious gaps in the space of imaginable configurations.

Embryologists and students of teratology study the range of variation that occurs across as many individuals of a population as possible. Humans have been paying attention to variation across individuals in our own species for centuries, with the result we have compiled an extremely rich record of the commonly observed, the very rare, and the apparently never observed configurations in people (Geoffroy de Saint-Hilaire, 1832; Gould and Pyle, 1897; Wilder, 1908; Bendersky, 2000; Blumberg, 2009; Held, Jr, 2009, 2021). These compilations show that humans can be born with three or four eyes, but that these eyes won't be just anywhere on the body. They will be arranged along the transverse plane at about the middle of the face, where eyes are normally found (Figure 1.3). They will be associated with two mouths and two noses. If there is just one eye, then the nose is a tubular proboscis above the eye. Many plausibly imaginable configurations don't seem to exist (Alberch, 1989). Some of these include four eyes but one nose and one mouth, one central eye and a normal nose below it, or three heads. This is certainly not a continuous field of variation. Instead, developmental potential is an object of study in its own right and a necessary consideration in explanations of organismal form.

Biology has no shortage of terminology, often with several labels for closely overlapping ideas; here are some considerations for why the term "developmental potential" is used throughout, especially in light of related concepts like evolvability and developmental bias. "Evolvability" (Hendrikse *et al.*, 2007) often refers to essentially the same idea as developmental potential: the range of variants that development can produce. The term is avoided here because it often seems to inspire intimations of value and purpose-ladenness in evolution. Examples include "the ability of a biological system to produce phenotypic variation that is. . .adaptive" (Payne and Wagner, 2019, p. 24) or "The capacity of a lineage to evolve" (Kirschner and Gerhart, 1998, p. 8420). These sorts of framings can reinforce a tendency in evolutionary biology to focus preferentially on lineages with many species and high morphological diversity, as though these were inherently more interesting or important than those with few species and low diversity. The "-ability" part of "evolvability" risks reinforcing this framing, suggesting that producing lots of variation is inherently good, when the true goal of evolutionary biology is to understand the causes of the whole span of diversity, from groups with low species number and morphological diversity to the highest.

"Developmental bias" is another near-synonym, as in the definition "developmental systems tend to generate some characters, or trait combinations, more readily than others" (Lala *et al.* 2024, p. 14). This is a useful and clear definition, but here I worry that it can misleadingly suggest that a single underlying mechanism is responsible for generating variation in certain directions. Evolutionary biology has many terms that seem like they might describe processes or entities that can be appealed to explain patterns, such as phenotypic integration, canalization, G matrices, phylogenetic inertia, or niche conservatism (see Section 5.7.11), to say nothing of "constraint" (Section 1.4), when they are just patterns that require explanation. It's helpful to use terminology that guides away from this pitfall.

"Developmental potential," "evolvability," and some senses of "developmental bias" mean essentially the same thing, but "developmental potential" elicits well the notion of an array of possibilities, without the possible imputation of beneficialness that "evolvability" or the risk of mistaking "developmental bias" for a process that can explain patterns rather than a pattern to be explained. Perhaps my worries are misplaced; in any case, what is important is whatever term is being used to try to define it and use it consistently. We will explore below many examples of developmental potential, methods for studying it, and how developmental potential participates in explanations of organismal form. We can then fold the study of developmental potential into the overall effort to build explanations of organismal form.

2.3 Adaptationist evo-devo case studies: layers of evidence, not "full proof" in a single study

There is a truly wide arsenal of approaches for exploring developmental potential. In reading the examples that follow, it's important not to get antsy for "full proof." We will carefully explore the structure of an explanation of organismal form in Chapter 5. But in the meantime, don't fall prey to the common but misguided expectation that a single study needs to, or even can, provide anything like "definitive proof" of adaptation, developmental impossibility, or indeed any evolutionary phenomenon. This is not how evolutionary explanation works. It builds gradually, by adducing many, many tiny pieces of evidence to shine a dim light on phenomena and events that happened in the distant past (Griffiths, 1996b; Larson and Losos, 1996; Pigliucci and Kaplan, 2000). Here are two categories of these small pieces of evidence that adaptationist evo-devo approaches often address and to keep an eye out for in the case studies.

Often, an exploration of developmental potential will simply show that *some* species or individual was able to make a given "empty space" configuration. This is a good example of how an explanation of organismal form is made up of many small pieces of evidence. Say in my clade of interest, I haven't found any species with a certain combination of traits. This is a classic "empty space" in developmental potential. One possible cause of this empty space is that, for whatever reason, organismal matter simply cannot make such a combination, a modification to a standard adaptation scenario if there ever was one. In exploring developmental potential, one day you discover that a species, living in a recondite and poorly studied habitat, does fall in "empty" space. These individuals demonstrate that such a combination is developmentally possible, that organismal matter can indeed produce such a configuration. This information leaves out a great deal of questions that biologists would like to have answered in coming up with explanations they regard as solid, such as: Can the "empty space" variant be produced in other species? Does the variation arise in normal populations of other species or as a teratology? If it arises in other species, is the variant heritable? Is it associated with lower or higher fitness than the commonly observed configurations? And so on. But it does settle one important point: this region of phenotype space is, at least under some conditions, occupiable. The variant is developmentally possible.

Another very common outcome in evo-devo adaptation studies is to find that empty space configurations can indeed be produced in development, often with little provocation, and that they have manifestly lower performance or fitness than the commonly produced ones. This finding is exactly what a standard adaptationist account would predict. Along these lines, biologists usually take possible-but-low-fitness as evidence that the commonly observed configurations are commonly observed because they are the ones favored by selection. We will explore this interpretation in more detail throughout.

Your job as an evo-devo adaptationist is to find ways to amass many layers of evidence into an explanation that you and the community of biologists consider satisfactory. Chapter 5 will cover in more detail the complementary sources of evidence that biologists use to build what they regard as a satisfactory explanation of organismal form. For now, the case studies that follow highlight some of the many clever and creative strategies scientists have used to explore what we've been calling "empty" space, configurations that seem possible in principle but are absent or only rarely observed in an available sample. Some studies test whether these variants can be produced at all, others examine how they perform once they exist, and a few do both.

2.3.1 *Arabidopsis* directed mutagenesis

Adaptationist evo-devo, with its exploration of developmental potential, comes in many guises. Camara *et al.* (2000, p. 1011) set out a beautiful expression of developmental thinking. They noted that it is important to study "the impact of mutations on morphological and phenological covariation...rather than on fitness per se. Different combinations of phenotypic traits may be adaptive in novel or changing environments and mutations considered detrimental in one set of conditions may well be advantageous in others." This is a concise summary of the central concern of adaptationist evo-devo. We are interested not just in the variants of highest fitness or in standing variation, but in the entire panoply of configurations that can be produced. Fitness will depend on selective context, so under the right conditions, in principle, any possible heritable variant could be favored as the common one. Standing variation doesn't begin to provide a window onto the possible (Wolf *et al.*, 2004). To explore the "different combinations of phenotypic traits" that might be possible and therefore "may be adaptive in novel environments," we need a way of producing variation.

Camara *et al.* produced such variation with mutagens. They studied the plant *Arabidopsis thaliana,* which is as inoffensive and uncomplaining in cultivation as it is unexciting to look at. The plant is nothing more than a tiny rosette of leaves, a bolt of an inflorescence with pitifully plain flowers and some filmy roots. This modesty translates into a marvelous laboratory subject. Its simplicity of structure means that fairly small sets of measurements can describe axes of variation of major functional significance for the species. Its short generations make it amenable to lab research.

These virtues played directly into Camara *et al.'s* work. They exposed seeds from a Michigan population of *Arabidopsis* to ethyl methanesulphonate, which causes mostly point mutations. That ethyl methanesulphonate produces mostly point mutations was central in Camara *et al.'s* choice of the mutagen. They wished to explore the regions of morphospace directly adjacent to the wild-type plants. Point mutations are both common and potentially evolutionarily important. Using them gave Camara *et al.* a way of scattering the population away from the wild-type morphospace, close enough to seem likely to parallel the sorts of things that might happen in a natural population.

The seeds thus rejiggered were grown and a series of traits was measured. Camara *et al.* picked traits that reflected crucial aspects of both vegetative and reproductive function in *Arabidopsis.* These included: the duration of the vegetative growth period; the duration of the flowering period; the number of rosette leaves at flowering, a proxy for investment in nonreproductive biomass; the height of the main flowering stalk; overall plant size, captured as the summed length of all flower stalks; the number of fruits produced and the presence of any unelongated inflorescences at the time the first fruit shed its seeds, interpreted as potential but unrealized reproductive allocation.

These traits together provided a multidimensional map of plant development and reproductive effort. Camara *et al.* were interested not so much in individual traits in isolation, but in how traits covary, how changes in one might predict or entail changes in another. This multivariate approach reflects the biological reality that organisms are integrated wholes: their parts draw on shared pools of materials and energy, and are often functionally or developmentally coordinated. Covariation can result, for example, from tradeoffs, such as beetles investing in large eyes or large horns but not both (Emlen, 2001), or from selection favoring some trait combinations, such as cryptic coloration in animals together with locomotory chariness, over others (Ioannou and Krause, 2009). Confronted with patterns of covariation between traits in wild-type *Arabidopsis,* Camara *et al.* asked whether the "empty spaces" in trait space reflected natural selection favoring certain combinations, or whether these areas were unoccupied because they are developmentally difficult to access, a very logical first empirical step. Their mutagenesis approach offered a way to find out.

Saying that "traits covary" also implies empty morphospace, which directs us to the question of the causes of empty space. Covariation means that traits don't vary independently of one another. Therefore, they tend to fall together in some parts of space but not in others. Those parts of space where the data don't fall require explanation just as much as the occupied space. If the wild-type covariation patterns are adaptive, then this would imply that at least some parts of empty space are accessible but not commonly observed because those variants in those areas have low fitness. If they are possible, mutational perturbation should scatter the population into the adjacent and usually empty morphospace. If, on the other hand, they are developmentally inaccessible or only of very difficult access, it is impossible or very unlikely that the point mutations inserted by Camara *et al.* would displace individuals into these areas. To distinguish between these possibilities, mutagenesis should provide many small shoves, along the direction of the wild relationships, perpendicular to them, and every way in between.

Camara *et al.* (2000) found that these shoves did indeed scatter the population into the morphospace surrounding the wild type. The means of most features changed little, but the variances—the ranges of variation in single traits—increased in almost all features, sometimes dramatically. The duration of the vegetative period, for example, went from 28.7 days in the control to 99.3 days in the group exposed to a moderately high dose of mutagen. The duration of the vegetative period is the time after germination and before the flowering stalk emerges. This is the period in which the plant amasses the reserves that will be invested in reproduction. A long vegetative period would be associated with a greater store of reserves directed to reproduction. However, in a cold temperate plant like *Arabidopsis*, dawdling too long before reproducing could lead to being buried under snow or ice before reproducing. For these reasons, the duration of the vegetative phase is a trait with clear connections to fitness. Camara *et al.* showed that under induced mutation, the duration of the vegetative phase can vary markedly, plenty of grist for the mill of selection. The relatively low variance seen in the wild type seems plausibly a result of selection favoring a vegetative phase short enough to reproduce before weather becomes prohibitive but long enough to provide a substantial investment of photosynthetically captured carbon in progeny. Traits do not vary independently, though, and changes in variance were associated with changes in the structure of the multivariate patterns of trait covariation.

Camara *et al.* then examined the ways that these multivariate patterns of covariation might differ from those between traits of the control. One way they did this was via the search for common principal components. This technique involves reducing the original datasets to a smaller number of orthogonal axes describing the main dimensions of variation in the hypervolume described by the values of the traits measured. They could then see whether the matrix of trait covariation patterns of the control group shared similar dimensions of variation with the treated plants. This technique revealed that, across their doses of mutagen, not only did the variances of most traits increase, but the patterns of covariation detected also altered in various directions, though some remained nearly constant. Duration of the vegetative period remained highly correlated with the number of leaves across all treatments. This meant that if a plant grew vegetatively for more time, it had more leaves, and vice versa (presumably this means that leaf size remained relatively similar across treatments, rather than longer vegetative periods being correlated with larger leaves). In contrast to the constancy of the vegetative phase-leaf number correlation, that between the duration of the vegetative phase and the height of the flowering stalk pitched about wildly across treatments, even reversing sign in some cases.

Camara *et al.* concluded that most of the possible relationships between traits have abundant possible developmental alternatives. This observation supports the interpretation that the relationships observed in the wild population are plausibly regarded as the products of selection with regard to the developmentally possible variation adjacent to the commonly observed configurations. However compelling the results of mutagen experiments might be, biologists are a fretful lot and worry that the conditions of such experiments might be too artificial. Soothing these scientists are techniques that rely on old-fashioned reproduction to move populations through morphospace.

2.3.2 Radish flower artificial selection

A particularly illuminating series of artificial selection studies emerges from the work of Jeff Conner and his laboratory in Michigan on wild radish flowers (Conner *et al.*, 2011). The relationship between plants and their pollinators is an important model for studying evolutionary interactions between species. The intimate specificity of some plant-pollinator systems, in which a plant species depends on a single pollinator, and the pollinator requires the plant for its survival, is a paragon of coevolution (Anderson and Johnson, 2007). Plant-pollinator relationships run the gamut from such specificity to the relative indifference of "generalist" plants that may be visited by dozens of species of pollinators. A key player in determining where a particular plant species is located along this gradient is the configuration of its flowers, and one of the main factors in determining how much pollen is deposited on a pollinator, and where on its body the deposition occurs, depends on the relative positions of the stamens, the male, pollen-bearing parts of the flower, and the petals.

Precisely because of this importance in pollination, Conner *et al.* focused on the proportionality between the stamens and the petals. In radish, the petals project vertically for the basal half of their length, and these bases are drawn amorously close to one another, forming a tube encasing the female part of the flower, as well as the stalks (known as filaments) that bear the pollen-bearing sacs (anthers; a filament plus its anther is a stamen). The upper, showier half of the petals flare outward, exposing their surfaces in a colorful display (Figure 2.4A). Radish flowers can vary a great deal in size, from plant to plant growing in the same spot, between plants growing in dry or wet places, and between flowers along the same flowering stem as the flush season turns lean. But across the range of variation in absolute flower size, the proportions of anther to petal length remain similar: the anthers just peek out of the tube (Figure 2.4B).

Presumably, anther-petal proportionality is important for pollination, and good evidence suggests that it is adaptive (Waterman *et al.*, 2023). Very short stamens would nestle well down within the tube (Figure 2.4C). Hidden in the tube, it seems plausible to imagine that little pollen would transfer to the body of the sweat bees that are the most common pollinators of wild radish, and so the mating success of such individuals would be low. By the same token, anthers waving around on very tall filaments might hold the pollen so high that the bees could walk all over the petals without getting a single pollen grain on them. Having the anthers project just out of the floral tube, however, ensures that small bees burying their faces in the tube opening would be smeared with plenty of pollen. This interpretation of the proportionality between stamen length and petal tube length is therefore an adaptive one. Overall size may fluctuate, but the proportionality observed in nature is the result of selection favoring maximal pollen transfer.

Crucially, this explanation involves assumptions regarding development. These assumptions are not only that alternative stamen-petal tube proportionalities are of lower fitness but also that they are developmentally possible. They may even be occasionally realized in populations, but they are inevitably selected against because of their low mating success. These assumptions are usually taken for granted to the point of not even mentioning them, part of the tacitness of assumptions of continuous variation that we biologists so often invoke. What makes Conner and colleagues' work so special is that not only did they articulate these assumptions but they also tested them compellingly.

Conner *et al.* tested their adaptive hypothesis in two steps. They first documented the pattern of association between stamen length and petal tube height across wild-type plants. They found that, as expected, changes in stamen length track changes in petal tube length closely (Figure 2.4B; Conner and Via, 1993; Conner, 2003). This gave them a necessary baseline against which to compare the results of their selection experiments. Their plots of filament length versus petal tube length (Figure 2.4B) reveal large blank areas above and below the allometric scaling line. As always, these blanks demand explanation, and Conner *et al.* asked whether they can be accessed under selection.

The second step of Conner *et al.'s* test was the selection experiment. They wished to select as far as possible from the allometric slope in Figure 2.4B. In the "up" line, this meant relatively longer stamens and shorter petals. This direction would lead to stamens thrusting well out of the petal tube (Figure 2.4C). In the "down" line, selection was for relatively short stamens and long petal tubes, leading to plants with the stamens withdrawn into the petal tube. In two lines and over eight to nine generations, they selected the most extreme morphologies, up or down, in each of twelve families (the cohort of brothers and sisters from the same cross). They maintained two lines as randomly mated controls. Eight or nine generations worth of breeding involves a lot of plants. Conner *et al.* (2011) measured nearly 5,000, and worked with many more, because not all plants flowered, and seedlings need to be cared for even though not all plants survive. All told, selection experiments are extremely labor-intensive, but in the end, eight or nine generations is very little compared to the time scales available in nature. What they found is, therefore, all the more remarkable.

What they found was that eight or nine generations was plenty to achieve differences in proportionalities between stamens and the petal tube easily perceptible to the naked eye (Figure 2.4C). The "up" line had stamens obviously exserted and the "down" line had markedly hidden stamens. When family mean values of filament length versus petal tube length were plotted against one another, they fell well within previously empty space, with no overlap with their ancestral trait combinations. In statistical expressions of variation, the standard deviation of the filament-petal tube lip distance in the wild-type starting population was just 0.53. By the time Conner *et al.* coaxed the plants through nine generations of up- and down-selection, the lines diverged 5.8 standard deviations.

Conner *et al.* note that their results parallel the floral diversity observed across the Mustard Family, to which radish belongs (Figure 2.5). Most species resemble wild radish, with anthers poised near the lip of the petal tube. Some, like the prince's plume

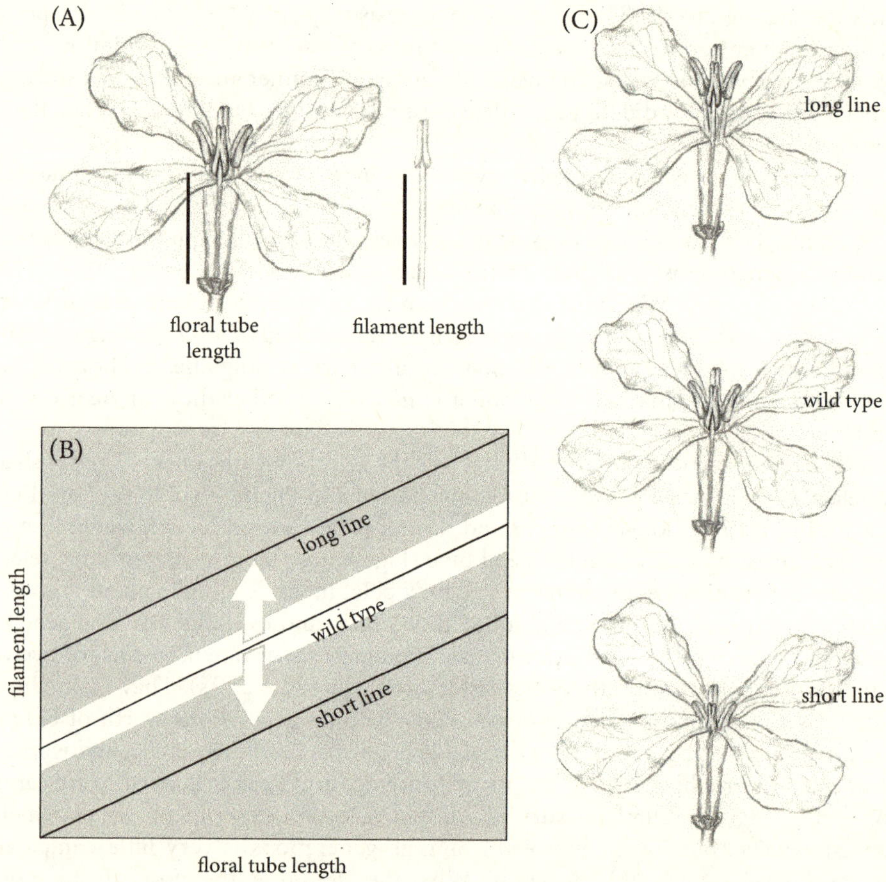

Figure 2.4 Radish flower artificial selection. (A) In wild-type wild radish, the male and female parts of the flower perch at the opening of the tube formed by the petals. The filaments are the stalks that hold the anthers (the male parts with the pollen) just outside the tube formed by the petals. (B) The larger the flower, the longer the petal tube and the longer the filaments are, with the result that the anthers remain just at the petal tube opening across the range of variation in flower length. The gray fields indicate the trait combinations not observed in normal populations. Conner *et al.* wondered whether they were developmentally inaccessible or whether they were accessible but of low fitness. After just nine generations of selection, the populations were dragged well into the formerly "empty" space around the wild-type scaling line. (C) The "long line" individuals above the line had much longer filaments than the wild type, with the male parts flaunting far outside the floral tube. The "short line" individuals below the line had much shorter filaments than the wild type, with the male parts nestled well into the floral tube.

Stanleya of the dryland western United States, have the stamens projecting well out of the petal tube. The stamens in other groups, like the Eurasian *Hesperis*, nestle well within the petal tube and look very much like the configuration of Conner *et al.*'s "down" line.

There is evidence all around us that artificial selection can move populations far from their original areas of morphospace. This is a happy property of living things, as illustrated by corn, beans, avocados, or practically any other domesticated food

Figure 2.5 Filament-petal proportionality across the Mustard Family Brassicaceae. The experiment of Conner *et al.* (2011) altered the proportionality between filaments and petals in ways that parallel diversity observed across Brassicaceae. (A) The wild radish *Raphanus sativus* has filament-petal proportions very similar to those of *R. raphanistrum* studied by Conner *et al.* In this flower, the anthers are positioned just at the entry to the floral tube. (B) In this wallflower (*Erysimum*), the female part of the flower projects out especially conspicuously, while in *Heliophila* (C), there is little tube and both male and female parts of the flower are exposed, projecting well beyond the petals. (D) This tendency reaches an extreme in the prince's plume (*Stanleya*), in which both the male and female parts jut far out of the tube formed by the petals. The finding that filament-petal proportionality can be so readily altered under selection is consistent with the idea that the diversity of filament-petal proportions across Brassicaceae species is adaptive.

organism, virtually all of which are gloriously more palatable than their all-but-inedible wild ancestors. Small changes in stamen-petal proportions might not be quite as dramatic in magnitude, but three considerations make the changes in Figure 2.4 outstanding. First, they were extraordinarily fast, just nine generations. Moreover, these changes mirror the variation seen in flowers across the family, so not only were these changes rapid, but they also tell us about the lability of traits that are manifestly of relevance in the interspecific diversification of the family. Finally, the selection response in radish was remarkable for being perpendicular to the allometric scaling line, the direction we would expect to have the slowest response to selection (Schluter, 1996). The steps that selection can take between generations, in the magnitude of phenotypic differences, can reasonably be expected to be proportional to the breadth of variation about the mean. Very low variances imply that steps should be small, whereas wide variances imply that steps can be large (Falconer & Mackay, 2009). Because there's a lot of variation parallel to the allometric scaling line, and relatively little perpendicular to it, it's reasonable to expect that evolutionary shifts in filament-petal tube scaling would be slow. And yet Conner *et al.* achieved functionally significant change about the Y axis in just a few generations. Their study suggests that if selection were to favor anther-petal proportionalities different from those observed in the wild, then this would surely be possible, and that the change between generations would be marked (Waterman *et al.*, 2023). That this selection mirrors an important axis of diversification across the Mustard Family suggests that this across-species diversity, too, is probably adaptive. Despite all of the advantages of laboratory selection, some biologists want even more generations of selection, and they turn to domesticated organisms.

2.3.3 Domesticated dogs

The morphological diversity of domestic dogs is the result of selection in directions as diverse as the needs and caprices of the human beings doing the selecting. Dogs have been subject to selection as domestic ornaments, like the many breeds of small house dogs, which are certainly diverse in appearance and also, admittedly, cute, a consideration that has certainly formed part of the selection criteria. Other dogs, not so fortunate cosmetically, have been bred for certain roles, like bulldogs, brought into being, as the name suggests, for the purpose of encouraging fights between bulls and dogs. The range spans swift running dogs, clever herding dogs, short-legged burrow-delving dogs, and an unhappy surplus of mean, powerfully built breeds of large dogs designed for nothing more than biting undesired people. This range not only attests to the practical needs of hunters and herders, human desire for servile companionship and their less tender impulses for power, but also, and most relevant for us, to the developmental possibilities offered by the dog lineage. This diversity can be translated into terms of developmental potential and therefore, to examine the questions regarding adaptation and the developmental assumptions hiding in these studies.

Drake and Klingenberg (2010) documented the way that domestic dogs fill morphospace and compared it to wild canids and a selection of carnivores at large. Focus on domestic dogs spared the authors the need to breed dogs themselves, and allowed them to address the issue of whether or not the morphology observed in wild canids is likely adaptive or not. They addressed this issue using geometric morphometric measurements of skulls. Geometric morphometrics is based on locating comparable points across the skulls of individuals measured. These "comparable points," such as the meeting point of two bones, are areas that are recognizable as being "the same" across all of the skulls examined (Zelditch, 2004). The scare quotes are because there is an important debate in biology, in the literature on homology, on what we mean as biologists when we say that two parts are "the same" (Assis and Brigandt, 2009; Ochoterena *et al.*, 2019). If it is difficult to pin down what "the same" means with reference to whole parts of organisms, it's even more difficult when it comes to single points. In any event, these points are carefully chosen to reflect the shape of the skull. Drake and Klingenberg were particularly interested in how different parts, especially the snout, could change relatively independently of the rest of the skull. They measured 677 skulls from 106 breeds of dogs, plus wolves and other wild canids, as well as samples from across Carnivora, the order that dogs belong to.

They found two major axes of variation in skull shape, which were largely independent of one another. One was how long or short the snout is (Figure 2.6). In domestic dogs, this axis ranges from short, pug-faced skulls on one extreme and elongate, murderous-looking skulls on the right. Remarkably, the range in snout length in domestic dogs exceeds by a very long shot what is observed not only in wild canids but in wild Carnivora as a whole. The other axis ranged from tall versus low braincases,

Figure 2.6 Artificial selection on dog skull morphology. Eduardo illustrates two dogs representing the striking morphological extremes uncovered in Drake and Klingenberg's study of domestic dog cranial variation. On the left is a brachycephalic breed, such as a Pekingese, with its dramatically shortened snout and tall, domed cranium. On the right is a dolichocephalic breed like the Borzoi, with an elongated, narrow snout and a low, sloping forehead. The skull shapes of domestic dogs span a range far beyond what is seen in ancestral wolves, or even in the entire order Carnivora. Most dog breeds diverge from the ancestral condition by having shorter snouts, but some have longer. Studies like this show how artificial selection can unlock latent variation in shape, revealing developmental possibilities that natural selection typically does not favor. Drawings by Eduardo Martínez Díaz.

and here domestic dogs again covered most of the diversity in wild dogs, though the canids had taller braincases than most carnivores. With these results in hand, Drake and Klingenberg provide a brief discussion very much in line with adaptationist evo-devo.

Drake and Klingenberg note that their data are in line with an adaptive explanation for wild canid morphology. Their data showed that a genuinely remarkable degree of developmental variation is clearly possible, and given the right environment, morphologies strikingly different from wild dog skulls can emerge. The "right environment" in this case is the extreme condition of manifold vectors of artificial selection. These authors attribute wild canid morphology to selection favoring high bite force. Skulls with longer, wolf-like mandibles should have an easier time of ripping tough sinew and cracking bones than the puny moment arm available to a pug or a shih tzu. Drake and Klingenberg say that selection should be most "permissive" (p. 298) in companion dogs. These dogs usually lounge on couches or ride in purses rather than running after sheep or cattle, let alone foraging for themselves in the woods. Therefore, the reasoning goes, the greatest degree of skull shape diversity should be found in the various lap and toy dogs bred for affection and amusement rather than for working.

Their results bore out their expectation, with the greatest disparity being found in companion dogs, less in working dogs, and less still in the wild ones. One of the great advantages of many adaptationist evo-devo approaches is that they bring unusual variants to hand, living and breathing, where variation in performance or fitness associated with morphological variation can be measured directly. Along these lines, plenty is known about the various aptitudes and ailments associated with different dog breeds, as well as, alas, which ones bite with greatest force and ferocity. Such indices of organismal "performance" (often taken as proxies for fitness) can be directly observed when "empty space" variants are brought into flesh-and-blood being.

Dogs illustrate quite dramatically that every species has not only a broad repertoire of developmental potential that it can produce, but also that much of this is, in one context or another, functional. Organisms don't simply produce their common variants plus a bunch of rare monstrosities, deformities, or teratologies doomed to die shortly after birth or germination. Instead, in every species there is a panoply of possible variants, latent potential that can be favored by selection. The next example, also from domestication, underscores just how remarkably wide this latent variation can be.

2.3.4 Domesticated goldfish

People have been keeping goldfish as pets for over 1,000 years (Smartt, 2001). In this time, breeders have produced a dazzling spectrum of forms starting from the fairly wild-type looking "common" goldfish, which have the usual fish shape of being sveltely laterally compressed and longer than tall. Common goldfish have trim finnage and single tails, and can swim relatively quickly. From this ancestral type, selective breeding has resulted in a bizarre array of breeds, with fish encumbered by egg-shaped bodies, bulging (even upward-pointing) eyes, extravagantly long fins, wobbling bags of fluid below the eyes, missing dorsal fins, or neoplasm-like "hoods" of tissue covering their faces (Figure 2.7) (Smartt, 2001; Sawatari *et al.*, 2009).

Of this array of morphologies unobserved in the wild type, we focus here on the big one: posterior bifurcation. Instead of the usual single tail, many varieties of domestic goldfish, such as the "fantail" that adorn so many fishbowls around the world, routinely and heritably produce two tails (Abe *et al.*, 2014). "Two tails" means that double-tailed goldfish exhibit a posterior bifurcation of skeletal elements. The spinal column itself doesn't bifurcate, but the elements just posterior to it do. Proceeding snout-to-tail, the first elements that double are bones called hypurals, which resemble short, sturdy versions of the long, delicate fin rays that support the fan-like part of the tail in teleost fish. The sturdiness of the hypurals concords with their role as anchoring elements of the tail and the site of high mechanical loading, where the forces of body undulations are transmitted to the tail during swimming. They also transmit the muscle inputs that allow many fish to make their entrancing top-to-bottom sinuous tail fin movements (Gosline, 1997). Continuing our survey, snout-to-tail, in "twin tail" goldfish, all of the skeletal elements posterior to the hypurals, the parahypurals, and the fin rays themselves, are also present in two copies, mirror-image across the animal midline. The net effect is a fish with two tails.

Figure 2.7 Goldfish diversity. Eduardo offers his take on the bewildering range of body shapes, fin sizes, head morphologies, and most strikingly, tail configurations that have been bred from the standard-issue ancestral goldfish. Clockwise, starting at lower left is a Bubble Eye goldfish, whose swimming is burdened by jiggling, fluid-filled sacs below its eyes, along with a double tail and no dorsal fin. At upper left, a smugly grinning wild-type goldfish, with a single tail, streamlined fins, and an aquadynamic body shape. To its right, a Telescope Eye goldfish viewed from the top, showing its protruding eyes and double tail. Next comes a Fantail, distinguished by its prominent dorsal fin and egg-shaped body, whose waddling swimming contrasts starkly with the wild type. To the right of the fantail is a Ryukin, with a dramatically hump-backed body shape and upright dorsal fin. At upper right, an Oranda, with its *wen*, a fleshy, raspberry-like mass covering much of its head. At lower right, a Celestial Eye, with upward-pointing eyes, a double tail, and no dorsal fin. Finally, a comet goldfish, with a single, deeply forked, flowing tail. The wild goldfish gives no hint of the extraordinary variation latent in its developmental potential. Drawings by Eduardo Martínez Díaz.

The example of two-tailed goldfish is remarkable because in the history of fishdom, there aren't any other fish known to have two tails. There are, and have been, a lot of fish, to say the least. Fish make up more than 30% of the total animal biomass on Earth, and something like 99% of the current biomass of wild vertebrates, and more than half of vertebrate species diversity (Bar-On *et al.*, 2018; Friedman, 2022). There are over 30,000 species of ray-finned fish, and that's just the contemporary ones. Ray-finned fish have been around for some 400 million years (Friedman, 2022). So, between the number of extant fish and their long history, fish have had a lot of opportunities to evolve a species with posterior bifurcation. However, no extant or extinct species of fish is known to be characterized by two tails.

This example illustrates again the point that all species harbor sprawling storehouses of developmental potential, most of which is never or rarely expressed. Tellingly, when goldfish of any variety are released into the wild, selection rapidly trims away extravagant morphologies, returning them within a few generations to the lean, single-tailed form of their ancestors. This selection process, however, doesn't mean that other developmental possibilities are erased. Instead, broad and mostly unstudied fields of potential outcomes remain available to every developing individual. While we don't know whether just any fish could be coaxed into forming two tails, it's likely that many species harbor latent capacities for dramatic departures from their typical morphology, possibilities that may never be revealed without the right developmental perturbation or selective environment.

What's striking is how unpredictable such variation can be. The kinds of variation adjacent to the wild-type form—slightly longer fins, slightly shorter fins—are unsurprising. But nothing in a tank full of common goldfish, or even across the 30,000+ species of extant fishes, would suggest that a tail might duplicate. Still less would genome analysis or surveys of common variants reveal this. Some ontogenetic possibilities are readily produced, others are vanishingly rare, and many are never seen at all, yet they remain, folded into the developmental architecture. This unpredictability is what makes developmental potential itself an object of study in the effort to explain organismal form.

2.3.5 Domesticated and feral pigeon legs

Biologists have a gift for finding data in the least-expected places, a talent illustrated beautifully by Daniel Sol's work on urban pigeons (Sol, 2008). While most people pay no mind at all to the pigeons pooping on monuments and squatting in the town squares of Catalonia, Sol saw a fascinating evolutionary story. He took advantage of a local situation in which some pigeons remained domesticated, while others had gone feral, no longer shaped by artificial selection and once again subject to the pressures of the wild.

In as pithy a prediction from an adaptationist evo-devo perspective as one could wish for, Sol reasoned that "Because artificial selection is likely to have displaced individuals from the adaptive peak for proper functioning in the wild, if natural selection is the driving force behind a phenotype-ecology association, we should expect a reconstitution of the wild ancestral phenotype when individuals are exposed to selective pressures similar to those that originated such association" (p. 657). His specific prediction had to do with the tarsus, the segment of a bird's foot above its toes, leading back to what looks like its backward-bending "knee" (Figure 2.8).

Sol noted that pigeons with longer tarsi have a harder time walking and foraging. This pattern would seem to jibe with everyday experience: balancing an orange on a short, squat base is easier than balancing it on a tall, skinny one. The taller and more slender the base, the easier it is for the orange's center of gravity to lie outside the radius of the base and to fall over. Sol reasoned that wild pigeons have short legs, allowing them to move their rotund bodies with relative efficiency (though this efficiency must surely be highly relative, given that the head-bobbing strut of pigeons suggests anything but a paragon of streamlined motion).

Under artificial selection, though, pigeon fanciers have tended to favor longer-legged birds (Figure 2.8). A legacy of this selection is that domestic pigeons, at least those thought to be the ancestors of the feral Catalonian populations, have markedly elongated tarsi compared to their wild relatives. With charming directness, Sol tells us that "Although the modified tarsus of domestic pigeons might decrease locomotory performance, this should not represent a serious problem in captivity, where individuals need only to sit and wait for people to bring them food" (p. 658). It could, however, make all the difference in the feral state, where individuals are likely near the metabolic knife's edge, where any excess energetic expenditure could affect survival. Sol's prediction, then, was that as tarsus length increases beyond the short-legged wild type, survivorship should decline.

Sol's prediction is easy to articulate in the terms we have been using throughout. The basic prediction is that if the tarsus-body size relationship found in wild pigeons reflects adaptation, then it should be fitter than alternative configurations. In domestic

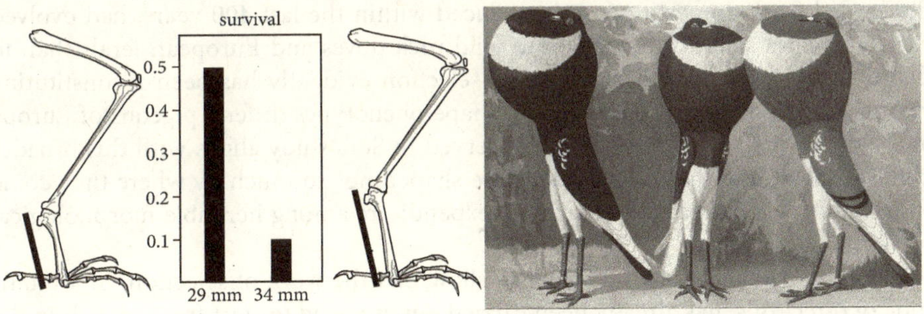

Figure 2.8 Pigeon tarsus length and survival. In his study of feral pigeons, Daniel Sol found that in Barcelona, where pigeons had to walk around to find food, individuals with tarsi (indicated by the black bars) that were "too long" had dramatically lower survivorship. Underscoring the point that domestic pigeons often have much longer legs than the wild type is the illustration at right of a French pouter pigeon, a domestic breed favored for its upright stance. Selection in feral conditions tends to return them to a wild-type tarsus length. Pigeon illustration from Schachtzabel (1906).

pigeons and their feral descendants, the tarsus is "too long" given body size as compared to wild progenitor rock doves. Artificial selection has dragged domestic pigeons far from the wild tarsus-body mass proportionality. If the wild tarsus-body mass proportionality of wild pigeons is adaptive, understood as the fittest variant culled from a wider set of developmental possibilities, then feral pigeons should be subject to selection favoring the short-legged individuals that most resemble the wild type. Finding pigeons to test these ideas was as simple as heading into the plazas of Barcelona and the nearby hill town of Moià.

Possibly to the bemusement of people sipping coffee in pleasant plazas, Sol's strategy was to capture feral birds, take measurements, and release them after tagging for observation. Over two years, he marked nearly 400 pigeons and tracked their fates, testing the prediction that individuals with shorter tarsi for their body mass would survive better. Pigeons with longer legs, presumably less efficient walkers, tended to favor a sit-and-wait strategy. In Barcelona, that approach was hit-or-miss: pigeons did loiter near café tables, but human generosity (or messy eating) didn't yield enough crumbs to live on, and many birds had to walk around to find their meals. In Moià, a quieter rural town, feeding was more structured. There, pigeons waited until workers at a grain factory finished their shift before starting theirs, swooping down to forage among the scattered seeds. With plenty of grain and no need to roam, the rural pigeons had it easier. Not so in the city: in Barcelona, where walking mattered, tarsus length predicted survivorship (Figure 2.8). Pigeons with moderately short tarsi fared best, while those with the longest legs had markedly lower survival. Sol concluded that the most likely explanation for the typical tarsus-body mass ratio was adaptive: longer-legged, less efficient walkers simply didn't survive as well.

Sol notes that this pattern of selection is consistent with a broad pattern of convergence observed by ornithologist Richard F. Johnston, who studied skeletons from wild rock doves, domestic pigeons, and feral populations in Europe and North America (Johnston, 1992). Remarkably, Johnston found that North American ferals, descended from domestic stock introduced within the last 400 years, had evolved skeletal shapes significantly closer to wild rock doves and European ferals than to domestics. He concluded that "natural selection evidently has been reconstituting reasonable facsimiles of wild size and shape phenotypes in feral pigeons of Europe and North America." The selection observed in Sol's study aligns with this broader pattern, suggesting that feral pigeons are shaped not so much by where they come from but by similar patterns of energy expenditure among heritable morphological variants.

Examples such as dogs and pigeons show that the work of selection, in a multitude of directions, has already been carried out, no need to start from scratch in the laboratory. Often, the characteristics being selected are documented, even as ideals of given breeds (Losos, 2023). Moreover, paintings, drawings, and photos help trace the history of breeds, helping track both selection criteria and responses. Whereas in the lab the labor-intensiveness of selection experiments allows breeding in one or two directions, people have often applied artificial selection in a multitude of directions, such as varieties of maize bred for sugar, starch, oil, color, cob length, kernel size, or popping (Mangelsdorf, 1974). The results of breeding reveal morphologies that are possible, and where, in relation to the possible, the common morphologies are found. Finding that they are a small subset of the possible is taken as consistent with the action of natural selection. In addition to within-species studies of domesticated organisms, similar reasoning can be applied in comparative studies of wild species, with an example in the next section.

2.3.6 The architecture of organismal circulatory systems

Organismal circulatory systems have similar architectures across species, and this conspicuous similarity demands explanation. Circulatory systems have narrow, highly branched conduits distally, which consolidate into wide ones proximally. More precisely, conduit surface area per fluid volume proportions are maximal at the distal ends of circulatory systems, and are minimal proximally. In human circulatory systems, the narrowest conduits are the distalmost capillaries, where conduit surface area is maximal for a given fluid volume. Proximally, toward the heart, arteries coalesce, becoming fewer in number and wider, until the aorta is reached, the widest artery in the entire system. The same happens in kidneys, where the highest surface areas are found in the glomeruli, where exchange occurs with the blood. Moving away from the glomeruli toward the collecting ducts and the ureter, the conduits become wider. From alveoli to bronchioles to bronchi and the trachea, a similar pattern of distal-to-proximal widening and coalescence is observed in the lungs.

Not only animals, but plants too share this distal-to-proximal architecture. Plants transport water from their roots to their leaves through conduits in a tissue called xylem. Unlike animal circulatory systems, which push blood by compression, plants suck sap under negative pressure driven by evaporation of water from the leaves. Despite this very different mode of fluid transport, just as in animal systems, conduit surface area per unit fluid volume is maximal at the distal end of plant circulatory systems. These distal ends are found in the leaves, where water exits conduits and diffuses out into the leaf cells and then into the air as transpiration. As in animal circulatory systems, proximal (toward the base) leaf conduits are fewer and wider than distal ones (Figure 2.9). The widest conduits in a leaf are found at the base of the petiole, the stalk holding a leaf. This basipetal (tip-to-base) widening continues down a plant's stem, even down the lengths of the world's tallest trees and into the roots (Williams *et al.*, 2019; Koçillari *et al.*, 2021). The West, Brown, and Enquist (WBE) optimality model posits an adaptive cause for this apparently universal narrow-to-wide pattern in organismal circulatory systems (West *et al.*, 1997, 1999). WBE hypothesizes that there are two main vectors of natural selection shaping organismal circulatory systems.

The first WBE vector of selection involves maximizing surface area relative to fluid volume at exchange interfaces. In conductive systems, the walls of conduits are the surfaces across which oxygen, carbon dioxide, or water diffuse. Because wall area scales with the square of conduit diameter, narrower conduits have much higher surface area relative to fluid volume. The rate of diffusion per unit fluid volume increases as this ratio increases, so WBE predicts narrower conduit diameters near exchange surfaces, where diffusion is concentrated.

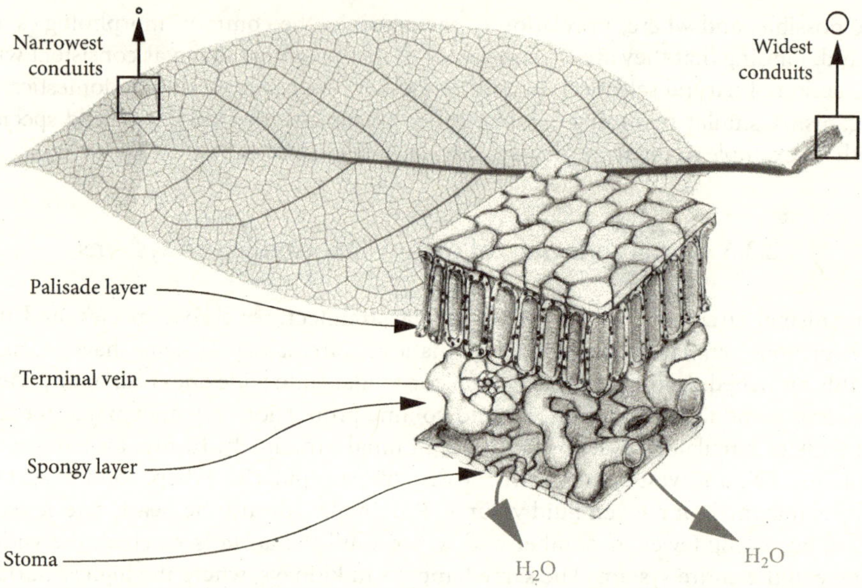

Figure 2.9 Leaf vein scaling. Natural selection favors very narrow conduits where water exits the conduits and enters the spongy tissue of the lower leaf surface. The diagram of a section of leaf at lower right shows the chloroplast-containing palisade layer, drawn slightly transparent to reveal the terminal vein in the middle of the leaf, with its spiral thickenings and surrounded by a parenchyma sheath. If the conduit surface area per unit fluid volume is very high, then when the stomata open (the tiny openings on the underside of the leaf), water will tend to rush out of the conduits, into the spongy, airy tissue that makes up the lower inside of a leaf, and then out into the atmosphere, thus driving transpiration. Narrow terminal "capillaries" maximize this surface area. Yet if conduits remained uniform in diameter from these narrow termini, then bigger leaves, with their longer conductive paths, would have higher resistance, thanks to the friction introduced by the conduit walls, meaning that conductance would steadily drop. In this context, selection favors conduits that widen tip-to-base, offsetting the resistance that would otherwise accrue, maintaining water supply to the terminal veins constant. The result of this process of selection is that conduit diameter scales with leaf size, with the widest conduits found at the base of the leaf, and larger leaves having wider conduits. Image of a cleared *Nyssa* leaf used with permission from the National Museum of Nature and Science, Japan, https://www.kahaku.go.jp/research/db/geology-paleontology/cleared_leaf

The other vector, toward the minimization of transport distances, opposes the first vector. Maximal surface area is achieved by a two-dimensional plane. A circulatory system that approximated a plane might consist of a single narrow capillary snaking densely back and forth across a flat sheet, forming a near-continuous exchange surface. While this would create an enormous surface for exchange, the metabolic cost of moving fluid through it would be commensurately high. Because of friction between fluid and the conduit walls, resistance to fluid movement increases linearly with conduit length (Vogel, 1994), meaning that a single long, snaking conduit would impose immense resistance. As a result, a long, narrow conduit might offer high exchange surface, but only at the cost of substantial energetic expense in transport. In contrast, the shape that minimizes distances between any two points in a given volume, and thus transport distances, as of blood or water, is a three-dimensional sphere.

Thus, WBE envisions one vector of selection pushing toward a plane (maximization of surface area) and the other toward a sphere (minimization of transport distances). What lies between the integer dimensions of 2 and 3 are fractional dimensions, geometries that fill space neither as planes nor as spheres. This is exactly what highly branched circulatory systems do. Trace an imaginary cube through a tree and see where the branches pass through the cube. A tree (or any other circulatory system) doesn't fill the cube fully, the way a 3D volume does. Instead, it fills it only partially, somewhere between dimensions 2 and 3, in other words, as a fractal.

In the context of this evolutionary tug-of-war, WBE also predicts the widening of conduits tip-to-base. The diameters of the terminal conduits at exchange surfaces—capillaries, alveoli, leaf veins—should remain constant across organismal sizes. The diffusion rates of key molecules such as water, CO_2, and O_2 remain constant as organisms increase in size. Given these constancies, the diameters of conduits where these molecules are being exchanged, the capillaries, do not increase systematically with body size. However, as organismal size increases, the circulatory system length increases, just as the distance between heart and capillary in a whale increases as an individual grows larger or the root-to-leaf distance increases as a tree grows taller. Because hydraulic resistance accrues linearly with conduit length, without a change somewhere in the system, conductance will drop predictably as an individual grows taller (Petit and Anfodillo, 2009). Because conductance only drops linearly with length increases but increases to the fourth power with diameter increases, small increases in diameter are, in principle, sufficient to maintain conductance constant as the circulatory system becomes longer. Accordingly, WBE also posits that conduits should widen tip-to-base in such a way that conductance remains constant as size increases.

WBE thus makes numerous adaptationist evo-devo predictions. One is that terminal conduit diameters should be invariant (i.e. not predicted systematically by body size) as circulatory systems get longer. Another is that conduits should widen predictably tip-to-base, such that larger individuals have proximally wider conduits (Figure 2.9). In addition, conduits should stretch from point A to point B in as short a distance as possible between exchange surfaces and the wide, proximal conduits, not ambling over routes that are longer than strictly necessary.

Just as importantly, by positing that the observed network geometries are the result of selection, WBE implies that a wider array of network geometries and conduit diameters is developmentally possible. For example, networks with distal conduits that are wider than proximal ones should be possible (Figure 2.10A). Conduits longer than strictly necessary, and that therefore don't minimize transport distances, should also be possible. Of this universe of possibilities, geometries that stray from those that minimize transport distance and maintain conductance constant should have lower fitness, at least under most conditions.

If selection favors a given network geometry "under most conditions," then one empirical strategy to test these predictions is to look for species that inhabit environments both unusual and likely to select for departures from the typical pattern. One such context is the lifestyle of tiny desert annuals that sprout after ephemeral rains. This is exactly the kind of exception to the "most conditions" assumption that WBE reasoning relies on. From the Atacama to the Sonoran Desert, small annuals in the genus *Cistanthe* produce fleshy leaves that gorge on water when it is briefly available, then shrink as drought returns and the plant flowers and fruits using the stored water.

In general, natural selection should favor veins that take the shortest path from the base of the leaf toward the "capillaries," minimizing the metabolic input of transporting fluids as well as minimizing the metabolic cost of building veins that aren't longer than they need to be (Durand, 2006). But most leaves don't shrink as much as *Cistanthe* leaves do. WBE-style veins that project unwaveringly into the leaf could buckle as the leaf shrinks, blocking water movement. Veins following undulating paths, however, could relax, concertina-like, as the leaf shrinks, preserving water movement. Squiggly veins are exactly what are found in some *Cistanthe* species (Figure 2.10D). They have veins that definitely do not go from point A to point B along the shortest path possible. These *Cistanthe* species growing in globally unusual conditions of natural selection show that alternative configurations to the commonly observed ones, in this case, the no-nonsense, shortest-path veins of most plants, are indeed developmentally possible. And not only are they developmentally possible, they are (probably) favored under some conditions. "Probably" because little is known about the working of *Cistanthe's* squiggly veins. Most species of the genus are rare and often live in remote places. But there is plenty enough known about vein structures in shrink-prone plant parts to make selection favoring collapsibility a very good bet for accounting for *Cistanthe* vein squiggliness (Landrum, 2006). Whatever the cause of their meandering paths, they are evidence consistent with the WBE prediction that developmental alternatives to the common direct-path veins should be possible, just not favored under most conditions.

Some clever authors, such as Brummer and Savage (2021), come at the "at least under most conditions" expectation in a different way. The central strategy of these authors, as for so many adaptationist evo-devo applications, is to search for situations in which deviation from the commonly favored configuration could be expected. But instead of looking for unusual conditions of selection in the wild, they turned to pathologies. Brummer and Savage reasoned that selection on circulatory system network geometries generally shapes characteristics at the whole-individual level. That is, selection favors intrapopulational variants whose network geometries meet the metabolic demands of an entire individual at the lowest metabolic cost. Cancer, though, represents a situation in which some cells break free of the usual harmony of coordination of resource allocation within individuals.

Brummer and Savage wondered whether cancerous tumors would therefore provide an arena in which network geometry would be gluttonous, overvascularizing tissues as compared to the austere economies of whole individuals. They examined lung tumors, focusing not only on network topology but also on the metabolism–body-mass relationship that different network geometries give rise to. Consistent with their expectations, they found radically altered relationships between conduit

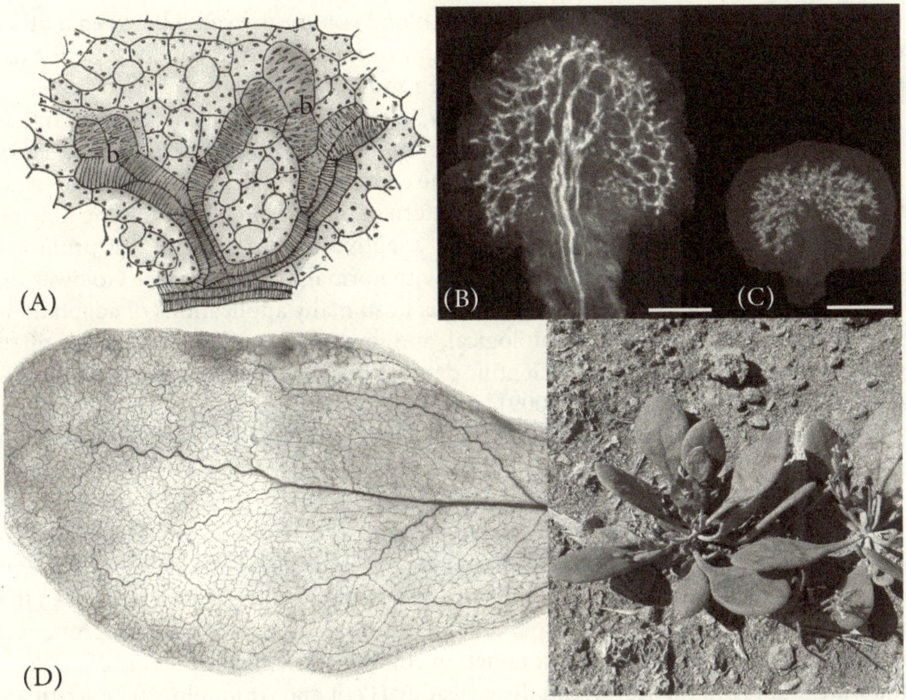

Figure 2.10 Rare but developmentally possible leaf vein geometries. Under some conditions, organisms depart from the usual pattern of tip-to-base widening in conductive systems. (A) Some plants have terminal conductive cells that, instead of being narrower than those lower down, are wider, as in these cells called water-storing tracheids, in *Euphorbia milii*, a succulent from the drylands of Madagascar. These cells cede water to surrounding cells, buffering transient moments when the conductive system doesn't keep pace with water demand. They show that the common pattern of distal conduits being narrowest is by no means a developmental inevitability. (B, C) Laboratory experiments show that even more extreme vein configurations are developmentally possible. In his lab, Enrico Scarpella perturbs hormone signaling pathways in *Arabidopsis* plants. This perturbation produces an array of leaf venation patterns unimaginable based on observed circulatory system geometries. In B, distal veins, toward the tip, are narrower than proximal ones, as in normal circulatory systems, but the distal veins are largely disconnected from the proximal ones. This leaf shows that the production of a circulatory system with disarticulated, unconnected veins is developmentally possible. The leaf in C is even more dramatic, with a circulatory system made up only of tiny, disconnected veins. That systems such as those induced by Enrico are not routinely produced in nature is almost certainly because of their low performance in supplying water. (D) The leaves of *Cistanthe maritima*, a desert annual, collected here in Baja California, Mexico, have squiggly veins that likely allow the fleshy leaves of this desert plant to contract when water is scarce. *Cistanthe* shows that veins that do not follow the shortest possible path are entirely possible developmentally, and can be favored under certain conditions.

Adapted from Stevens (1910). B and C adapted from Verna et al. (2019). Coordination of tissue cell polarity by auxin transport and signaling. *Elife*, 8, p.e51061, licensed under CC BY 4.0. creativecommons.org/licenses/by/4.0/

diameters and lengths. In normal tissue, the blood volume delivered between mother and daughter blood vessels remains constant. In metabolite-greedy lung tumor tissue, blood vessel volumes *increased* between mother and daughter blood vessels, a pattern essentially unheard of in circulatory systems globally. But such a pattern has to be possible if the usual decrease between mothers and daughters is the result of selection favoring tip-to-base widening in the context of a wider field of developmental possibility. The authors interpret this pattern of increase in blood volume flowing from mother to daughter blood vessels as "adaptive" in the sense that it promotes tumor proliferation as compared to tumors with normal vasculature (see Goldwasser, 2023). Brummer and Savage illustrate that as in so many applications of adaptationist evo-devo, the unusual, the teratological, and the frankly pathological, while often personal tragedies, offer crucial scientific data for building scientific explanations (as do Galis *et al.*, 2006; Blumberg, 2009).

Other authors provoke variation themselves, via directed mutagenesis, with Rodriguez-Zaccaro *et al.* (2021) providing a vivid example in poplar trees. The leaves of trees use copious amounts of water in photosynthesis, the process of fixing carbon from the atmosphere that is then invested in metabolic respiration and in the building of the plant body. A given area of leaf uses the same amount of water whether it sits atop a 1 m tall stem or a 100 m tall one. As a tree grows, the distance between the roots and leaves increases. Recall that because resistance increases as a linear function of conduit length, if plant water-conducting conduits remained the same diameters, the water supply to the leaf would drop steadily (Petit and Anfodillo, 2009; Koçillari *et al.*, 2021). As in leaves, plant water-conducting conduits widen tip-to-base in stems. The result is that taller plants have predictably wider conduits at the bases of their stems, as a function of how much downstream distance there is to the leaves. If this conduit diameter-plant height proportionality is one shaped by selection, as WBE suggests, then developmental alternatives should be possible, and these alternatives should affect performance and plausibly fitness. Plants tend to adjust their conduit diameters as they grow, producing wider or narrower conduits as the stem lengthens or (as after pruning) shortens. So, within any given species, the range in conduit diameter given stem length observed in a wild species is fairly narrow. Rodriguez-Zaccaro *et al.* (2021) wondered whether the narrow range of variation in conduit diameter given stem length diagnosed limited developmental possibility.

To see whether normally unobserved conduit diameter-stem length proportionalities were developmentally possible, Rodriguez-Zaccaro *et al.* (2021) subjected poplar pollen to gamma radiation, provoking mutations. They then pollinated poplar flowers with the irradiated pollen and measured the conduit diameter-stem length proportionalities in the resulting progeny, expressing height-corrected conduit diameter as conduit diameter-height residuals. The ranges of the residuals ballooned in both directions, with conduits that were significantly wider than those normally observed as well as narrower. Given the scaling of conductance as the fourth power of conduit diameter (Sutera and Skalak, 1993; Vogel, 1994), these represented highly functionally significant alterations. Standard quantitative genetic estimates of heritability, moreover, showed that height-corrected conduit diameter was heritable and could respond to selection, exactly what scenarios such as WBE predict (Koçillari *et al.*, 2021).

From the natural variation in *Cistanthe* veins or in cancerous tumors, to the manufactured variation provoked by radiation, there is abundant developmentally possible variation in organismal vascular systems. Essentially universal patterns in circulatory systems include narrow, profusely branched conduits at terminal exchange surfaces, scaling of conduit diameter with distance from the termini, and veins that traverse the source-to-sink distances as directly as possible. The developmentally possible variation is so wide that it seems certain that the placement of the normal variants—network topology, conduit widening exponents, etc.—is the work of selection culling from a much wider field of developmental possibility.

To illustrate the breadth of developmental possibilities in organismal circulatory systems, as well as the many avenues for research remaining in their study, we close this section with striking images from Enrico Scarpella's lab. Enrico and his colleagues manipulate hormone transport systems involved in vein development in *Arabidopsis*. In doing so, they show that mutants within a single species are developmentally capable of producing a sprawling range of venation patterns, far exceeding those observed in wild species (Figure 2.10C) (Verna *et al.*, 2019; Linh and Scarpella, 2022). The stunning traceries of venation in Enrico's experimental plants have probably never seen the light of day before, probably never been expressed by living plants. If they have, their tenure was brief—Enrico says that these variants have a hard time surviving even in the coddled environment of the growth chamber and remain tiny. And yet Enrico's experiments show incontrovertibly that they are perfectly possible developmentally. With their existence, these never-before-seen venations prove themselves to be just as much a part of organismal developmental heritage as the normal, usually observed vein configurations, and a reminder of the prodigious store of developmental potential available to each individual, if so rarely expressed.

Sometimes biologists want to study a trait not as amenable to perturbation as leaf veins. Sometimes, biologists are just plain impatient. These biologists have another option: the knife.

2.3.7 Fly microsurgery

The fastest way to produce an "empty space" configuration is via surgical manipulation. Surgical manipulation can involve adding structures (Basolo, 1990a), but more often it involves cutting, sanding, or scraping off the structure to be investigated. The performance of the bereft individual can then be compared to controls that have not been likewise peeved. Polak and Rashed (2010) turned their attention to the many spines and prickles that so often encrust the genital areas of male insects, spiders, reptiles, and other animals with intromittent organs. These genital fishhooks would seem clear impediments to coitus interruptus, and presumably, selection favoring enhanced mating success is what has encouraged their maintenance in so many groups. Polak and Rashed focused on a pair of cruel-looking spikes that characterize male *Drosophila bipectinata* (Figure 2.11). Presumably, the species epithet refers to the two spectacularly unpleasant-looking "claw-like projections from the male ventral cercal lobes" (Polak and Rashed, 2010, p. 1). The cercal lobes (making up a zone

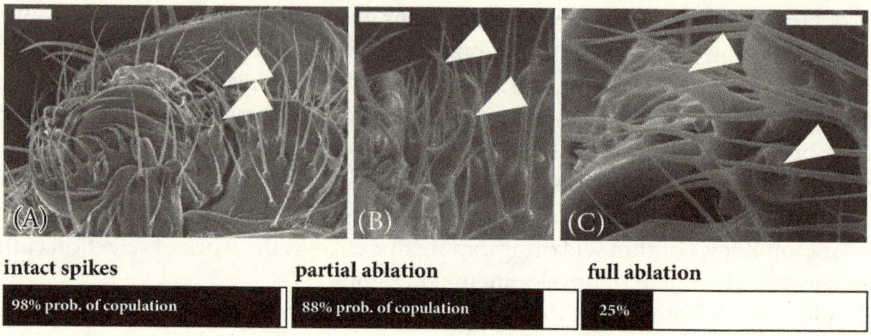

Figure 2.11 *Drosophila* genital spike microsurgery. (A–C) are scanning electron microscope (SEM) images of the genital spikes of *Drosophila bipectinata.* In (A), the two white arrows show intact spikes. In (B), the upper arrow points out an intact spike, and the lower one a partially ablated spike. In (C), the upper arrow indicates an intact spike, and the lower one a fully ablated one. At the bottom, the effects of ablation, with black bars representing successful copulation and white unsuccessful. Intact or partially ablated spines (A, B) were statistically indistinguishable and were associated with high probabilities of copulation. In contrast, full removal of the spikes was associated with a drastic drop in mating success. Scale bars = 20μm. SEM images kindly provided by Michal Polak, used with permission.

that Rice *et al.* (2019) tell us was formerly known as the "anal plate") appear to be the bits near the orifice from which the fly extends its spinulose penis.

If that image were not bad enough, the authors assault us with further detail: "We refer to these genital spikes as intromittent because they insert into female external genitalia during copulation, and not because they insert into the reproductive tract" (Polak and Rashed, 2010, p. 1; see also Polak and McEvey, 2022). Well, that's a relief. The authors never make clear what adversities in life led them to their muscogenital fixation, but one's heart goes out to them (the authors, as well as the female flies). One's heart also goes out to them (again, the authors) because variation in spike length presumably would have fitness consequences, but marked variation was missing in Polak and Rashed's *D. bipectinata.*

To produce variation in spike length, thereby obtaining variants whose performance could be compared with the wild type, Polak and Rashed resorted to laser microsurgery. Laser microsurgery focuses a tiny laser beam through a microscope objective. This allowed very precise cutting, and allowed the authors to cut just the genital claws and leave unaffected the forest of bristles projecting out of the fly terminalium (I am not making up this term: Rice *et al.*, 2019). In addition to controls, they produced two manipulated groups of 50–60 flies each, one with the claws completely removed, and another with the distal half of each claw removed. They allowed the flies to rest up for a few days after surgery and then presented the females. Males with shortened spines showed no significant differences in the probability of copulation as compared to control males, but spineless males had a decidedly tough time, with copulation probability plummeting to a fraction of control males (Figure 2.11). The authors conclude that "spineless males were unable to lock into the female genitalia with their own genitalia, thus they failed to achieve genital union...the folding action of the spines over each other draws the genitalia together" (p. 5).

Images of flies jimmying each other's terminalia notwithstanding, Polak and Rashed's study (as well as their Hotzy *et al.*, 2012) illustrates brilliantly the reasoning of surgical manipulation approaches. Many instances of low population-level variation in a trait are interpreted as past selection fixing a single variant and the consumption of the available variation. Low standing variation gives little traction for selection experiments or exploration of teratologies. This low variation makes it hard to test the hypothesis that the current commonly observed morphology is the fittest of the developmentally possible variants from which it was selected, the basic prediction of the hypothesis that the common morphology is an adaptive one. Surgical manipulation can mimic variation that biologists think may have been present at one time in a population, or that could be possible given enough time under selection. Finding that the created, and thus uncommon, variants have lower performance or fitness than the normal variants is consistent with the notion that selection favors the commonly observed ones.

2.3.8 Mammalian long bone scaling

Let's return to our standard adaptationist account of mammalian long bone scaling from Section 1.11.1 to examine another strategy for exploring empty space. In that example, Christiansen (1999) documented the remarkable ability of body mass to predict femur length across 79 species of terrestrial mammals ($r^2 = 0.97$!). The explanation offered in that chapter was natural selection acting in similar ways independently in each species. The scenario posits that keeping body mass constant, longer and shorter bones should be developmentally possible. Terrestrial mammals with bones that are very long and slender given body mass likely have low fitness because of bone breakage. Those with bones that are very short and stocky given body mass likely have low fitness because of the high locomotory and construction costs of relatively massive bones. Within each species, selection therefore favors, the scenario goes, bone length-body mass proportions that maximize stiffness and strength given construction and locomotory costs.

In that scenario, the "empty space" above the scaling line is assumed to be developmentally accessible but of low fitness. This is the totally reasonable, or at least perfectly pragmatic, assumption tacitly invoked in adaptation scenarios. It is tacitly invoked whenever biologists do not examine directly the developmental accessibility of these "empty" spaces. In the case of long bone scaling, the assumption implies that any given species should be able to produce individuals with bones significantly longer or shorter than the species' mean value (Figure 1.2). It might be that long- and short-bone variants appear in every generation, only to die or not reproduce and thus not contribute to the next generation. It could be that significantly long- or short-boned individuals are only born occasionally, either as teratologies or produced in pathological conditions (Petit and Adamec, 2005). It could also be that selection has extinguished the possibility of producing these variants, but that they were producible at some time in the past. Whatever the case, if long-bone-body-mass scaling is the result of selection, then it must be that variation about the scaling line is (or at least once was) possible.

The developmental accessibility of the empty space around the scaling line was not tested by Christiansen (1999). Instead, the explanation in Section 1.11.1 invokes the developmental accessibility of empty space as an assumption. Examining the assumption directly is a clear empirical priority in testing the scenario in Section 1.11.1. In addition to techniques such as selective breeding in the lab or under domestication, surgical manipulation, or mutagenesis, another key adaptationist evo-devo approach uses diversity exhibited across species as its source of information regarding what is developmentally possible.

Like Christiansen, Swartz and Middleton (2008) demonstrated a remarkable pattern of long bone scaling with body mass (Figure 2.12A). Across nearly 200 terrestrial mammal species, metacarpal length scales remarkably with body mass ($r^2 = 0.89$, slope = 0.41, intercept = 1.21). Swartz and Middleton reasoned that gravity is a potent factor of natural selection acting on all organisms. However, it is experienced differently by mammals that walk, run, jump, and climb on their metacarpal bones versus those that fly through the air on wings involving metacarpal bones.

So, Swartz and Middleton measured the metacarpal-length–body-mass relationship across bats. Their results were dramatic—metacarpal length scales with body mass across bats just as strikingly as across terrestrial mammals (Figure 2.12B). But the bat data fall far above the terrestrial mammal scaling and with a much shallower slope (slope = 0.24, intercept = 2.01). This result is consistent with the notion that the "empty space" above the terrestrial mammal bone-length-body-mass scaling line is ontogenetically accessible: at least some mammals can indeed produce bones that are much longer for a given body mass than is common in terrestrial mammals. This observation rules out the possibility that for some reason mammalian developmental systems are incapable of producing anything but the common terrestrial configurations. The results become even more dramatic when adding data from aquatic mammals: metacarpal length-body mass relations fall in the space well *below* the terrestrial mammal scaling line (Figure 2.12C), consistent with the notion that this area, too, is eminently accessible by selection in the right environmental context (Olson, 2012a). It is likely that the spaces of teratologies or bone dimensions available under artificial selection would be wider still (Figure 2.12D).

This comparative pattern provides key information in explaining terrestrial mammal long-bone scaling. The comparative pattern showing that bats and aquatic mammals fall in terrestrial mammal "empty space" is not a direct demonstration that terrestrial mammals can produce metacarpals as long and thin as bats or as short and thick as whales. However, it is consistent with, and complementary to, what is observed across terrestrial mammals. The delicate, thin bones of bat hands that allow them to fly would certainly be subject to breakage under the punishment of repeated pounding against the ground or tree branches that terrestrial mammals would expose them to. As discussed in Section 1.11.1, this expectation is consistent with what is observed within species, with thin-bone teratologies being associated with high rates of breakage, and, unfortunately, lowered survival rates (Folkestad *et al.*, 2016). No single source of empirical evidence provides a complete explanation of organismal form.

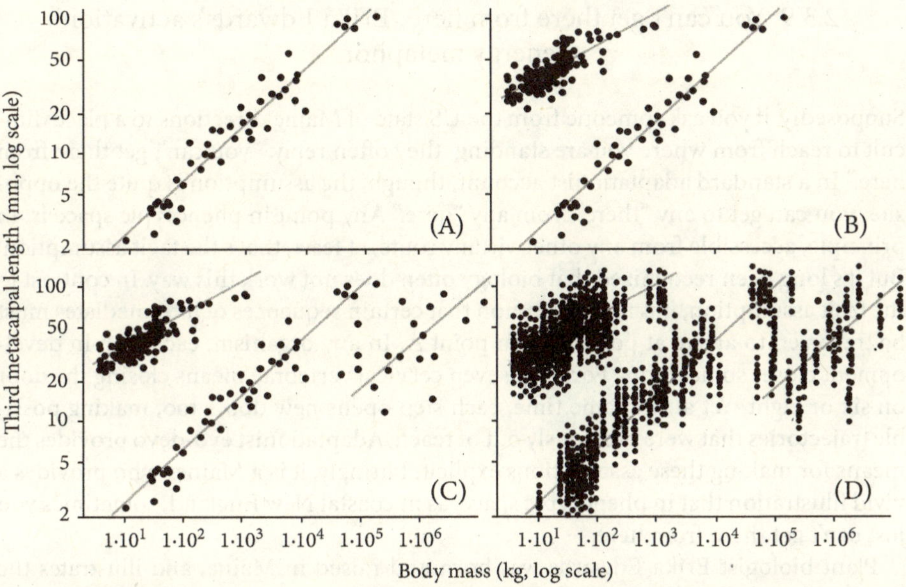

Figure 2.12 Accessible "empty" spaces above and below terrestrial mammal long bone length-body mass scaling. (A) Across terrestrial mammals, bone length, in this case the third metacarpal, scales with body mass with a very tight fit. Such tight relationships often inspire talk of "allometric constraint" or mysterious "physical laws" that somehow obligate biological variables to covary obediently. Such statements make predictions about developmental potential, for example, that the "empty spaces" above and below the scaling line should be inaccessible. (B) Swartz and Middleton added data from bats, finding that the space above the terrestrial mammal scaling line is obviously accessible. (C) Adding data from cetaceans and other aquatic mammals shows that the space below is accessible as well. Moreover, these occupations above and below the terrestrial mammal line make adaptive sense, in that they are not arbitrary with regard to habitat and lifestyle. Bones much longer for a given body mass, and thus lighter for a given length, correspond to bats, with their aerial lifestyle. Bones much thicker for a given length, and thus much heavier for a given length, correspond to aquatic mammals, which are supported by the water. (D) The space of teratologies is surely much larger still. This evidence regarding the accessibility of the space surrounding the terrestrial mammal scaling line is consistent with the notion that the slope and intercept of the terrestrial scaling line are shaped by selection in the context of a wide field of developmental alternatives. Figure adapted with permission from Olson, M.E. 2012a. The developmental renaissance in adaptationism. *Trends in Ecology & Evolution* 27: 278–287.

Well-supported explanations only emerge from the piecing together of evidence from diverse sources, and in this case, Swartz and Middleton showcase the crucial contribution of comparative data. But moving from one part of morphospace to another is not always as seemingly straightforward as it is in bone-body-mass space, as the next example illustrates.

2.3.9 You can't get there from here: Erika Edwards's activation energy metaphor

Supposedly, if you ask someone from the US state of Maine directions to a place difficult to reach from where you are standing, they often reply, "you can't get there from here." In a standard adaptationist account, though, the assumption is quite the opposite: you can get to any "there" from any "here." Any point in phenotypic space is, in principle, accessible from any other via any route, at least, that's the tacit assumption. But it's long been recognized that biology often does not work this way. In contrast to the tacit assumption, it is more common that certain sequences of intermediates must be traversed to arrive at point B from point A. In any organism, each step in development closes some doors. Forming seven cervical vertebrae means closing the door on six or eight. Yet at the same time, each step opens new doors too, making possible trajectories that were previously out of reach. Adaptationist evo-devo provides the means for making these assumptions explicit. Fittingly, it is a Mainer who provides a vivid illustration that in phenotypic space, as in coastal New England, sometimes you just can't get there from here.

Plant biologist Erika Edwards was born and raised in Maine, and illustrates the "you can't get there from here" principle with the evolution of C_4 photosynthesis in flowering plants. Most plants engage in C_3 photosynthesis, which involves simply waiting around for the sun to come up and opening the tiny pores on their leaves called stomata. Once the stomata are open, the warmth of the sun evaporates water from the leaves, generating the negative pressures that pull water up from the roots. In most habitats worldwide where plants find themselves, plenty of CO_2 drifts into the stomata and diffuses into mesophyll cells, which make up, as their name suggests, the middle of a leaf. In the mesophyll cells, the enzyme ribulose bisphosphate carboxylase (RuBisCo) catalyzes a reaction that captures CO_2 in the form of a 3-carbon molecule, hence the name "C_3" photosynthesis. This 3-carbon molecule then proceeds along the Calvin cycle, fueled by ATP and NADH supplied by the light-powered reactions of photosynthesis. The Calvin cycle spits out the building blocks for the sugars that form the basis of plant growth and metabolism, and indeed net primary productivity on Earth.

With their moderate growing season temperatures and relatively abundant moisture, most habitat conditions on Earth, from tundras to temperate forests to rainforests, are conducive to C_3 photosynthesis. However, some habitats impose sufficiently frantic conditions that C_3 photosynthesis is subverted via a tragic Achilles heel in RuBisCo. If we can be charitable for a moment, and view them from the point of view of RuBisCo, CO_2 and O_2 are structurally similar. They are both small, linear molecules with highly electronegative oxygen atoms dangling provocatively at either end. Higher temperatures lower the specificity of the active site of RuBisCo, with the result that instead of fixing CO_2 and producing sugars, RuBisCo catalyzes the fixation of oxygen. The resulting molecules are essentially nuisance products, whose recycling squanders ATP and NADH without resulting in carbohydrate synthesis. To make things worse, the blazing sun of tropical dry habitats means that the light reactions often proceed at full tilt, pumping even more O_2 into the inner environment of the leaf. At the same time, any CO_2 present is promptly removed to the Calvin cycle, further lowering the CO_2-O_2 ratio and making things even harder for the already heat-beleaguered RuBisCo. This fixation of O_2 instead of CO_2 under hot and often seasonally dry conditions is known as photorespiration.

Given the metabolic cost of photorespiration, a "you can't get there from here" mode of photosynthesis, known as C_4, has evolved dozens of times within the flowering plants from the ancestral C_3 photosynthesis (Figure 2.13). As it happens, in C_3 plants, there is an enzyme already present that has a CO_2 affinity much higher than that of RuBisCo, known as phosphoenolpyruvate carboxylase (PEPC). In C_3 plants, PEPC is off minding its own business in a wide array of cellular processes (discussed in Section 1.9.2 on the notion of biological "function"), none of which are directly connected with the usual C_3 carbon fixation pathway. The greater affinity of PEPC for CO_2 as compared with RuBisCo is coopted evolutionarily in C_4 plants. In C_4 plants, instead of RuBisCo being at the front lines of carbon fixation in the mesophyll, PEPC is the enzyme that catalyzes the fixation of CO_2.

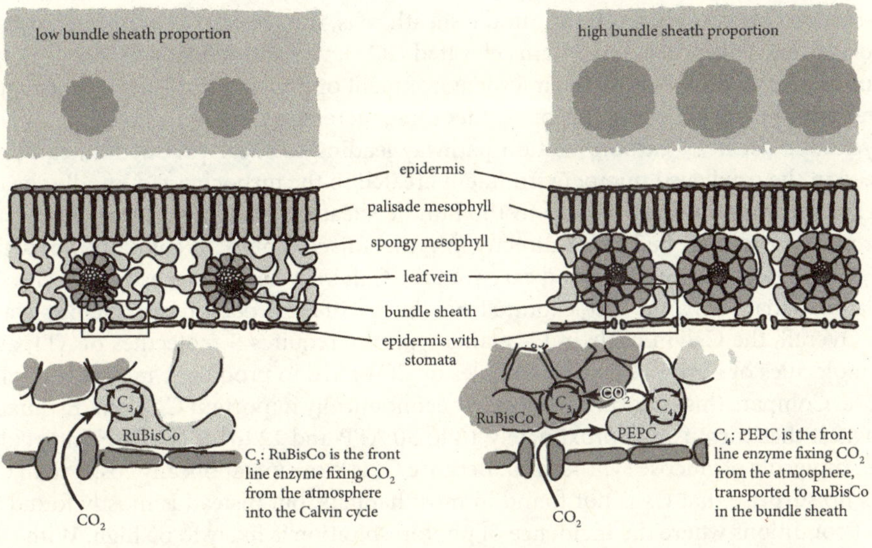

Figure 2.13 Leaf anatomy of C_3 and C_4 plants. The figure shows schematics of leaf cross sections, showing the palisade mesophyll layer, rich in chloroplasts, and the spongy mesophyll layer. The spongy layer has lots of air spaces; water diffuses out of the veins into the air, and then out of the stomata. The veins are surrounded by parenchyma cells, making up the bundle sheath. The black rectangles indicate the areas enlarged below each leaf section. In C_3 plants, RuBisCo is found in mesophyll cells, where it fixes CO_2 from the atmosphere, directing it into the Calvin cycle. The resulting sugars are loaded into the leaf veins, where they are transported to the rest of the plant. In C_4 leaves, which operate under conditions that are hostile to RuBisCo, PEPC is the front-line mesophyll cell enzyme that fixes CO_2 from the atmosphere. The resulting 4-carbon compounds are moved into the bundle sheath, where CO_2 is liberated. The CO_2 concentration in the bundle sheath cells is thus elevated, and the RuBisCo there can work efficiently.

As you might have imagined, instead of the 3-carbon molecule that is the first stable product of carbon fixation in C_3 plants, the first stable compound in carbon fixation in C_4 plants, often malate, has four carbons. Unlike in C_3 plants, where RuBisCo catalyzes carbon fixation in mesophyll cells, C_4 plants have PEPC in the mesophyll cells. With its low oxygen affinity and high CO_2 affinity, PEPC can slowly but surely capture whatever CO_2 might be available in the relatively low CO_2 conditions in the mesophyll of a plant baking in tropical heat. The higher affinity of PEPC for CO_2 and its very low O_2 affinity mean that its presence, as compared with RuBisCo, essentially obviates the possibility of photorespiration in the front-line mesophyll cells.

The 4-carbon compound fixed by PEPC is then actively transported to a different part of the leaf, where RuBisCo is waiting (Figure 2.13). In what amounts to a plant CO_2 turbocharger, after carbon fixation occurring in mesophyll cells, active transport mechanisms shuttle 4-carbon compounds to what are known as bundle sheath cells. In plant parlance, "bundles" are the veins that include both the conduits transporting water from the root to the leaves, but also the conduits, known as phloem, that move sugars produced in the leaves to the rest of the plant, where they fuel cellular metabolism. The "bundle sheath" cells surround these veins or bundles. Once the 4-carbon compounds reach the bundle sheath cells, they release CO_2 through decarboxylation. This process maintains elevated CO_2 levels and reduced O_2 levels in the bundle sheath cells, exactly the microenvironment optimizing RuBisCo performance and minimizing photorespiration. Under these increased CO_2 conditions, the Calvin cycle, the usual C_3 carbon fixation pathway leading to sugar production, can proceed in the privileged microenvironment created in the turbocharged bundle sheath cells. Locating the Calvin cycle in the bundle sheath is ideal, given that the sugars produced can then be loaded directly into the adjacent phloem conduits. So, while C_3 plants both capture CO_2 and carry out the Calvin cycle in the mesophyll cells, C_4 plants spatially separate CO_2 capture from the rest of the process of producing sugars.

Overall, the Calvin cycle in C_3 plants typically requires 9 molecules of ATP and 6 molecules of NADPH per 3 molecules of CO_2 fixed to produce 1 molecule of glucose. Compare this to corn, an extremely economically important C_4 plant. Estimates suggest that corn uses approximately 18 to 30 ATP and 12 to 18 NADPH molecules per molecule of glucose synthesized. Because C_4 is more metabolically costly than C_3, it's no surprise that C_4 is not found in most habitats, but instead is mostly found in hot conditions where the incidence of photorespiration is likely to be high. With this information in hand, we can now turn to why the C_4 photosynthesis is an example of evolutionary "you can't get there from here."

Using the evocative metaphor of "activation energy," Edwards (2019) explains that a specific combination of anatomical leaf traits must evolve first before C_4 metabolism can evolve. To get to the "there" of C_4 photosynthesis, a C_3 lineage must first evolve a high ratio of bundle sheath volume to mesophyll volume. C_4 does not generally evolve in species with low proportions of bundle sheath to mesophyll, with such configurations leaving little room for sugar production. Also, more mesophyll means longer transport distances to the bundle sheath. Edwards notes that, based on comparative studies across hundreds of species, the evolution of the appropriate bundle-sheath-mesophyll ratio is relatively rare, but once it evolves, evolution of C_4 occurs commonly.

She depicts the likelihood of evolutionary transitions by likening them to the activation energy of chemical reactions (Figure 2.14). In enzyme-catalyzed reactions, the activation energy is the energy threshold that must be overcome for the reaction to proceed. The activation energy associated with the evolution of high bundle sheath-mesophyll proportions is much higher than that associated with the transition from C_3 to C_4 (Figure 2.14). The appropriate leaf anatomy arises relatively rarely. In the absence of C_4 photosynthesis, why it might arise at all is not clear. One idea is that large bundle sheath cells store water that can be ceded to water-conducting conduits on hot days when the usual root-shoot-leaf pathway can't keep up with water demand. Whatever the reason for its evolution, once the "there" of high bundle sheath proportion evolves, then C_4 evolution is common. Through the lens of Edwards's activation energy metaphor, the evolution of high bundle sheath proportion is a high-activation energy, rate-limiting step. So, you can't get to the "there" of C_4 photosynthesis directly

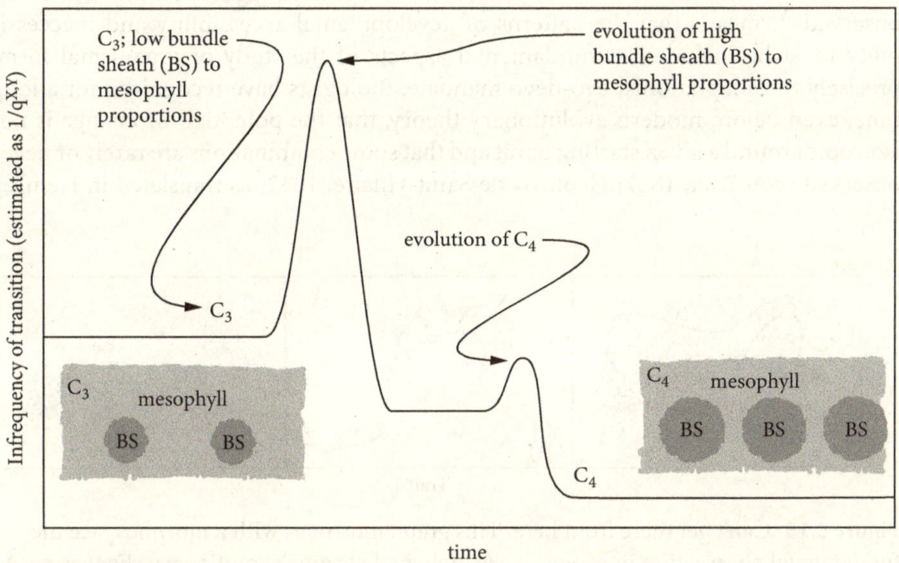

Figure 2.14 Erika Edwards's activation energy metaphor. This metaphor draws on chemical reaction kinetics to illustrate that the evolution of one state opens up previously inaccessible evolutionary possibilities. The peaks between states in a diagram of chemical reaction kinetics represent energetic hurdles, with their heights corresponding to how improbable a transition is under given conditions. Translated to evolutionary terms, the peaks correspond to evolutionary shifts, and their heights correspond to the relative rarity of those transitions. These were estimated based on transition rates ($q^{x,y}$) from an empirical study on grasses (Christin *et al.*, 2013). Plants can't get to the "there" of C_4 photosynthesis from the "here" of standard C_3 photosynthesis and a low proportion of bundle sheath (BS) to mesophyll tissue, represented by the diagram at the bottom left. The evolution of such proportionalities is rare, as indicated by the high peak. Once a high proportion of BS to mesophyll evolves, then C_4 photosynthesis evolves readily, represented by the much lower peak.

from just any point in phenotypic space. In at least 67 times that C_4 has evolved, it has practically always involved first acquiring the appropriate anatomy. As depicted in Erika Edwards's expressive "activation energy" metaphor, C_4 photosynthesis illustrates the way developmental possibility, in the form of which anatomies can arise in ontogeny, interacts with selection in shaping the range of variants available to selection, and thus evolutionary trajectories, including whether a lineage can reach a given phenotypic "there" from its current "here" (Figure 2.15).

The evolution of C_4 photosynthesis illustrates the adaptationist evo-devo quality of biological evolution in general, the inextricable interaction between developmental potential and selection. Precisely because any given developmental step closes some doors and opens others, any given evolutionary transition also precludes some possibilities while making others possible. If it is assumed that any point can be reached from any other, then the fact that a given configuration is unobserved is likely due to selection. Finding that it is developmentally inaccessible fundamentally changes the explanation for its absence and for the presence of the configuration that is actually observed. It means that the patterns of developmental accessibility and inaccessibility must be included as fundamental aspects of the study of organismal form, precisely the adaptationist evo-devo mandate. Biologists have recognized for a long time, even before modern evolutionary theory, that the potential for change is not isotropic around a given starting point and that some combinations are rarely or never observed (von Baer, 1827; Geoffroy de Saint-Hilaire, 1832; as translated in Henfrey

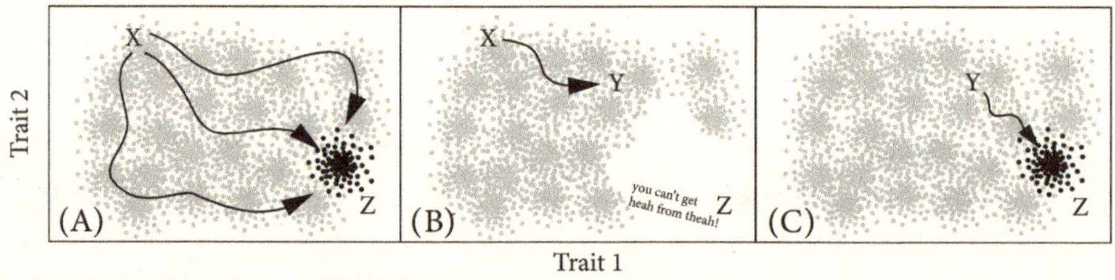

Figure 2.15 Can't get there from here. This graph illustrates with a morphospace the fundamental observation in evolutionary biology that some evolutionary changes open up previously inaccessible possibilities. (A) Standard scenarios of adaptation, ones that don't take into account developmental potential, assume, tacitly or as a starting point, that evolutionarily relevant variation is distributed continuously. Each cloud of points represents a population with trait values approximating bivariate normal distributions. Unless shown otherwise, any point Z is accessible from any point X via any route. (B) Erika's metaphor illustrates that in many cases, a given point Z isn't accessible from X until reaching Y. This inaccessible area is represented by the empty space at the lower right. (C) Once Y is reached, Z becomes accessible. In the example of C_4 photosynthesis, X represents the normal C_3 leaf anatomy, with a low proportion of bundle sheath tissue to mesophyll. Y represents C_3 photosynthesis with a high proportion of bundle sheath to mesophyll. Once such a condition is reached, the possibility of evolving C_4 photosynthesis, here represented by Z, opens up.

and Huxley, 1853; Bateson, 1894; Gould and Pyle, 1897; Wilder, 1908; Wright, 1932; see also the history traced by Arnold, 1992). Adaptationist evo-devo provides a means to bring these often tacit assumptions into the light, and, as with Erika's metaphor, give name to them.

2.3.10 *Drosophila* flight performance maneuvering around panadaptationism

A favorite criticism of adaptation studies is that they slip too easily into panadaptationism. Panadaptationism is the practice of explaining all of organismal form by appeal to natural selection. In their critique of adaptationism (about which more in Section 5.10), Gould and Lewontin target precisely this practice: "An organism is atomized into 'traits' and these traits are explained as structures optimally designed by natural selection for their functions. . . Organisms are integrated entities, not collections of discrete objects. . ." (Gould and Lewontin, 1979, p. 585). The problem they highlight is the implicit assumption that each trait is driven independently to some global possible optimum, independently of all other traits. Gould and Lewontin's panadaptationist biologist would assume that selection can increase fox ears strictly as a function of their heat-radiating quality, unfettered by any other factor, such as snagging on the underbrush or flopping down over the eye (Figure 2.16).

Even if a biologist accepts that traits do not evolve in isolation and tries to explain form as the result of multiple selection pressures acting in different directions—for example, that large ears help radiate heat while small ears reduce snagging—Gould and Lewontin criticize that too. They object that "any suboptimality of a part is explained as its contribution to the best possible design for the whole," arguing that this is also panadaptationist (Gould and Lewontin, 1979, p. 585), because it still treats natural selection as the sole sculptor, even when it's juggling multiple tasks. So the problem isn't just assuming that each trait is free to respond to selection independently of any other; it's also the impulse to explain any trait falling short of its maximum possible performance as the result of selection balancing competing demands. And that leaves the biologist in a bind: even if they acknowledge the organism's integration and appeal to selection acting in multiple directions, they're still panadaptationist. By exploring what is possible and not in development, adaptationist evo-devo shows why these critiques are confused, providing much-needed empirical perspectives and conceptual order, as admirably illustrated by the example of flight maneuverability in the study of Ray *et al.* (2016).

Adaptationist evo-devo approaches allow probing of the central assumption that selection favors the "least bad" trait combination, as Ray *et al.* (2016) did for *Drosophila* wing shape. *Drosophila* wing shape presumably reflects the effects of selection favoring multiple aspects. Aerodynamics are clearly important for volant animals, and wing shape has profound effects on aerodynamics. This is especially true in insects, which, unlike birds or bats, can't markedly adjust individual wing surface area or curvature (Combes and Daniel, 2003). Wings that are short (from root to tip) for their chord (from leading to trailing edge) tend to confer maneuverability, whereas those that are long for their chord confer better soaring or long-distance flying. High

Figure 2.16 Developmental quasi-independence and selection on organisms as integrated wholes. Arctic foxes (*Vulpes lagopus*), left, have very small ears for their body size, whereas desert kit foxes (*Vulpes macrotis*), center, have ears that are relatively larger than those of most wild canids for their body size. These different proportionalities are usually explained by adaptation: selection favors heat-retentive, frostbite-resistant small ears in cold places, large, heat-radiating ones in hot deserts. Such evolutionary changes in proportions can only occur if there is some degree of developmental independence between parts. In this case, ear size can clearly vary markedly with respect to body size. This quasi-independence is necessarily bounded, though: indefinite enlargement of the ears is impossible because of finite resources in development. Also, at some point ear size is presumably acted against by selection (excessive conspicuousness, tearing on vegetation, etc.), as in the imaginary fox at right. Thus, while there is *sufficient* independence between "parts" to observe the obvious and striking diversity of the natural world, for many reasons this independence is only partial.

body weight per wing area is associated with faster flight and greater resistance to turbulence, whereas lower weight per wing area provides slower flight speeds and higher maneuverability. Wingtip shape can affect the vortices that are produced when the relatively low air pressure on the upper surface of the wing sucks air from below, reducing lift. In long, slender wings with pointed tips, the wingtip vortices produced during flight tend to be smaller and weaker compared to species with broader or more rounded wingtips (Norberg, 1989; Dudley, 2018). These and other associations give reason to think that selection should mold wing shape as a part of the *D. melanogaster* aerial lifestyle.

But additional factors of selection likely influence wing shape as well. In addition to flight, *Drosophila* wings are involved in courtship (Menezes *et al.*, 2013). Male flies vibrate their wings to produce what Menezes *et al.* refer to, surely generously, as a "love song." In at least some *Drosophila* species, wing shape and patterning are involved in visual courtship displays. Ray *et al.* note that the wing features that are likely to make *D. melanogaster* individuals maximally maneuverable in flight are virtually certainly not those that make them maximally appealing to females. Therefore, selection on

D. melanogaster wing features should represent the "least bad" combination of opposing vectors of selection involving, at a minimum, flight performance, visual display, and love songs.

Ray *et al.* (2016) employed an adaptationist evo-devo approach that manipulated the space of developmental possibility with exceptional precision. Surgical manipulation of wing shape and area injures the wing, potentially degrading performance. Artificial selection on wing shape almost certainly affects other variables, such as flight muscles, making it hard to attribute differences in performance to wing shape per se. Ray *et al.* (2016) turned to the Gal4-UAS system, which is used in fruit fly research to control gene expression. This system takes advantage of the yeast transcription factor Gal4 and its corresponding upstream activating sequence (UAS) to drive the expression of target genes. Gal4 is a transcription activator protein that binds specifically to UAS sequences, activating transcription. For their part, UASs are DNA sequences recognized by the Gal4 protein. When Gal4 binds to a UAS, it initiates transcription of the downstream gene. By breeding flies in just the right way, progeny are produced in which Gal4 is produced, binds to the UAS sequences, and activates the expression of the gene of interest.

Ray *et al.* (2016) implemented this system such that Gal4 production was restricted to the developing wing, and was negligible in the rest of the fly body. In the developing wing, Gal4 affected the expression of the *narrow* gene, which affects wing shape but not venation and other wing variables. With careful crosses, Ray *et al.* (2016) produced progeny that expressed a range of wing length-chord relations (aspect ratios) that went from normal to much higher than observed in natural populations. Because the changes were restricted to wing shape, and not in other factors that affect flight performance, such as the wing hinge, the flight muscles, or the characteristics of the halteres, Ray *et al.'s* experiment admirably examines the effect of variation in wing shape all else being equal.

With a set of *Drosophila* individuals with a wider-than-normal array of wing aspect ratios in hand, Ray *et al.* (2016) could then compare their flying performance. They examined turn radius in flight, a maneuverability metric relevant most obviously to predator avoidance, with tighter turns meaning that a given fly can see-and-avoid more quickly than one with a wider turn radius. In activities such as foraging, a tighter turn radius would also contribute to the possibility of exploring the resource landscape on a finer grain before alighting than wider turn radii would allow. For our purposes, the most important result of their performance tests was that some aspect ratio variants had tighter turn radii, meaning that they were *more* maneuverable than wild-type flies. If selection favored traits independently, then maneuverability should be maximal: the experiment of Ray *et al.* (2016) shows that *Drosophila* wing shapes that confer greater maneuverability are, after all, developmentally possible. That such wing shapes are developmentally possible is a necessary consequence of the supposition that wing shape reflects multiple, opposing vectors of selection, which include maneuverability but also include other factors as well, such as sexual selection on "love songs" and visual displays.

Let's now return to the criticisms of Gould and Lewontin with Ray *et al.'s* (2016) results in mind. The adaptationist evo-devo approach reveals that Gould and Lewontin and similar critiques confuse *trait* optimality with *fitness* optimality. If developmental potential is sufficiently wide that other, functionally significantly different configurations are possible, then in general, the farther an individual is in phenotypic space from the mean value, the lower its fitness should be. That the area of commonly occupied phenotypic space corresponds to the highest fitness is what is meant by "fitness optimality." The position in phenotypic space corresponding to the highest fitness reflects the fundamental fact that organisms function as integrated wholes. Assuming that developmental potential is wide, selection in one direction on one trait will always be halted by selection acting in another direction, either on the same trait or on a different one linked to the first one in some way. This means that displacement away from the region of highest individual-level fitness will correspond to some traits having higher or lower performance values. *Pace* Gould and Lewontin, these statements are totally non-controversial (Arnold, 1983; Garland and Losos, 1994).

Supposedly, the hot conditions in which desert foxes live favor large ears because these provide a high surface area from which body heat is radiated. If this is the case, then there is a vector of selection that pushes toward larger ears. Presumably, there is some limit to the benefit provided by large ears, say when they snag excessively on shrubs or flop over the eyes, or when selection favoring energetic investment in, say, the immune system or musculature opposes selection on ear size (Figure 2.16). If we somehow relaxed selection favoring investment in the immune system or musculature, it would be no surprise if fox ears became larger. It would be no surprise if larger ears relative to body size radiate even more heat. In this imaginary experiment, when we relaxed selection along the opposing vector, the performance of ears as radiators increased when ear size increased.

In organisms as multifunctional wholes, individual *traits*, like ear size, are not "optimal" in the sense of being at the trait values (like ear surface area) that make the performance criterion (like heat radiation) maximal. Instead, selection maximizes individual-level fitness given the many competing vectors of selection, in the context of what is and is not possible in development. So, despite criticisms such as those of Gould and Lewontin, it's not panadaptationist naiveté to regard both decreases in performance, such as those encountered in *Drosophila* genital spike removal, or increases in performance, such as the increase in maneuverability found by Ray *et al.* (2016) as consistent with basic adaptationist expectations. In fact, expectations regarding both trait optimality and fitness optimality are entirely testable with adaptationist evo-devo tools. Even the possibility that trait distributions are shaped by factors other than selection is addressed by adaptationist evo-devo. The next example continues the theme of insect wings and maneuverability, illustrating the way that not just the performance of individual structures but fitness assumptions can be tested as well.

2.3.11 Butterfly wing proportionalities

One example in which biologists examined a component of fitness directly comes from studies of artificial selection on butterfly fore- and hind-wing allometry and the mating success of the resulting unusual variants. For nearly 30 years, a group of researchers, originally in Leiden and now scattered in various parts of Europe and the United States, have worked on the small African butterfly *Bicyclus anynana* (Brakefield *et al.*, 2009). With fast-cycling populations, markedly different color morphs between seasons, and an obliging inclination to laboratory life, this butterfly has provided decades of data on the causes of morphological disparity, the developmental mechanisms underlying diversity in behavior and color, and the causes of allometry, to name a few (Beldade *et al.*, 2002; Frankino *et al.*, 2007, 2010). It is to this last point, allometry, which we will direct our attention now.

Allometry is the proportionality relationship between organismal attributes, such as the examples of bone-length-body mass proportions or the relative lengths of stamens and petals in wild radish (Sections 1.11.1, 2.3.2, and 2.3.8). Frankino *et al.* (2007) examined the proportionality relationship between fore- and hind-wing area in *B. anynana*, in which larger hindwing sizes predict larger forewing sizes (Figure 2.17A). The scaling line leaves marked empty areas above and below it: we would like to explain both why the line has the slope and elevation that it does, as well as being able to explain why we do not observe individuals with morphologies corresponding to the empty areas above and below the empirical scaling line. Different explanations seem equally plausible.

On the one hand, there might be abundant variation in fore- and hind-wing proportionality as in Figure 2.17B. In any clutch or litter or brood or whatever term is used for butterflies, there might be some individuals with slightly larger forewings for their hindwing size, others with slightly smaller forewings, and still others with the usual proportionality. The relationship between wing area and body weight is a crucial one aerodynamically. Lighter bodies with relatively large wings produce configurations that are difficult to maneuver and at the mercy of air currents, neither of which would likely prove beneficial to a small flying insect. However, relatively large wings given weight provide good slow flight characteristics. Slow flight would presumably be of benefit to a butterfly maneuvering through the woods looking for food, or even in landing without damage. The opposite situation, relatively small wings given body weight, would allow faster, more maneuverable flight, excellent for zipping through the woods quickly, finding widely spaced food sources, and evading predators. Faster flying speeds permit greater directional control in the wind. The downside is that small wings are more vulnerable to separation of laminar flow, a phenomenon known as stalling, which results in a loss of lift and, if there is insufficient height from which to recover, in the insect falling to the ground (Betts and Wootton, 1988). Because of the intimate relationship between wing size and body weight, there is reason to expect wing area to change with body size.

There is also reason to think that within any given wing size–body size relationship, not just any proportionality between fore- and hind-wings will do. Butterflies, at least often, fly with their forewings slightly separated from their hindwings. When large jets have to fly slowly, as at landing and takeoff, they deploy leading-edge slats. These slats help direct air smoothly over the surface of the wing and lower the airplane's stall speed, mercifully allowing for safe maneuvering at lower speeds. Butterflies may be doing something similar in slow flight. By slightly separating the wings, the forewing

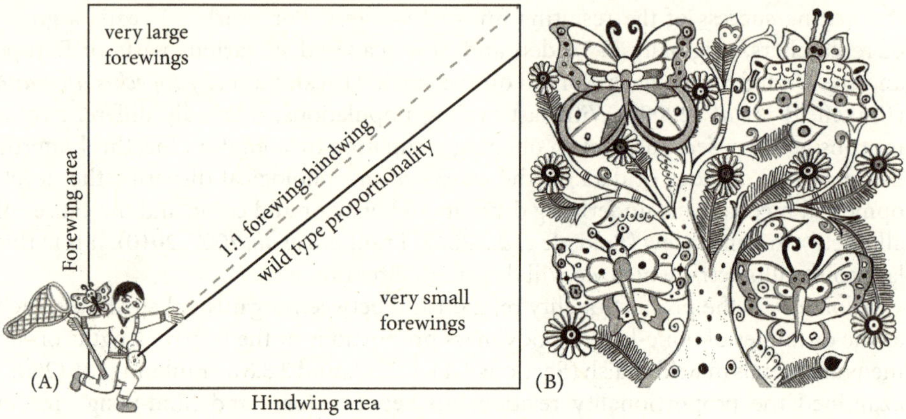

Figure 2.17 "Empty" spaces above and below *Bicyclus* fore- and hind-wing allometry. (A) This graph following Frankino *et al.* (2007) shows the relationship between forewing and hindwing area across some 700 adult female *Bicyclus anynana* butterflies. Hindwing area predicts forewing area very well, with a slope of 0.94, with the confidence interval not including a 1:1 slope. The empty spaces above and below the scaling line demand explanation—are they empty because they are developmentally inaccessible or are they developmentally accessible but of low fitness? (B) Eduardo's take on the occupants of these different areas, including 1:1 forewing to hindwing proportionalities, very large forewings relative to hindwings, and very small forewings relative to hindwings. The empirical wild-type pattern does not reveal whether these variants away from the line are developmentally possible or not. Drawings by Eduardo Martínez Díaz.

directs air over the hindwing, making its flow of air less likely to become turbulent (Betts and Wootton, 1988; see also Lindhe Norberg, 2002 and references therein). Not just any fore- and hind-wing proportionality would be expected to provide the ideal combination of laminar flow and maneuverability. There would seem to be a good reason, then, to expect that hind- and fore-wing sizes across butterflies of different body size should show a distinct proportionality and that this proportionality should be maintained by selection in current populations.

Even though it's easy to come up with adaptationist explanations for the proportionality between fore- and hind-wings in insects, there may simply be no developmental alternative. Fore- and hind-wings appear to be serially homologous structures, structures derived from the duplication of an ancestral feature (Fusco, 2022 provides a helpful exploration of serial homology). Serial homology is widespread in animals, appearing in structures that repeat along the body axis, such as vertebrae in mammals or limb pairs in arthropods. In many cases, these repeated parts evolve into regionally distinct structures with different roles. Mammalian vertebrae, for example, have diversified to such a degree that we distinguish cervical, thoracic, and lumbar regions, and even name individual elements like the atlas and sacrum. Even though serially homologous structures may diverge to lesser or larger degrees in their structure and function, they still possess features that identify them as betokening the "same" sort of structure (Patterson, 1982). The shared evolutionary origin of serially repeated structures may mean that they share regulatory bases so fundamental that truly marked developmental independence is impossible (e.g. Cullinane, 2000; Young and Hallgrímsson, 2005). The proportionality between butterfly fore- and hind-wings could therefore be due to the sharing of essential developmental pathways, even though selection would favor other proportionalities if they were possible. This interpretation leads to the expectation that major, functionally different developmental alternatives to the usual scaling pattern are developmentally impossible.

Confronted by distinct explanations for the same scaling pattern (Figure 2.17A), Frankino *et al.* (2007) turned to artificial selection. Starting with the minute variation in laboratory populations, they bred the individuals with the largest forewings for a given hindwing size with one another in one line. This line was intended to see if the population could be moved into the empty area above the scaling line. They also set aside the individuals with the largest hindwings for a given forewing size and bred those with one another. This line would test the accessibility of the empty space below the wild-type scaling line. They made a point of selecting individuals all along the absolute size range of the butterflies. If they hadn't done this, they might have simply picked the largest or smallest individuals in absolute size, which would have altered the slope of the line rather than shifting the proportional relationship upward or downward.

They began their experiment by measuring butterflies, hundreds of them. From the 750 measured, they selected for each line, the up, down, and control lines, the 30 most extreme females for the next round of breeding (the males were selected at random). With each round of selection, the authors measured all of the 150 to 220 females produced and again selected the 30 most extreme ones. Measuring wing area on butterflies that then go on to breed obviously requires a delicate touch—their technique was obviously gentle enough that it allowed them to cross these individuals with randomly selected males from the same lineage. After 12 generations—this must have been a great deal of work—the authors compared their results with their hypothesis.

Frankino *et al.* specified a clear prediction regarding the results they should find if fore- and hind-wing allometry was caused by a lack of developmental alternatives. They reasoned that if artificial selection could readily move the population in fore- versus hind-wing space, then this would indicate abundant, heritable developmental variation on which selection could act. On the other hand, if under repeated bouts of selection the population refused to budge, then this would suggest that heritable variation of the sort required to observe selection was, for some reason, lacking, and open the door to understanding the cause of this lack of variation. They found clear evidence that ontogenetic alternatives were indeed possible (Figure 2.18A). The slope of the relationship remained essentially unchanged; what shifted was the y-intercept, revealing that the wild-type proportionality was just one of many developmental possibilities. The authors therefore concluded that the wild-type proportionality is most likely the fittest of the many developmental possibilities available.

Interestingly, the line in which selection favored relatively large hindwings largely stopped responding to selection after about six generations, and despite the repeated application of selection, the relative hindwing increase stopped. The authors investigated the cause of this apparent barrier to movement in morphospace with additional experiments, and found evidence for a lack of heritable variation. The cause of this bottoming out clearly deserves further attention. However, for the purposes of their hypothesis, it is clear that abundant developmental potential exists such that y-intercepts alternative to the commonly observed ones could be occupied, and a sufficient swath of morphospace is thus manifestly occupiable such that even the slope of the allometric proportionality could be different.

Having created large-hindwing and large-forewing variants, the authors could now compare the performance of these variants to the wild type (Figure 2.18B). The hypothesis that the wild-type scaling line falls where it does because of selection is to say that the other possible positions for the line represent morphological configurations that are lower in fitness. Frankino *et al.* focused on mating success. They released males of their variant butterflies into a free-flight greenhouse, where large forewing, large hindwing, and control males could interact with wild-type females. The fitness of the variant males, as measured by mating success, was abysmal (Figure 2.18B). For every variant male that mated with a female, three normal males mated. The reasons for this disparity remain to be investigated. It is unclear whether female choice is involved, say females preferring males with a certain proportionality. Or butterfly aerodynamics could be raising its head, with clumsy flight performance putting variant males at a disadvantage in alighting next to appropriate females. Whatever the case, the fitness difference between variants produced in so few generations of selection reveals just how swiftly fitness can unravel when a population ventures into "empty" regions of developmental possibility.

Before leaving the *Bicyclus anynana* study system, we can turn to an example of presence-absence features. With examples of wild radish stamen-petal allometry and butterfly fore- and hind-wing proportionality, we have examined examples of continuous traits. To be sure, a major component of morphological diversity across clades is continuous variation, manifest as variation in the sizes and proportions between parts. However, the presence or absence of some features, think the spots of the jaguar versus the solid tawny of the lion, also participates in describing the arc of diversity. In investigating the developmental potential and selection underlying diversity in qualitative traits across species, authors have also turned to selection experiments in *Bicyclus*. Beldade *et al.* (2002) studied color patterns on the wings of *B. anynana*. The genus name refers to the circular eyespots often present on the wings. These eyespots are often regarded as having some sort of antipredator effect (e.g. Cordero, 2001). The name *Bicyclus* suggests that there must always be two spots, and most species indeed do have two. But across the forewings of the eighty or so species of *Bicyclus*, there can be two spots, one spot, or no spot. When there are two spots, they are in characteristic positions, one anterior and one posterior. The spots can be both large, both small, or a combination. The possibilities of presence or absence, large or small, anterior or posterior define a range of possible combinations (Figure 2.19). Some of these are not observed in any *Bicyclus* species, such as the pattern of lacking an anterior spot in combination with a large posterior spot.

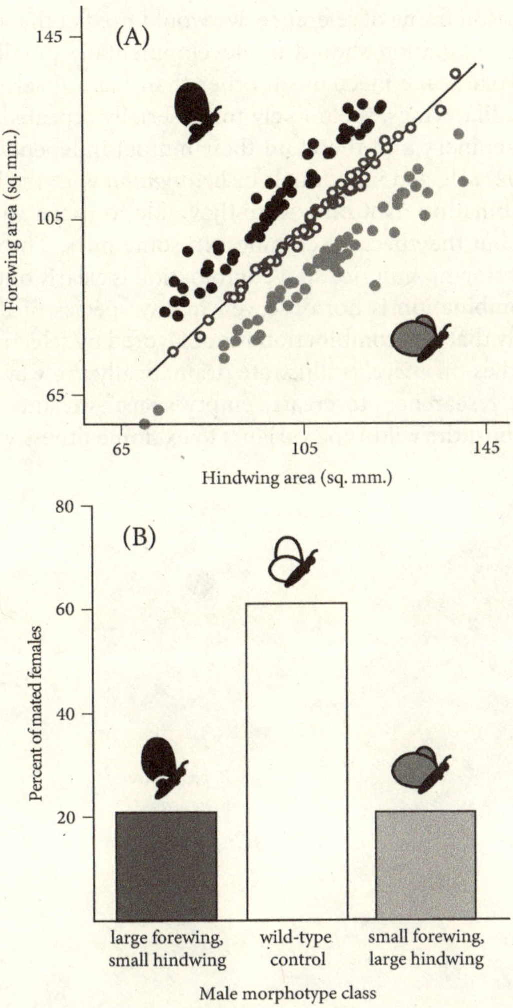

Figure 2.18 Drastic reduction in the mating success of male "empty space" *Bicyclus* variants. (A) The wild type is represented by hollow circles, and the line shows the wild-type allometric relationship between forewing and hindwing area. Solid dots represent "empty space" variants that Frankino *et al.* (2007) generated through artificial selection. The black dots are males with forewings "too large" for their hindwings, and the gray dots are those with forewings "too small." The graph dramatically shows that selection produced variants far outside what is observed in wild-type populations. (B) This variation, moreover, is strongly associated with fitness differences. The bars show the percentage of females that mated with males of three categories, which were wild type, forewings "too large," and forewings "too small." The mating success of both types of "empty space" males is markedly lower than that of the wild type. Frankino *et al.* (2007) provide a dramatic case study showing that manipulation experiments, including artificial selection, can generate living, breathing "empty space" variants, forms never seen in standing variation, and that once in hand, their performance and fitness can be directly compared to the wild type. Figure adapted with permission from Frankino *et al.*, 2007. Internal and external constraints in the evolution of morphological allometries in a butterfly. *Evolution* 61: 2958–2970.

From the adaptation frame of reference, we would predict that the large-posterior-no-anterior spot combination should be developmentally possible. Otherwise, we would need to invoke some mechanism other than natural selection to explain its absence. Eyespots, like wings, seem likely to be serially repeated structures sharing developmental machinery and therefore their mutual independence might be low. Beldade *et al.* (2002) selected individuals of *B. anynana* with the large-posterior-no-anterior spot combination. Not only were they able to breed variants with exactly this combination, but they became common in some lines. The authors concluded that the large-posterior-no-anterior spot combination is clearly ontogenetically accessible. That this combination is not observed in any species of *Bicyclus* in the wild makes it seem likely that this combination is not favored by selection under wild conditions. These studies on *Bicyclus* illustrate dramatically the way that manipulation experiments allow researchers to create "empty space" variants, and with them in hand, pit them against the wild-type variants to examine fitness variation directly.

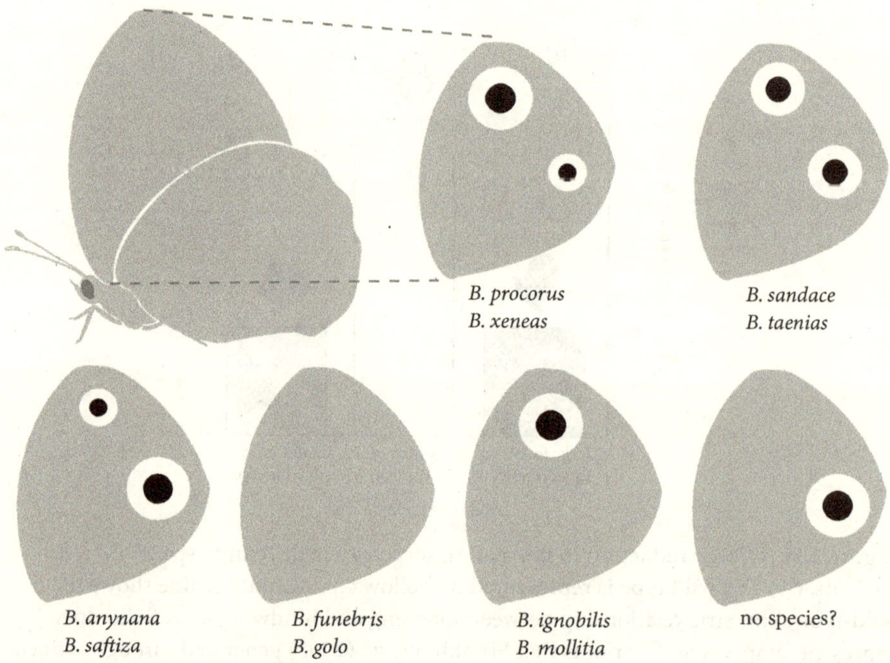

Figure 2.19 *Bicyclus* forewing eyespot morphospace. Many combinations of forewing eyespots are observed across *Bicyclus* species, including only a leading-edge eyespot, as well as no eyespots. The plausibly imaginable condition of no leading-edge eyespot and a large trailing-edge eyespot is, apparently, not observed. A natural first step in empirical research is to find out whether an unobserved condition is possible but not favored or developmentally inaccessible, and if so, why. Beldade *et al.* (2002) found that the combination is readily produced under artificial selection. Figure based on Beldade *et al.* (2002).

2.3.12 Eppendorf tube flowers, 3D printed insects, and virtual morphospaces

In many cases, even under the most painstaking artificial selection or the most delicate surgery, the necessary variation is not present. In these cases, biologists can break out the clay, the paint, and the computer to construct models, physical and *in silico*. Plant biologists have wondered for a long time whether differences in inflorescence structure, the ways flowers are arranged, affect the movement of insect pollinators from flower to flower. However, across species, different inflorescence structures come bundled with a host of confounding variables, such as flower type, color, number, nectar content, and more. To minimize this variation, Jordan and Harder (2006) turned to constructed models (Figure 2.20). Flowers were represented by painted Eppendorf tubes, which they mounted on supports designed to mimic different inflorescence types, each with the same number of flowers. They even filled the tubes with a standardized sugary "nectar." This level of uniformity, unthinkable in natural systems, allowed them to isolate the ways that differences in flower positioning alone affected the movement patterns of various bumblebee species (including a colony they purchased from a supplier; I had no idea that making a living selling bumblebee colonies was an option). Because all other features were held constant—flower color, nectar content, flower size, and so on—they could confidently attribute behavioral differences to inflorescence architecture itself. Their results showed that flower arrangement significantly influenced the consistency of foraging paths among flowers, which in turn shaped patterns of pollen import and export and reduced self-pollination. By using constructed variation, Jordan and Harder made it far more likely that what they observed in pollinator behavior truly reflected responses to spatial flower arrangement, rather than being tangled up with the confounding variation that inevitably accompanies natural systems.

In an animal example with as clear an adaptationist evo-devo motivation as one could hope for, Taylor *et al.* (2025) lament "the difficulty of assessing the fitness of phenotypes absent among extant species." Their focus was Batesian mimicry. In Batesian mimicry, harmless mimics resemble poisonous or otherwise harmful models. Just as predators avoid the dangerous models, they also tend to avoid the mimics. A long-standing question in this area is how similar the mimic must be to the model in order for selection to favor mimicry as a predator deterrence strategy. This is clearly a question that hinges on the unique perceptual systems of different predator species. Natural variation among insects isn't fine-grained enough to determine what degree of resemblance counts as successful mimicry, nor to tease apart which features fool which types of predators. Ingeniously, Taylor and colleagues turned to 3D printing. They scanned hoverfly mimics and their target wasps, which are avoided by birds because of their stings, then interpolated intermediate shapes, sizes, and colors to produce full-color printed models from key points along the spectrum. Because these were artificial models, the researchers could vary shape, size, color, and pattern independently, allowing them to disentangle the effects of each. They then presented

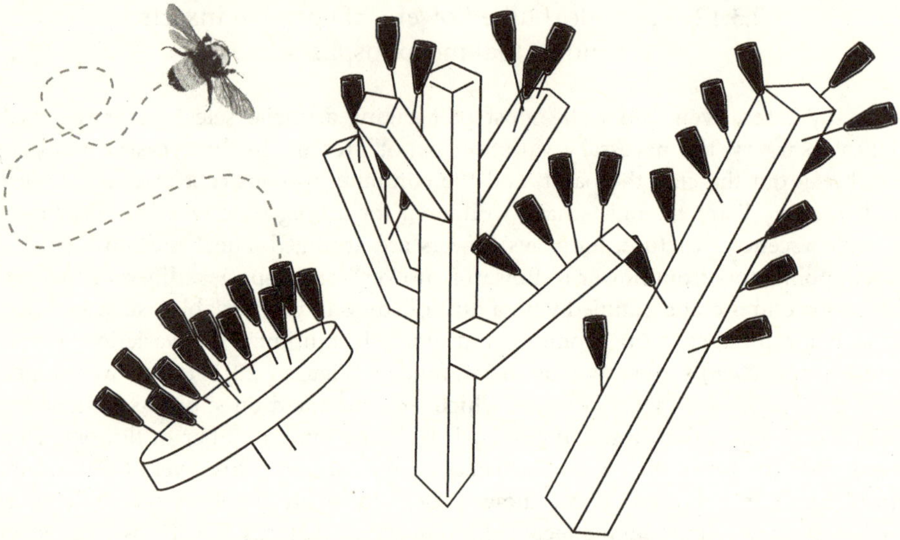

Figure 2.20 Constructed variation. Different arrangements of flowers are expected to produce different movement patterns by pollinators, and therefore different patterns of pollen exchange and fertilization. But these differences are usually observed across species, which vary not just in inflorescence architecture but also in flower size, color, scent, nectar content, and more. This makes it difficult to isolate the effects of spatial arrangement alone. Jordan and Harder (2006) addressed this by building artificial inflorescences, each with twelve identical flowers matched in color, size, and nectar volume and arranged them in three configurations (left to right): umbel, panicle, and raceme. This allowed them to show that bumblebees do, in fact, follow distinct flower-to-flower foraging paths solely in response to inflorescence structure. Bumblebee from Lovell (1918).

the models to bird predators, which behaved just as predicted: the birds most readily attacked the least wasp-like models and became increasingly hesitant as the models became more wasp-like. The authors concluded that their results show that predators can discriminate even tiny differences between mimics and models. They found that color was especially important in predator discrimination, and ended by stating that testing their hypotheses "would not be possible using only real insect specimens" (p. 7).

In addition to physical models, theoretical morphospaces offer a structured lens through which to investigate potential forms within a group. These spaces are defined by selecting variables that are shared, ideally, across all members of the group in question. As with empirical morphospaces, this requirement imposes a major limitation: it is impossible to plot individuals that lack those variables within the same coordinate system. That limitation aside, theoretical morphospaces allow researchers to generate complete, a priori maps of form by systematically varying key traits (McGhee, 2007). Classic examples include David Raup's pioneering work in the 1960s on mollusc shells. He defined a three-dimensional morphospace using coiling parameters: the whorl expansion rate (how quickly the coil grows), the translation rate (how far the shell displaces along its vertical axis), and the distance of the generating curve from the coiling axis (how far the tube forming the shell is offset with each coil). Later versions sometimes added a fourth parameter for the shape of the generating curve itself. Raup compared these spaces with those actually occupied; findings included that real ammonite shells tended to cluster in regions associated with minimization of shell material for a given strength and space for the body, as well as hydrodynamic efficiency, while large regions of the morphospace remained unoccupied (Clements *et al.*, 2008; McGhee, 2007; Raup, 1961, 1966, 1967; Raup and Michelson, 1965). In plants, Karl Niklas used variables such as stem taper, length, and branching angles to create a multidimensional space encompassing most of the diversity of plant architectures, from whisk fern-like forms to Christmas tree shapes and flat-topped acacias (Niklas, 2009). Strikingly, every form his model generated matched a real plant species, living or extinct, suggesting that what may seem inviable in one context (like a tree that collapses under its own weight) may be perfectly viable in another (like a liana that relies on external support). Theoretical morphospaces thus allow scientists to construct entirely filled spaces, and with these explicit combinations, they can explore where empirical limits lie and why.

2.3.13 Seven cervical vertebrae

A standout example will introduce a key empirical adaptationist evo-devo theme, which is the inextricability of adaptation and "constraint," coming from Frietson Galis and her colleagues' (2006, 2018, 2021) work on cervical vertebrae in humans. A celebrated example of limited developmental potential is the pattern for (virtually) all mammals to have seven cervical vertebrae (Bateson, 1894; Narita and Kuratani, 2005; Hautier *et al.*, 2010; Varela-Lasheras *et al.*, 2011; Galis *et al.*, 2022). This example is celebrated and conspicuous because seven vertebrae seem so clearly impossible to be the optimal number in situations as different as, say, whales, humans, and giraffes (Uller *et al.*, 2018).

These extremely different lifestyles would reasonably seem to befit differing numbers of vertebrae. Indeed, in cases in which neck length relative to body size is similar to that of the giraffe, many extant animals, such as birds, and extinct ones, such as sauropods, have much higher numbers of cervical vertebrae (Figure 2.21). By contrast, mammals seem to maintain a strangely fixed vertebral count, despite the wide range of neck lengths and functions found across the group. This breadth of functional variation across the clade would seem to demand commensurate variation in cervical vertebra number. The low variation is thus an example of "limited" developmental potential, again because of adaptationist expectations. If development weren't limited somehow, surely selection would favor higher or lower numbers of cervical vertebrae in different situations.

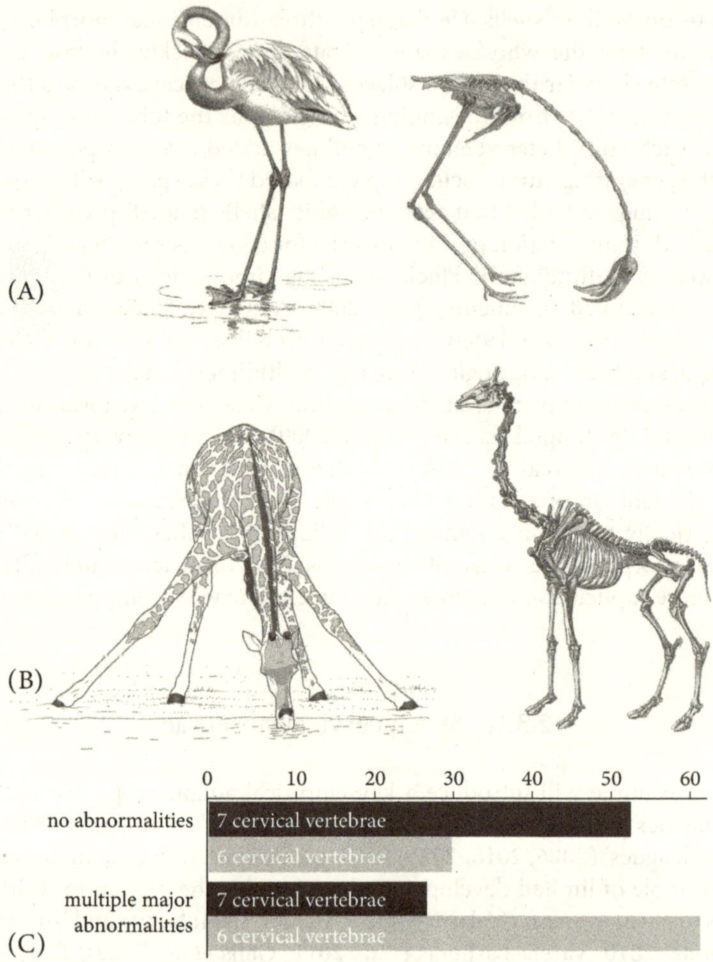

Figure 2.21 Seven cervical vertebrae: "constraint" or "selection"? (A) That birds with long necks, such as this flamingo, routinely have more than seven cervical vertebrae shows that vertebrates are developmentally capable of producing higher numbers of vertebrae. (B) Just seven cervical vertebrae in giraffes are associated with a stiff neck, requiring the trademark awkward, vulnerable splay in reaching water. With their more numerous vertebrae, flamingoes reach the water readily. (C) The Y-axis in this graph shows two categories of human fetuses from the study of Galis *et al.* (2006), those with no abnormalities and those with multiple, major abnormalities. The higher the number of abnormalities, the lower the likelihood of survival. The X-axis gives the percentages of fetuses within each category with the usual seven versus six cervical vertebrae. Within the normal fetuses, the majority had seven cervical vertebrae; remember that these were fetuses that had died, so the proportion of seven versus six cervical vertebrae in the general population is even higher. The fetuses with the highest numbers of abnormalities were consistently associated with only six cervical vertebrae. This situation could be considered to reflect "constraint" in that viable fetuses with only six cervical vertebrae are very unlikely to be viable and survive to adulthood and thus not part of the pool of

If the widespread pattern of seven cervical vertebrae across nearly all mammals were the result of natural selection acting on a wide array of developmental possibilities, the biological world would be an elaborate theater of coincidences. If the explanation for why so many mammals have seven cervical vertebrae were a standard selection scenario, then every species—whales, giraffes, humans, bats—would need to exhibit heritable intraspecific variation in vertebral number (say, five to ten neck vertebrae per species). These conditions seem possible and reasonably plausible. The part that beggars belief is that selection would sift through this variation for entirely different, species-specific reasons, yet somehow always arrive at seven. In blue whales, for instance, calves with more than seven neck vertebrae might jackknife tragically while diving, their heads snapping back with fatal fluid-dynamic consequences. Those with fewer than seven might swim like stiff-backed torpedoes, unable to make the fine head adjustments needed to feed, mate, or dodge orcas. But with seven, suddenly the whale is a marvel of aquatic grace: its neck flexes just enough to aid streamlining, but not so much as to be structurally compromised. Meanwhile, giraffes with only six cervical vertebrae would be too stubby to reach the upper branches or deliver a decisive blow in combat with other males. Giraffes with eight or more vertebrae might have long and elegant necks, but tragically whippy ones, vulnerable to wind shear or sudden neck entanglement during dominance battles. Seven cervical vertebrae provide giraffes with the perfect balance between combat and cuisine: stiff enough to deliver bone-shattering blows in courtship duels, but limber enough to weave between acacia thorns. And in humans, we must imagine similarly bespoke selective forces at work. Individuals with more than seven cervical vertebrae might have been prone to over-rotation of the head, leading to existential dizziness, frequent fainting during philosophical contemplation, or an inability to hold down a desk job. Those with fewer than seven might walk with the tragic rigidity of someone trying not to spill soup. But seven? That, of course, allows for the perfect nod, enough to signal agreement across a crowded room, but not so much that you sprain your neck when checking your blind spot. If all of this were true, it would imply that in species after species, across every imaginable environment and lifestyle, natural selection has independently discovered that seven is not just good, but perfect, every single time. This scenario is so implausible that no biologist entertains it. Instead, in seven cervical vertebrae they see a perfect place to look for the clearly non-adaptive causes of limited developmental potential.

Figure 2.21 *(continued)* coexisting variants. By the same token, this situation could be interpreted as representing selection because the variant of six cervical vertebrae can be produced, but is simply lower in fitness than the variants with seven. Pronouncing in favor of one or the other in fact adds no clarity to Galis *et al.*'s account, illustrating the inextricability of "constraint," however interpreted, and adaptation in any explanation of organismal form. Figure adapted from Galis *et al.* (2006); the percentages in the categories don't quite sum to 100 because Galis *et al.* included conditions in addition to just the seven versus six cervical vertebrae shown here. A and C are reproduced with permission from Galis *et al.* (2022).

When detailed empirical information is in hand, though, it becomes difficult—indeed, seems unnecessary—to declare that adaptation or limited developmental potential as the chief architect of the pattern of seven cervical vertebrae in mammals. This empirical information comes from the study of Galis *et al.* (2006; also Ten Broek *et al.*, 2012). The authors realized that a hospital in Amsterdam took and saved X-rays of fetuses and babies that died there, and autopsy reports were also available. The authors pored over nearly 600 radiographs in their initial study, later increased to 1400 (Ten Broek *et al.*, 2012), checking for the presence of cervical ribs. Typical cervical vertebrae are not associated with ribs, whereas thoracic ones are. So, a baby with cervical ribs at one vertebra would effectively have just six cervical vertebrae. They also checked for the absence of the first (uppermost) thoracic rib. When the uppermost thoracic vertebra lacks ribs, then effectively the neck has eight cervical vertebrae. Together, such cases show without a doubt that in addition to the usual seven, six and eight cervical vertebrae are indeed developmentally possible. Galis *et al.* then scored data from the autopsy reports for congenital and cytogenetic abnormalities.

What they found was striking, not just because of the strong statistical association, but because the results so defy easy classification into "constraint" or "adaptation" categories. As summarized in Figure 2.21, babies with severe abnormalities had a much higher likelihood of having cervical ribs as compared to babies with no abnormalities. The authors note that "at least 78% of the individuals in recognized pregnancies with a cervical rib die before birth and 83% before one year...For individuals without cervical ribs 9% are dead by one year of age" (p. 2647). This low viability of babies with cervical ribs was not due to the rib presence per se, but rather associated conditions, such as spina bifida, cyclopia, anencephaly, and a host of other developmental tragedies. Galis *et al.* (2006, 2018, 2021) note that the identity of cervical vertebrae is determined during early organogenesis with head-to-tail patterning involving *Hox* genes, when cell division, migration, shape changes, segmentation, and left-right symmetry maintenance are all tightly coordinated across the embryo. Because of this intense, system-wide coupling, teratogenic disturbances often induce cervical ribs along with other malformations, frequently marked by asymmetry.

The establishment of vertebra identity (e.g. whether a vertebra has or lacks ribs) is one such crucial early developmental moment, and the results of Galis *et al.* are consistent with variation between individuals in the establishment of vertebra identity at this early stage being associated with profound differences in survivorship. In animals with many cervical vertebrae, such as birds, vertebra identity is determined after this intensely interactive phase (Galis *et al.*, 2021). Given this developmental timing, and in contrast to mammals, changes in cervical vertebra number in birds do not set off as many deleterious changes elsewhere in the embryo. And talk about putting the "evo" in evo-devo: the authors even found that late Pleistocene mammals, such as Irish Elk, woolly rhinoceros, and woolly mammoth, started showing a high incidence of cervical ribs as they declined into extinction (Reumer *et al.*, 2014; Van Der Geer and Galis, 2017; Cuxart-Erruz *et al.*, 2024). These data suggest that eight or six cervical vertebrae should be developmentally possible in mammals at large, but that in most species these conditions should be associated with low survivorship (Varela-Lasheras *et al.*, 2011).

Both "constraint" and "adaptation" advocates will claim this situation as their own. "Constraint" advocates will point to the low viability of non-seven cervical vertebra individuals. Individuals with six cervical vertebrae only rarely reach adulthood. So, they might argue, from the perspective of natural selection, it is as though the "six" condition were impossible to produce. "Constraint" can usefully describe a situation in which variation in a trait is low within a population due to limited developmental potential. Because it's so unlikely that individuals with six or eight cervical vertebrae will be present in adult populations, this could be considered a situation of limited developmental potential and thus "constraint." But adaptationists will point to variation in survivorship as a key aspect of natural selection. In the case of cervical vertebrae, variation in cervical vertebra number is developmentally possible, as the hospital X-rays and the data of Galis *et al.* document. This developmentally possible variation is simply associated with survivorship differences. Differences in survivorship between developmentally possible variants are just a part of natural selection. Moreover, individuals with six cervical vertebrae that do survive to adulthood seem to have a propensity to spontaneous abortion of their fetuses, meaning that these individuals might have lower fecundity than those with seven cervical vertebrae (Schut *et al.*, 2020). Because they can point to variation in survivorship, and perhaps fecundity, associated with variation in cervical vertebra number, adaptationists can construct a case to defend the posture that the low variation in cervical number in mammals is due to selection. Both these "constraint" and "adaptation" interpretations have their merits, and there really isn't a way to select one or the other as being the correct one.

The study of Galis *et al.* is a particularly vivid illustration of a principle that applies in all explanations of organismal form. In all cases, when sufficient detail is in hand, it is impossible to adjudicate unambiguously for "constraint" or "adaptation" as being a complete and satisfying explanation for any given pattern. Instead, in all cases, selection is the shaping of development, and developmental potential circumscribes selection's Lebensraum, such that all explanations must include this situation in all its inextricability (Schwenk and Wagner, 2004).

2.3.14 *Drosophila* sex combs

Once the biological mechanisms underlying the production of different sets of developmental variants are elucidated, the need for vague terms such as "constraint" evaporates, because we have in hand a precise description of the mechanism.[1] A stellar example of this inextricability is provided by the evolution of sex combs in *Drosophila* by Malagón *et al.* (2014). The legs of both male and female *Drosophila* species are covered with bristles, which are arranged in rows that are more or less transverse with respect to the proximo-distal axis of the leg. In male *Drosophila*, some of these bristle

[1] adapted from Olson (2019a). Overcoming the constraint-adaptation dichotomy: long live the constraint-adaptation dichotomy. In: Fusco G, ed. Perspectives on evolutionary and developmental biology. Padova, Italy: University of Padova Press, 78–94. With permission from University of Padova Press.

rows are arranged longitudinally and, rather than fly-like and unkempt, are of sufficiently regular shape, length, and even of dark pigmentation, so as to look like small black combs arrayed with their long axes parallel to the limb main axis. These play some ineffable rôle in titillating or at least maneuvering female flies (Ng and Kopp, 2008; True, 2008; Malagón *et al.*, 2014). In *D. melanogaster*, the sex combs form transversely and rotate in development to their mature longitudinal position. The number of teeth in a sex comb, and therefore how long it is, varies markedly and responds readily to selection. The longer a comb, the more likely it is in its rotation to blunder up against one of the transverse rows of normal bristles. Producing longer combs does not automatically mean that the transverse bristles make way for the sex comb as it rotates in the pupa to its mature, longitudinal position. This obstinacy on the part of the transverse bristles means that the proximal ends of longer combs become distorted as they crowd against the transverse rows. This distortion leads to combs that presumably have lowered functionality.

As in the example of seven cervical vertebrae, Malagón *et al.* describe a situation that can be seen both from the viewpoint of selection as well as "constraint" beyond selection. In this example, adaptationists will see the all-powerful hand of selection at work: of the developmentally possible variants, some "work" better than others and are favored. Fans of constraint will instead see the crucial role of development in driving evolution: the course of development, indifferent to function, involves a competition for developmental resources, in this case, space, and a bent sex comb grudgingly abutting and distorted by a transverse bristle row is the result. Declaring the pattern of *Drosophila* sex comb diversity and distortion as the result of selection or constraint adds nothing to the detailed description that Malagón *et al.* (2014) offer. Malagón *et al.* and the other case studies in this section illustrate that events in development are directly relevant for studying adaptation and for studies of organismal form in general. They give empirical expression to the widely recognized but often tacit assumptions about developmental accessibility or inaccessibility of functionally significant different variants adjacent to the commonly observed ones. Moreover, they illustrate that the field of the possible is an object of study in itself. One final case study will drive home this point.

2.3.15 Carotenoid-based coloration in birds: selection acting in a finite space of possibilities

Alex Badyaev's work on bird coloration is enough to convince even the most diopter-challenged biologist that these twitchy, hard-to-spot animals are more than just a punishment for masochistic early risers. His apparent patience with their maddening habit of flying away has paid off handsomely, yielding one of the clearest examples available of the inextricability of developmental potential and selection. Much of Badyaev's work is on passerines (Landeen and Badyaev, 2012; Badyaev *et al.*, 2017), small, fluttery songbirds that, to the civilian observer, look more or less all the same at first glance. His work delivers the pleasant surprise that they are, in fact, varied in the yellows, oranges, and reds they display on their feathers, colors that are based on carotenoids, carbon-rich pigments with chains of alternating double bonds, whose molecular structure produces vivid yellows, oranges, and reds. Sexual selection is a major driver in the evolution of this coloration in birds. Redder hues, most often expressed in males, are costly and frequently favored by females during mate choice, and signal condition through traits such as immune function and food intake (Delhey *et al.*, 2023).

By this point, dear reader, you will know that the "no free lunch" principle is coming, and in this case, despite the importance of carotenoids in passerine lives, birds at large (not just passerines) cannot synthesize carotenoids and must obtain them from their diet. Yellow carotenoids are more common in the environment and can be enzymatically converted into red ones (Delhey *et al.*, 2023). The enzymatic transformations possible for a given dietary carotenoid are limited, though. Getting to a given final product from a given dietary starting point must go through a certain series of enzymatic transformations and intermediates. While a few compounds can only be reached from a single dietary entry point, most of the carotenoids in the network can be reached from any starting point. However, the distances between them, understood as the number of enzymatic transformations, vary greatly, from just one step to, at least in principle, over ten (Figure 2.22).

The space of carotenoids that selection can favor in birds is therefore highly structured, in various ways (Morrison and Badyaev, 2016, 2017; Badyaev *et al.*, 2019). For one, certain carotenoids require ingestion of certain precursors, so the enzymatic transformations that are possible limit what compounds can be favored by virtue of the availability of precursors in bird diets. Moreover, even though most compounds can be reached from most starting points, enzymatic transformation is associated with energetic investment, meaning that more distant transformations should generally be more costly. Therefore, of the limited set of possible transformations, selection is likely to favor minimization of transformations because it minimizes metabolic investment. However, sexual selection could favor the costly carotenoids, often red ones, with the healthiest males being able to tolerate the costs of these more elaborate transformations (McGraw *et al.*, 2005). Moreover, in at least some cases, the yellow-to-red transformation is even associated with oxidative stress (García-de Blas *et al.*, 2016; Delhey *et al.*, 2023), with males that are able not only to invest in more enzymatic steps but also tolerate oxidative stress being favored by females. In some species, carotenoid-dependent coloration is associated with reduced male parental care, suggesting a trade-off in which males invest more in attracting mates than in offspring care (Rincón-Rubio *et al.*, 2023).

As with the previous two cases, both selection-focused and "constraint"-focused interpretations can plausibly claim Badyaev's case as their own, which precisely illustrates the inextricability of selection and developmental structure in explaining organismal form. Those who see selection as the main shaper will point to the structured yet chromatically very diverse array of possible carotenoids available, with sexual selection favoring certain pigments, balanced by countervailing pressures favoring lower energetic costs, simpler transformations, or reliance on broadly acting enzymes. Meanwhile, those who view "constraint" as the primary force shaping form will highlight the regimented pathways of biochemical transformation, the limited dietary inputs into the network, and the obvious fact that only certain compounds are available from any given starting point. Both views are defensible because

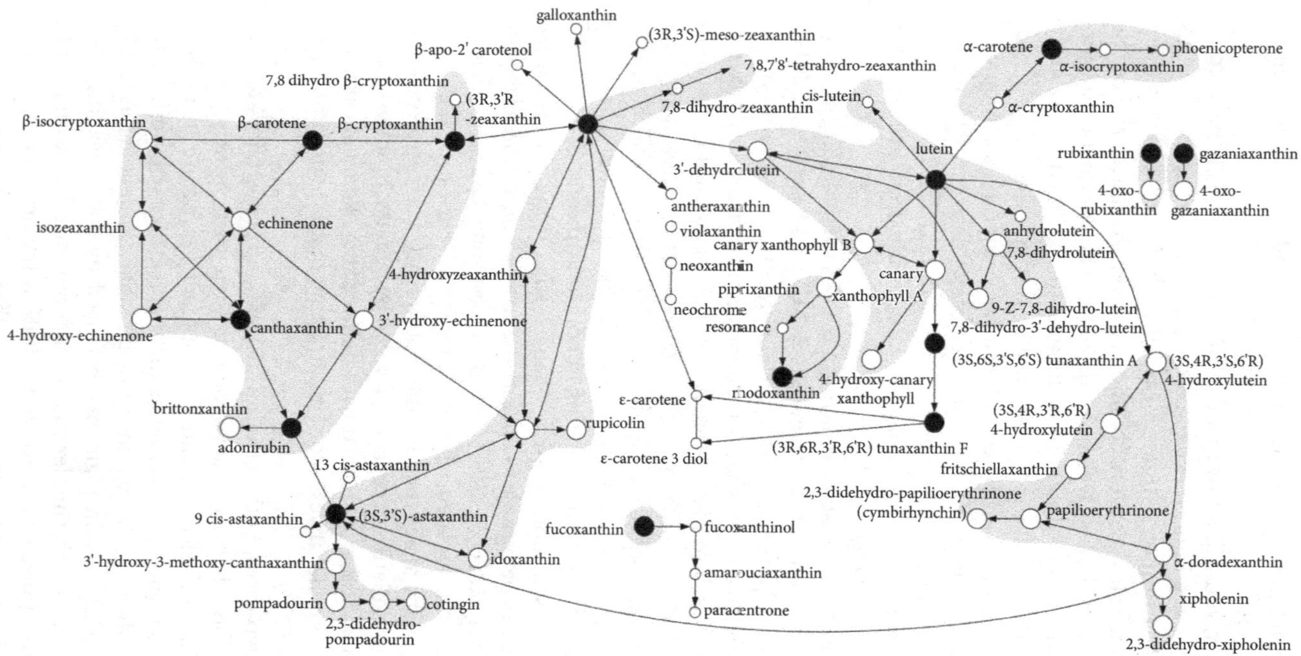

Figure 2.22 Highly structured field of potential variation: the avian carotenoid network. This network shows dietary carotenoids as black circles and metabolic derivatives as white ones. The arrows represent enzymatic transformations between compounds. The gray fields designate what the authors identified as "functional modules," groups of compounds with more enzymatic transformations (black lines) among themselves than with compounds outside the group. The first thing to notice in this network is that not all products are possible from all starting points. Therefore, the dietary availability of carotenoids is one obvious factor that can limit or direct the options that can be favored by selection. Based on data from 250 species across 11 bird orders, from Morrison and Badyaev (2016). Structuring evolution: biochemical networks and metabolic diversification in birds. *BMC Evolutionary Biology* 16: 168. Reproduced under the terms of the CC BY 4.0 license (creativecommons.org/licenses/by/4.0). Colors slightly modified for black and white reproduction and some labels removed.

developmental possibility—broadly construed—shapes the variants that selection can favor, and is itself shaped by selection.

2.4 Toward complementarity

Each case study in Section 2.3 illustrates just how painstaking evolutionary biology is. Each study represents time, patience, and resources, and together they represent thousands upon thousands of measurements. And yet, any single study contributes just a brick or two—valuable and admirable bricks, to be sure—in the larger edifice of evolutionary explanation. Showing that bats and whales occupy the "empty space" in the bone-body mass scaling pattern of terrestrial mammals demonstrates that such combinations are possible in mammals, but it doesn't prove that *terrestrial* mammals can produce them, nor that heritable intrapopulation variation with fitness consequences exists within species. Studies of single species, from dogs and pigeons to radishes, goldfish, and *Arabidopsis*, open rich windows onto variation and population-level processes, but say nothing directly about other species. This is, it seems, the best we can hope for as individual scientists: to spend our careers adding a few well-placed bricks, often specializing, in fact, in just one kind of brick. This one-brick-at-a-time quality has profound implications for how biologists build explanations of organismal form, a topic we take up in Chapter 5.

First, though, the reader may have noticed that the authors of the studies discussed so far rarely found it necessary to invoke specific genes. If genes are the ultimate directors of development, as is often claimed, then why aren't they more central in these evolutionary explanations? This question leads directly to the view of genes as components of broader developmental systems. Taking this systems perspective opens up an adaptationist evo-devo treasure chest, or at the very least, an engaging and revealing debate. We now turn to it.

3

Evolution and developmental systems

Watching an organism develop is always mysterious. The way a tiny mass, whether it's a frog embryo or the first stirrings of a leaf primordium, somehow turns into a larger and more complex individual or organ is striking for many reasons. Particularly relevant for us here is that, in general, in normal development, this transformation outwardly appears so similar across individuals of a species. Each developing individual or organ, as tiny, frail, delicate, and vulnerable as they are, seems confidently to know what it's doing. It does often appear that development is the inexorable reading out of some mysterious program, the fulfillment of some hidden blueprint, the manifestation of something preordained (Corning *et al.*, 2023 provides readings on such apparently goal-directed processes). Often, these programs and blueprints are conceived of as being in the genes.

From the view that development is directed by the genes, evolutionary biologists can be forgiven if they see little of evolutionary interest in ontogeny. If development is the mere reading out of instructions specified in the genome, then development really is an irrelevant black box. The evolutionary causes are in the genome, and the evolutionarily significant organism-environment interactions take place between the phenotypes produced by this reading-out process and the environments in which they develop. Along these lines, biologists use terms like "developmental pathways" or "developmental trajectories" that make it seem as though there's a straightforward route between the embryonic and the mature. If this route is just an inevitable expression of a blueprint, then it cannot be a significant locus for evolutionary change. Instead, ontogeny is not only a locus for the evolution of form, it is, as we will see, the only locus of form.

Understanding why ontogeny is the very source of evolutionary change requires some understanding of the way development occurs. For most of us evolutionary biologists, ecologists, physiologists, and others, developmental biology is unfamiliar territory, which makes it difficult to gain an understanding of its evolutionarily most salient aspects. The way development works is a subject so large that it fills shelves of books, each rich in the details of your particular group, whether you're drawn to imaginal discs, auxin gradients, or mesenchymal signaling. In this book, I've tried to touch on development in ways that stay accessible to biologists of any taxonomic allegiance. The goal is to open a good window onto developmental mechanisms, one that doesn't require everyone to be an expert in every group, but still lets us discuss examples and ideas that speak across taxa. In this chapter, we'll start in familiar territory: with genes.

The notion that genes, understood as DNA loci, "cause" organismal traits, remains very widespread. Proponents of the Human Genome Project promised us that it would reveal "what it means to be human," or give us the "book of man" (Morse, 1998; Noble, 2015; Baverstock, 2021). In his book *How Life Works*, Philip Ball (2023)

Adaptationist Evo-Devo. Mark E. Olson, Oxford University Press. © Oxford University Press (2026).
DOI: 10.1093/9780197831533.003.0004

presents a very entertainingly written parade of some of these claims. Similar promises were made here in Mexico when the "Mexican genome" was sequenced (López Beltrán, 2011). It is very common for biologists to talk about the "genetic basis" of a trait, or the "genotype-phenotype map." They talk about genes "coding for" traits. The term "genetic blueprint" is common in popular speak, and is even used by scientists, as are notions of "genetic programs" or phenotypes emerging from "genotype by environment interaction." DNA is regarded as containing the "instructions" for building an organism; all of these ideas are compatible with genes being ultimate specifiers of organismal features. If they are ultimate directors, then development is the expression of the information in the genes, and evolution is genetic change. This "genes cause phenotypes" language coexists with the uncontroversial recognition that DNA on its own can't make a new individual, and thus can't, all by itself, actually cause ontogeny. Two conspicuous reactions to this coexistence bubble to the surface in biology.

One reaction is to conclude that genes are manifestly the basis of inheritance. We can point to considerations as obvious as cases in which simple allelic differences are associated with clear differences in phenotype (Nachman *et al.*, 2003; Colosimo *et al.*, 2005; Hoekstra *et al.*, 2006; Reed *et al.*, 2011). We can point to mechanisms such as meiosis, with its elaborate dance in which homologous chromosomes align and undergo recombination, followed by careful segregation of chromosomes through two rounds of division, and involving hundreds of proteins. Surely such metabolically costly processes would be eliminated by selection if DNA were not utterly central. We can point to DNA's lengthy chains of nucleotide bases, and describe them with terms like "information" or "instructions." Moreover, mutations, insertions, deletions, and rearrangements in DNA provide by far the richest source of information for reconstructing phylogenies of organismal groups—no comparably detailed record has been found in morphology, biochemistry, behavior, environment, or any other organismal attribute. By this reasoning, DNA holds a special and privileged role and must be the locus of inheritance (Noble, 2013). This is one reaction.

On the other hand are biologists and others who excitedly argue that genes alone can't possibly be the sole basis for inheritance. They point to evidence as obvious as the fact that DNA, on its own, is inert and incapable of causing anything to happen. Instead, whole cells and environments are necessary for development. Therefore, their reasoning goes, these extragenomic factors are also evolutionarily necessary and therefore important too. I don't want to dwell on any factionalism in biology too much, and I don't want to strawman any positions, but it's undeniable that there are biologists who worry that the conception that the unit of inheritance is the gene and that DNA provides the instructions for development risks misrepresenting biology (Lewontin, 2000a, 2000b; Müller and Newman, 2003; Oyama *et al.*, 2003; Robert, 2004; Jablonka and Lamb, 2005; Pigliucci and Kaplan, 2006; Barker, 2007; Noble, 2008; Blumberg, 2009; Pigliucci and Müller, 2010; Noble, 2012, 2013; Gilbert and Bard, 2014; Laland *et al.*, 2013, 2015; Bourrat *et al.*, 2017; Bonduriansky and Day, 2018; Fábregas-Tejeda and Vergara-Silva, 2018; Akbay, 2022; Archer and Lavie, 2023; Smocovitis, 2023; Lala *et al.*, 2024). These genes-aren't-everything arguments often provoke other biologists to reassert the primacy of genes, arguing that nongenetic factors merely modulate or facilitate rather than drive inheritance and evolution (e.g. Coyne, 2009).

So, while these different positions do exist, fundamental disagreement doesn't seem quite necessary. It is not necessary because the genes-aren't-everything folks actually base their arguments on observations that are entirely noncontroversial and known to all biologists. Its proponents simply connect the dots from these noncontroversial observations to arrive at a different conclusion from the genes-are-everything folks (Gawne *et al.*, 2018). Because this debate intersects with key adaptationist evo-devo concerns, this chapter examines genes-aren't-everything reasoning. Accepting a genes-aren't-everything perspective is not a requirement for being an evo-devo adaptationist. Even if you don't buy genes-aren't-everything reasoning, you can still appreciate the value of exploring developmental potential as the heart of adaptationist evo-devo.

If you have an expansive notion of inheritance beyond DNA, great, but probably the majority of us biologists do invoke at least an implicit notion that genes are fairly straightforwardly found in the DNA, and that genes essentially cause phenotypes. Therefore, taking a moment to at least entertain the "genes-aren't-everything" view is helpful. It is helpful because genes-are-everything versus not-everything views lead to different ideas about issues as central as the genotype-phenotype relationship, the unit of inheritance, the sources of evolutionarily relevant variation, how natural selection acts, the importance of development in evolution, all the way to the definition of evolution itself. So, this line of reasoning can provide enriching inspiration for biological practice. One goal of this chapter is to explore that reasoning, not as a mandate, but as a potentially useful perspective.

The other thing I want to do with this chapter is suggest that there is a lot more convergence in views than is apparent from debates in the literature. The main theme that will emerge later on, in Chapter 5, is that if there is one thing standing in the way of robust evolutionary explanations, it is factionalism within biology. Recognizing factionalism and finding ways around it is, therefore, an important effort for biologists. In the debate over the role of genes, the disagreement may also be largely unnecessary, or irrelevant, for a more practical reason: regardless of terminology, there is reason to think that natural world is compelling biologists toward convergence on similar views. It certainly highlights large tracts of common ground. With this sense of shared ground in mind, this chapter explores the "genes-aren't-everything" perspective by starting from an uncontroversial observation.

3.1 Everyone knows that genes alone aren't sufficient for inheritance

All biologists know that DNA alone is not enough to build a new individual. This is entirely noncontroversial. DNA is inherently inert and requires cellular machinery to have any role whatsoever in heredity and development. In fact, DNA can even crystallize in vitro, demonstrating its complete lack of intrinsic activity without a cellular context. It crystallizes so obligingly, in fact, that it is intensively explored not for any role in "information" content but simply as a versatile and non-reactive scaffold for custom building nanotech structures (Kong *et al.*, 2023; Islas *et al.*, 2024). Surely, if there is something uncontroversial to say about DNA it is that it can't build a new organism or organismal part all on its own.

It is also uncontroversial that not just any cellular context will do. Epigenetic modifications, transcription factors, and signaling pathways can drive dramatically divergent developmental trajectories even with identical DNA sequences (Jaenisch and Bird, 2003; Feil and Fraga, 2012; Allis and Jenuwein, 2016; Lu, 2023). Beyond the cell, proper development requires an extracellular context with the right temperature, pH, and availability of water and nutrients (Gilbert, 2000). Cordeiro and Tanaka (2020) build a compelling case for regarding even environmental oxygen as a "key modulator of development and evolution."

Biologists have known for a long time that protein synthesis requires a great deal beyond DNA. In 1961, J. H. Matthaei and M. W. Nirenberg published groundbreaking experiments demonstrating that synthetic RNA sequences can participate in protein synthesis in a "cell-free system" (Matthaei and Nirenberg, 1961; Nirenberg and Matthaei, 1961). Their "cell-free system," though, required ribosomes, tRNAs, and enzymes required for translation. They obtained these key ingredients from *E. coli* cell extracts. They also included all 20 amino acids in the reaction, ATP and GTP, magnesium ions at a precise concentration, a buffer system, and, crucially, water too. Their "cell free system" required carefully mimicking cellular conditions, with components mostly extracted from living cells. From experience with everything from Jurassic Park to Dolly the sheep, everyone knows that cloning requires not only nuclear DNA but also an enucleated egg cell of the correct species, with the necessary cytoplasmic factors and environment for development (Wilmut *et al.*, 1997; Egli *et al.*, 2008).

Identifying the factors that are necessary and sufficient for proper development has special urgency in therapeutic contexts, where scientists try to coax stem cell development into treatments for everything from heart attacks to neurodegenerative diseases. Under such pressing conditions, these researchers can't afford the luxury of dogmatic positions. And their literature is filled with extragenomic considerations, as in their discussions of the importance of growth factors (signaling proteins that help regulate proliferation and differentiation) and cell-cell contacts, which facilitate communication and coordination between neighboring cells, even genetically identical ones (Discher *et al.*, 2009). Additionally, interactions with the extracellular matrix through cell-matrix adhesions play a crucial role in anchoring cells, transmitting mechanical forces, and influencing intracellular signaling pathways that guide lineage specification and tissue organization (Dibus *et al.*, 2024). In their pursuit of viable embryos, in vitro fertilization specialists also don't have time for dogma. You can get them ranting for hours on the advantages of "good centrosomes." In vitro fertilization research shows that different centrosomes produce meaningfully different embryo viabilities, and that centrosome development is affected by everything from pH to temperature (Vidair *et al.*, 1993; Xie *et al.*, 2023). All of these examples, plus so many others, serve to illustrate that it is well known in biology that DNA alone is not sufficient for development.

All of this adds up to a point no biologist disputes: nucleic acids cannot do anything on their own. DNA is chemically inert outside a functioning cell and cannot assemble, build, or instruct without the right molecular and environmental context. But even within the right context, any relationship that there might be between genes and traits is anything but straightforward, and that, too, has long been recognized.

3.2 Everyone knows that there is no straightforward connection between genotype and phenotype

Close on the heels of the rediscovery of Mendel's work in 1900, biologists recognized that if genes produce traits, the relationship between them was unlikely to be a simple matching game. Attributing a simplistic view of the genotype-phenotype relation to biologists is therefore inconsistent with the evolutionary literature, where complexity (i.e. not simple gene-phenotype matching) and interaction have long been important themes. When biologists criticize one another as though entire fields or movements held monolithic views, these critiques often fail to persuade (Wray *et al.*, 2014; Welch, 2017). For example, Stephen Jay Gould (1980a) argued that the Modern Synthesis initially welcomed developmental perspectives, but then in the 1940s "hardened" around the notion that genes cause phenotypes and that adaptation is the predominant cause of evolutionary change. Gould took care to qualify that he was trying to identify broad tendencies and not paint all biologists with the same brush (see also Gould, 1986), but his essays nevertheless provoked protests from biologists. These protests involved biologists pointing to statements in the literature that contradict Gould's generalizations (Orzack, 1981; see also De Jong, 2007).

Perhaps the real problem lies not in the reaction of mainstream evolutionary biologists, but in the rhetorical strategy often adopted by advocates of an "expanded" evolutionary biology. By characterizing the Modern Synthesis as narrowly gene-centric and rigidly reductionist, they invite defensiveness and derail the conversation into historical accounting. We end up combing through the literature to argue that "so-and-so already said this" or "actually, they didn't" instead of discussing how we would like evolutionary biology to be practiced and spoken about today. The potential importance of "genes-aren't-everything" thinking isn't that it somehow exposes everyone during the Modern Synthesis as misguided. Instead, it is an important and interesting scientific debate that overlaps with the adaptationist evo-devo focus on development. So, let's take a moment to acknowledge long-standing recognition, which continues to the present, among evolutionary biologists, that any relationship between genotype and phenotype that there might be is not a trivial one.

Very early on, biologists realized that there was no one-to-one map between genotype and phenotype, and this remains true today. As an example, George Udny Yule's scientific career spanned pre-Mendelian "blending inheritance" and the rediscovery of Mendel's laws. He quickly saw that if genes cause phenotypes, the relationship is not straightforward. For example, he thought that many genes likely contribute to any given feature (Yule, 1902). His work shows that practically immediately after the rediscovery of Mendel's laws in 1900, the relationship between genotype and phenotype was not regarded as a simple one gene-one trait correspondence (Depew and Weber, 1997; Tabery, 2004; Akbay, 2022). Also, soon after the rediscovery of Mendel, and based on his work on *Daphnia*, Richard Woltereck (1913) introduced the concept of the reaction norm, which describes the range of phenotypes that a single genotype can produce in response to different environmental conditions. Woltereck's work provides another early example that biologists have recognized for a long time that phenotype is not solely determined by genotype. Evo-devo biologists often trace their views to works such as those of Waddington (1942, 1957, 1961) and Schmalhausen (1949), who set out forceful notions in the mid-20th century that genotype alone is not sufficient to predict phenotypic outcomes. Off and preoccupied with concerns such as epigenetics, plasticity, and Lamarckian-sounding inheritance, workers such as Waddington and Schmalhausen are often considered peripheral to the Modern Synthesis (Depew and Weber, 1997; Jablonka and Lamb, 2005). It isn't necessary, though, to look to them to find abundant recognition of the complexity of the relationship between genotype and phenotype.

Even the architects of the Modern Synthesis were well aware that any relationship between genotype and phenotype wasn't a simplistic one. For example, R. A. Fisher (1930, p. 54), a central figure in the foundation of population genetics, and often regarded as the most gene-centric of the architects of the Modern Synthesis, wrote "The fashion of speaking of a given factor, or gene substitution, as causing a given somatic change, which was prevalent among the earlier geneticists, has largely given way to a realization that the change, although genetically determined, may be influenced or governed either by the environment in which the substitution is examined or by the other elements in the genetic composition." Likewise, Theodosius Dobzhansky, whose writings were central to the development of the Modern Synthesis, noted that "It by no means follows that every gene produces just a single character or trait. This is a misconception refuted by the striking manifold (pleiotropic) effects of many genes. . . The process of the development of an organism should not be misinterpreted as a gradual accumulation and superposition of independent contributions of its genes. Actually, the development is a complex network of processes. In these processes the gene products play, of course, the leading roles; taken all together, these processes are, however, integrated into harmonious systems capable of being alive. In other words, the genetic materials are aggregations of particulate or atomistic genes; the development is a unified network of interrelated events" (Dobzhansky, 1970, p. 35). Sewall Wright consistently emphasized points such as "modes of response to external conditions. . . can be related to the genetics of individuals only as statistical consequences of the latter" (Wright, 1931, p. 98), or "development is an epigenetic process. The genes cannot stand in the simple one to one relation to morphological characters of a preformationist theory" (Wright, 1934, p. 30), clear examples that we can point to in showing that even architects of the Modern Synthesis regarded the genotype-phenotype relationship as anything but straightforward.

The perspective that any relationship between genotype and phenotype is complex is more emphatic today than ever. Scientists today use metaphors such as the "genotype-phenotype map" (e.g. Cotterell and Sharpe, 2013; Snell-Rood and Ehlman, 2023). The map metaphor does depict the genome causing the phenotype, but the biologists using it emphasize that the "map" is both complicated and something to be determined empirically rather than simply assumed to have a certain structure (Brodie III, 2005; e.g. Stadler and Stadler, 2006; Gjuvsland *et al.*, 2013; Svensson, 2023; Sappington and Mohanty, 2025). Clinical biologists work together under the umbrella of the genotype-phenotype "map" metaphor to identify associations of medical significance (Bastarache *et al.*, 2023). Rather than simple gene-trait associations, they emphasize a highly indirect correspondence between genotype and phenotype via gene regulatory networks (Cotterell and Sharpe, 2013; Wagner, 2014).

Even in popularized science, the metaphors have become less simplistic. In *The Selfish Gene*, Richard Dawkins depicted the DNA as "a set of instructions" or like a blueprint ("an architect's plans"), both of which imply a straightforward relationship between genotype and phenotype (Dawkins, 1976). In *The Extended Phenotype* (Dawkins, 1990) as well as his more recent *Genetic Book of the Dead*, he invokes "recipe" metaphors (Dawkins, 2024). A recipe provides a set of instructions for creating a dish, but the final outcome depends on how the recipe is executed, the ingredients used, the cooking techniques, and the environmental conditions (e.g. oven temperature). The recipe metaphor depicts genes providing guidelines for development, but the actual outcome depends on context and interactions with other factors.

So, I hope it is clear that biologists can point to a rich tradition of thinking that contradicts accusations of naive gene-centrism, which is the reliance on a simplistic model of genes directly causing phenotypes while not even recognizing the importance of development. Implying that whole movements, such as the Modern Synthesis, were naively gene-centric can't be the center of the bullseye that some "genes-aren't-everything" biologists are trying to hit with their invitations to "expand" evolutionary theory.

3.3 The contradiction that sticks in the craw of the genes-aren't-everythingers

The more relevant critique would seem to be not that biologists have been universally naive about genes and development, but that, even while recognizing that genes alone don't determine outcomes, we still often slip, sometimes subtly, sometimes overtly, into describing genes as if they do. Despite all of the above considerations, that DNA on its own can't do anything, that making a new individual that looks like its parents requires whole organisms in whole ecosystems, and that no one has found anything like a really direct relationship between genotype and phenotype, biologists routinely turn around and talk about genes causing phenotypes (Noble, 2015; Akbay, 2022). These biologists are mostly ones not directly studying DNA and gene expression; the literature of those biologists is very cautious (e.g. Gerstein *et al.*, 2007). But in fields that don't often come into contact with the actual biological mechanisms of gene expression, like most of evolutionary biology, ecology, physiology, and related fields, unburdened by the discipline imposed by molecular detail, our imaginations float free.

And our imaginations lead us more often than not to language that references genes "causing" phenotypes, as well as "underlying," "underpinning," "determining," "driving," "encoding," "specifying," "directing," "governing," "programming," and "master-regulating" them. This language is very common in biology, from basic science to outreach. As an example of the prevalence of gene-speak in evolutionary biology, a little over 60% of papers published recently in the journal *Evolution*, on my reading, make reference, either in very strong language or with at least gestures, to "genetic" causes or "underpinnings," or use "genotype" when they actually mean a different phenotype. Most of the other 40%—papers that did not mention considerations of heritability—were comparative and just used DNA as phylogenetic markers. In the papers that did mention heritability, in no case were authors so agnostic about the causes of inheritance that they simply used "heritable" without ever mentioning "genetic" (cf. Archer and Lavie, 2023). In other words, when heritability was mentioned, it was always tied to genetics. No authors used terms like "developmental systems," about which more below.

This tendency shows up not only in academic writing but also in popular outreach. Science popularizers like Richard Dawkins continue to talk about genes causing phenotypes, as in "every aspect of an animal's survival machinery was bequeathed via its genes," or development being "programmed" by genes (Dawkins, 2024). Siddhartha Mukherjee says in *Song of the Cell*, "Genes provide instructions—code, if you will—to build proteins...": an RNA "strand carries the message, or code, to build a protein. You might imagine it as a set of instructions; a Morse code stretched along a tape" (Mukherjee, 2022). This isn't a complaint about simplification; these are gifted communicators, and simplification is part of the task. It's that simplification always leans in one direction: toward genes as commanders, instructors, or source codes.

What these examples reveal is that across biology and its public-facing counterparts, even though biologists recognize that genes cannot act on their own and are not literal recipes or instructions for life, the language biologists use and the design of their research so often depict genes as if they were the sole determinants of biological outcomes. They do so when they imply that genetic material alone is the basis of inheritance. They do so when they talk about the need to discover the "genetic basis" for a trait. They do so when they define evolution as "change in allele frequency" or when they talk about development emerging from "genotype-phenotype maps," or "genotype by environment interaction," where genes are the causal anchor, the reference point that interacts with a varying or unspecified world. They do so when they imply that the relevant inheritance across generations is genetic material. They do so in countless models of evolution that depict genes causing phenotypes. They certainly do so when they say that the genome provides "instructions" or is a "blueprint." This leads us to the point that if genes really were the causes, codes, recipes, or instructors, then it would be possible to know what a gene is, define "gene" and then recognize genes on the basis of that definition.

3.4 If genes are causal agents, it should be possible to recognize them

If DNA contains autonomous causal agents, bosses sitting in the boss's chair, the causes of phenotypes connected to developmental outcomes via a "map," then we should be able to pick them out when looking at just DNA sequences. When hearing this argument, biologists often protest that it is well-known that genes participate in developmental processes in many different ways, and that it's not always possible to know where the genes lie by just looking at the DNA sequence. Of course, these biologists are correct; this is exactly the point in this section.

The "genes-aren't-everything" view is that it is contradictory to say that genes cannot be identified by sequence alone and at the same time act as though we can look solely to DNA to find the "genetic basis" of traits, or claim that genes "cause" phenotypes, "specify" traits, or "underlie" development. If gene function depends on cellular and developmental context, then genes are not autonomous causes, they are context-dependent participants. Something cannot function as a sole, independent cause and simultaneously require context to be recognized as such. If genes truly caused phenotypes all by themselves, little bosses sending instructions from their little boss's chairs, they would have intrinsic, defining features that could be recognized directly from the sequence. But if, as biologists readily recognize, genes can only be understood through their interactions with cellular processes, then their causal status also depends on those processes. And if that's the case, then DNA alone cannot "cause," "underlie," or "specify" phenotypes.

Whatever your take on these questions, one thing becomes clear in the examples that follow: the recognition that DNA is not a blueprint but a participant is yet another insight that draws development to the very center of evolutionary biology. Even if you're skeptical of "genes-aren't-everything" reasoning, or unconvinced by the argument that true causal agents ought to be identifiable by sequence alone, you can still gain insight from the following examples into how development actually unfolds, or at the very least, revisit your admiration for the intricacy of ontogeny.

With that in mind, here are some conventions used in the diagrams in Sections 3.4.1–3.4.4. The horizontal solid line represents double-stranded DNA. Exons are shown as boxes: gray for coding exons (transcribed and translated into protein), and white for noncoding exons (transcribed into mRNA but not translated) (Aspden *et al.*, 2023). Solid line segments between exons are introns, which are transcribed into pre-mRNA but spliced out and excluded from the mature mRNA.

3.4.1 Mitochondrial ATPase subunits 6 and 8 in humans and mice

The human genome contains some 3.2 billion base pairs and around 20,000 protein-coding genes (Amaral *et al.*, 2023). That works out to roughly one gene per 160,000 nucleotides—plenty of space, it would seem, to keep genes neatly separated. And yet, thousands of overlapping genes have been identified in prokaryotes, eukaryotes, and viruses (Johnson and Chisholm, 2004; Veeramachaneni *et al.*, 2004), though they appear to be relatively uncommon in plants (Schwarz *et al.*, 1981; Song *et al.*, 2001). "Overlapping genes" are cases in which two or more products are generated from shared portions of the same nucleotide sequence, either by being transcribed from opposite strands or by different reading frames on the same strand (Wright *et al.*, 2022).

In the mitochondrial genome of humans and mice, for instance, positions 8366–8572 correspond to the nucleotide sequence incorporated into the mRNA for ATPase subunit 8, while positions 8527–9207 are incorporated into the mRNA for ATPase subunit 6 (Chrzanowska-Lightowlers *et al.*, 2004; Dautant *et al.*, 2018). These regions overlap by 46 base pairs (Figure 3.1). Although this shared stretch of DNA is transcribed into both mRNAs, each transcript is read in a different frame, leading to the production of distinct proteins. In addition, the mRNA involved

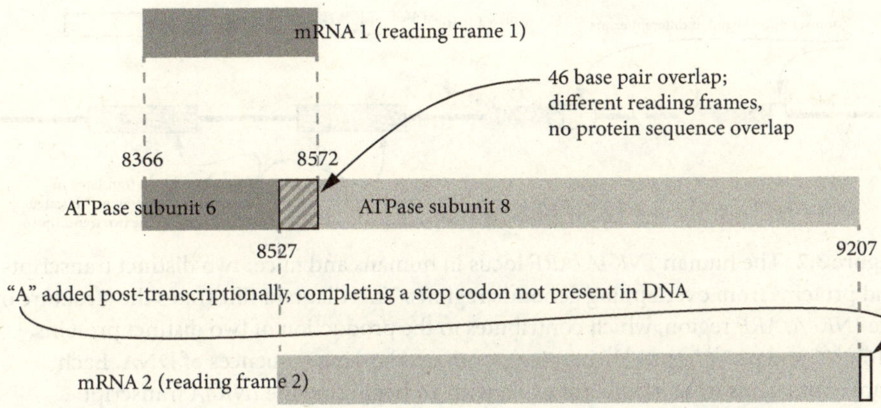

Figure 3.1 Overlapping genes in the human mitochondrial genome. ATPase subunits 8 and 6 contribute to the production of distinct mRNAs despite a shared genomic sequence. The human mitochondrial genome is involved in the production of ATPase subunit 8 from positions 8366–8572 (top), and ATPase subunit 6 from positions 8527–9207 (bottom). These regions overlap by 46 base pairs, shown hatched in the boxed region, but are translated in different reading frames, resulting in different proteins. The mRNA regions are shaded to match their corresponding genomic sequences, and a posttranscriptionally added "A" nucleotide completes the stop codon for ATPase subunit 6, which otherwise ends in a TA dinucleotide.

in the production of ATPase subunit 6 ends with an incomplete stop codon, a TA dinucleotide instead of the canonical TAA. A complete stop codon is formed only after posttranscriptional polyadenylation. Posttranscriptional polyadenylation, common in mitochondrial gene expression, is necessary for proper translation termination. The added poly(A) tail also protects the mRNA from degradation and stimulates ribosome binding (Anderson *et al.*, 1981; Chrzanowska-Lightowlers *et al.*, 2004; Dautant *et al.*, 2018).

The upshot is that the genomic region spanning positions 8366–9207 contributes to the production of two distinct proteins, despite overlapping sequences and the presence of an incomplete stop codon in one transcript. With these considerations in mind, according to your working definition of what constitutes a gene, how many genes are there in this region? The widespread occurrence of overlapping genes is a good illustration that identifying "genes" can only be done by observing actual gene expression and not by sequence alone.

3.4.2 Human and mouse *INK4A/ARF*

Like the example above, the *INK4A/ARF* region in humans and mice is also involved in the production of two distinct polypeptides (Quelle *et al.*, 1995; Sherr, 2001; Safwan-Zaiter *et al.*, 2022) (Figure 3.2). Two different mRNAs are transcribed from this region, each initiated at a different promoter and beginning with a distinct first

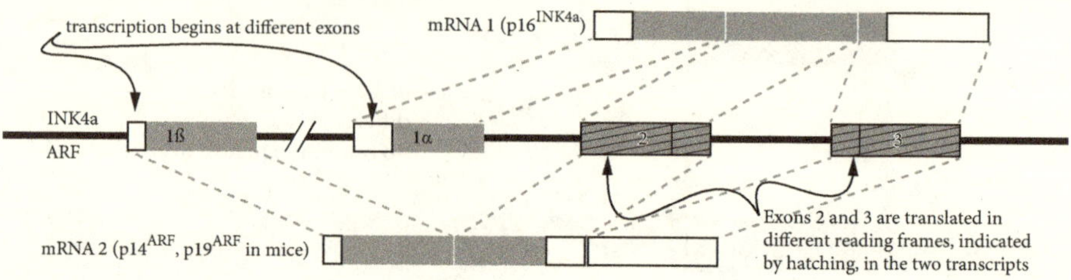

Figure 3.2 The human *INK4A/ARF* locus in humans and mice: two distinct transcripts and proteins from overlapping genomic regions. This diagram illustrates the structure of the *INK4A/ARF* region, which contributes to the production of two distinct proteins, p16^{INK4a} and p14^{ARF} (p19^{ARF} in mice), from overlapping sequences of DNA. Each transcript begins from a different exon, with 1α beginning the *INK4A* transcript (mRNA 1), and exon 1β beginning the *ARF* transcript (mRNA 2). Despite differing in their initial exon, the two transcripts share exons 2 and 3 at the DNA level. However, these exons are translated in different reading frames in each transcript, represented by hatching. As a result, the two proteins share no amino acid sequence despite overlapping genomic sequences. The interruption between 1β and 1α highlights that they are separated by a monster intron spanning more than 15 kb.

exon (1α for *INK4A*, 1β for *ARF*). These transcripts, in turn, are involved in the production of two different proteins, p16^{INK4a} and p14^{ARF} (p19^{ARF} in mice). Although the two transcripts share exons 2 and 3, their distinct first exons shift the reading frame. As a result, the shared exons are translated differently, producing entirely distinct amino acid sequences, so despite overlapping genomic regions, the proteins have no sequence similarity. The "ARF" in the name stands for "alternate reading frame," reflecting this offset in reading frames. And in addition to all this fancy genetic footwork, exons 1α and 1β aren't even close together: they're separated by over 15,000 base pairs (not to get too far into the weeds, but the introns also seem to have functional impacts, Harland *et al.* (2005)). The proteins, moreover, participate in different regulatory pathways: p16^{INK4a} plays a role in the retinoblastoma (RB) pathway, while p14^{ARF} is involved in the p53 pathway (Sharpless, 2005). According to your definition of "gene," does this situation represent one gene or two genes?

3.4.3 *IP259/DUb80* in *Drosophila*

The previous examples helped think about gene definitions that include stipulations like "a sequence of DNA that codes for a protein or RNA..." or "a segment of DNA that is transcribed into RNA..." They help illustrate that the way a given stretch actually participates in the production of RNAs or proteins depends on much more than the DNA sequence itself. The *IP259/DUb80* sequence complex in *Drosophila* stretches even further definitions that specify "a" coding sequence (Figure 3.3). Within a large intron of the *DUb80* gene lies an open reading frame involved in the production of IP259, a 259-amino acid protein conserved across eukaryotes from yeast to mammals (Barrio *et al.*, 1994; Mottus *et al.*, 1997; Stotz and Griffiths, 2004). *DUb80* includes a

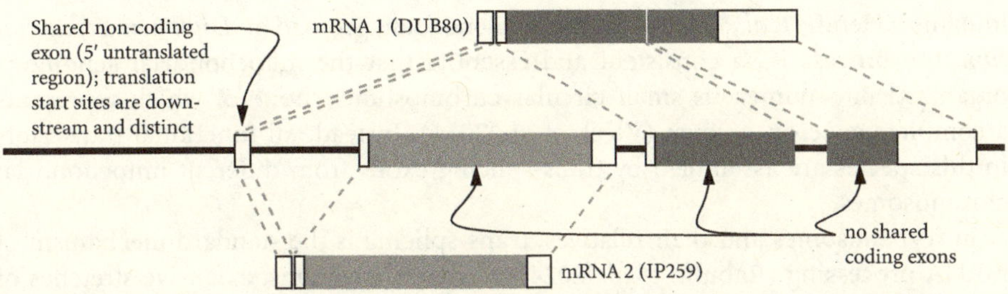

Figure 3.3 Genomic organization and alternative splicing of *DUb80* and *IP259* in *Drosophila.* The diagram illustrates the shared 5' untranslated exon that serves as the translation initiation site for both DUb80 and IP259 mRNAs. Two mRNAs are transcribed from the same start site but undergo alternative splicing, leading to the production of two distinct mature mRNAs with different coding regions that are translated in different reading frames and are involved in the production of proteins with differing sequences and structures.

sequence that contributes to the production of ubiquitin, a small protein involved in protein regulation, including posttranslational modification (Herhaus and Dikic, 2015). Transcription of both *DUb80* and *IP259* mRNAs begins at the same start site and includes a shared exon that forms part of the 5' untranslated region and does not contribute to the translated sequence of either protein. The primary transcript is processed in two different ways through alternative cis-splicing, in which exons from a single pre-mRNA molecule are joined together in different ways following the removal of introns (though in cases like this, one transcript's intron is another's exon). In this case, alternative cis-splicing leads to the production of two mature mRNAs. These mRNAs, while sharing untranslated sequences at their 5' ends, do not overlap in their coding regions and are translated in different reading frames. The result is two structurally unrelated proteins. According to your definition of "gene," how many genes are there in this example?

3.4.4 Gene expression in trypanosomes

In addition to cis-splicing, trans-splicing can also occur. In trans-splicing, exons from two separate pre-mRNA molecules, sometimes transcribed from entirely different locations in the genome, are spliced together to form a single, chimeric mRNA. This process can occur between transcripts originating from the same chromosome (intrachromosomal trans-splicing) or, even more remarkably, between transcripts from different chromosomes (interchromosomal trans-splicing). If intrachromosomal splicing requires precise coordination of transcript processing and availability, then intrachromosomal splicing would seem to demand an even more extraordinary level of regulatory synchronization and RNA mobility. Interchromosomal trans-splicing has been documented in a wide array of organisms, including humans (Herai *et al.*, 2018). In the euglenozoan protozoan *Diplonema papillatum*, the process is so consistent and essential that the mitochondrial genome is organized into numerous small circular chromosomes, none of which represents a complete gene on its own (Vlček *et al.*, 2011). Instead, all functional transcripts in this species are assembled by trans-splicing exons from different mitochondrial chromosomes.

In trypanosomes and their relatives, trans-splicing is the standard mechanism of mRNA processing (Rabuffo *et al.*, 2024). In these organisms, extensive stretches of DNA are transcribed into polycistronic transcripts, single RNA molecules that contain multiple transcribed sequences in long chains. These polycistronic RNAs are then rapidly cleaved and processed into individual, mature mRNAs. Importantly, before translation, each of these mRNAs must first receive a short sequence known as the spliced leader (SL). The SL is a universal 5' leader sequence trans-spliced onto the beginning of every nuclear-encoded mRNA (Kooter *et al.*, 1984; Laird, 1989; Liang *et al.*, 2003). Like the 5' ends of mRNAs in many eukaryotes, the SL is "capped" (Agabian, 1990), meaning that it carries a modified guanine nucleotide at its 5' end. This cap protects the mRNA from degradation by exonucleases, assists in mRNA export from the nucleus to the cytoplasm, and promotes recognition by the ribosome during translation. In this way, a single mature mRNA molecule can be assembled from exons that derive from multiple chromosomes (Figure 3.4). Again, hold up your definition of "gene" to this situation. What does it say about how many genes there are in Figure 3.4?

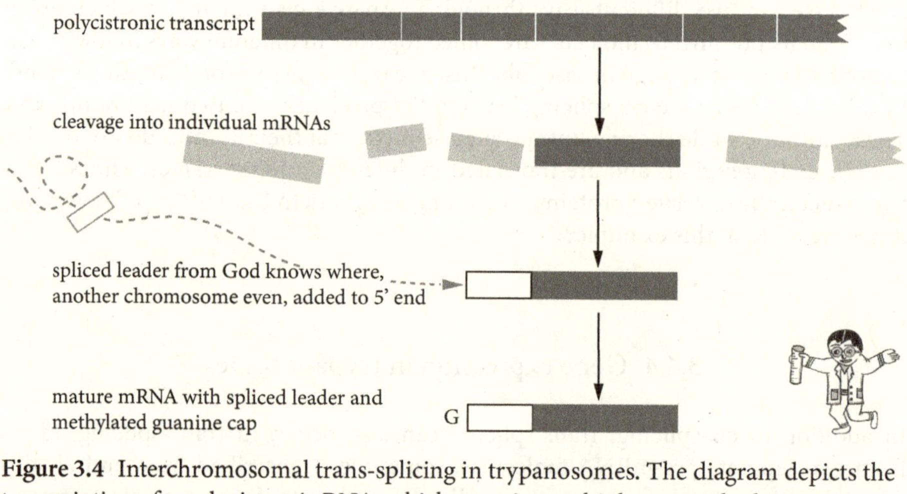

Figure 3.4 Interchromosomal trans-splicing in trypanosomes. The diagram depicts the transcription of a polycistronic RNA, which contains multiple transcribed sequences in a single long transcript. After cleavage into individual mRNAs, each molecule undergoes trans-splicing, with a spliced leader added to the 5' end. This spliced leader can derive from anywhere in the genome, often from another chromosome. The spliced leader is then capped, promoting mRNA stability, export, and translation. Drawing of a bemused molecular geneticist by Eduardo Martínez Díaz.

3.4.5 *Slowpoke* in chickens, *Dscam* in fruit flies

Most protein-coding regions in metazoans are involved in alternative splicing (Shin and Manley, 2004), and the number of protein isoforms involved can be mind-boggling. In chickens, the *slowpoke* gene gives rise to an astonishing 576 distinct protein forms, and even more have been found in humans and mice (Rosenblatt *et al.*, 1997; Black, 1998; Jablonka and Lamb, 2005). This staggering diversity arises through alternative RNA splicing, allowing a single region of DNA to be involved in the production of a whole library of RNA variants and, in turn, a wide array of proteins. The *slowpoke* gene is involved in the production of proteins that help determine how cells respond to different sound frequencies, shaping an animal's hearing. In the chick cochlea, for instance, hair cells express different splice variants of *slowpoke* mRNA depending on the position of each hair, which contributes to each cell's unique electrical resonance frequency. Because individuals grow up in environments filled with different ambient sounds, this flexible splicing system likely supports a kind of auditory fine-tuning. It seems plausible that being especially attuned to the dominant sounds of one's environment could offer a clear functional advantage. In species that rely heavily on vocal communication, like chickens, mice, and humans, distinguishing the calls of one's own kind from background noise could mean better chances of survival and reproduction. Likewise, detecting local predators or food sources through sound would also be favored.

If *slowpoke* weren't enough, the *Dscam* locus in *Drosophila melanogaster* is associated with over 38,000 isoforms through mutually exclusive splicing of exons. Different isoform profiles are involved in establishing cell identity. In a conclusion that must have taken an exhausting amount of work to come to, Neves *et al.* (2004) note that "the *Dscam* repertoire of each cell is different from those of its neighbors, providing a potential mechanism for generating unique cell identity in the nervous system and elsewhere." Because different *Dscam* protein isoforms can result in differing cell type identities, it could be argued that these products strain the idea of a "coherent set of products from the same DNA region," a clause in one of the most permissive definitions of "gene" I could find (Gerstein *et al.*, 2007).

With so many hundreds of proteins associated with stretches of DNA like chicken *slowpoke* or *Dscam*, and according to your definition of "gene," how many genes are there in these examples?

3.4.6 The point of these examples

The aim of these examples is not to debate this or that definition of "gene" or to come to some firm position on how many genes there are in each example, but to underscore the point that it is impossible to say much about how a given stretch of DNA participates in cellular processes without actually examining those processes and knowing something about the products (Falk, 2010). Taking the literature at face value tends to show a gradient across subfields within biology of how literally biologists seem to invoke the notion that genes cause phenotypes in any straightforward way. In ecology, evolutionary biology, organism-level physiology, and other fields far removed from cellular innards, "DNA causing phenotypes" language is very common. These biologists ask whether traits have a "genetic basis," whether variation is "genetic" or merely plastic, and express notions such as the gene being "a discrete unit of hereditary information consisting of a specific nucleotide sequence in DNA (or RNA, in some viruses)" in a prominent basic biology textbook (Urry *et al.*, 2021). It is very common for models of evolution, even today, to focus exclusively on genes and allele frequency variation, based on the assumption that genes directly specify phenotypes. These expressions have in common that they are compatible with the view that DNA contains more or less discrete units that, all by themselves, cause phenotypes to come into being.

In contrast, molecular geneticists, genomicists, and developmental molecular biologists, those who directly study how DNA participates in cellular processes, operate with pretty different understandings. For them, DNA is a crucial participant in dizzyingly intricate cellular processes, not an autonomous source of form. Gene definitions in these fields reflect the extreme diversity of ways that DNA is involved in cellular processes (Griffiths and Stotz, 2006, 2013; Gerstein *et al.*, 2007; Hopkin, 2009). One recent definition, for instance, describes a gene as "DNA or RNA that encodes a coherent set of potentially overlapping functional product molecules, either protein or RNA—that influences phenotype in ways we may or may not be able to measure" (Hopkin, 2009, p. 931). That final, perhaps slightly harried, caveat—"in ways we may or may not be able to measure"—captures just how far removed the working concept of a gene is from the picture often assumed in other parts of biology.

The above examples help illustrate how development unfolds, and why there are always some biologists arguing that the genome cannot be "the" cause of the phenotype. Even the brief examples above illustrate that what biologists call "genes" are often fragmented across the genome (even across chromosomes), interspersed with regulatory sequences, and subject to alternative splicing, RNA editing, and posttranscriptional modification. In all of the examples, what is clearly missing is any evidence of DNA sequences directing development all on their own. In the ATPase case, for example, identifying two distinct coding regions required observing not the DNA alone but how that stretch of DNA participates in developmental processes. The so-called "instructions" in the genome have no functional meaning apart from the cellular context that determines which sequences are used and how. It's clear that identifying "genes" requires knowing something about their roles in cellular processes (Gerstein *et al.*, 2007; Hopkin, 2009; Aspden *et al.*, 2023). If identifying genes depends on understanding their roles within larger systems, the genes-aren't-everything argument goes, then DNA cannot be the ultimate repository of developmental information.

The argument is that a gene identified by what it does, rather than what it is, cannot be said to determine phenotypes on its own. If an entity in nature is truly a foundational cause, then we should be able to establish some criteria for recognizing it. I can recognize other human beings and more or less reliably distinguish them from chimpanzees, which was helpful back in my dating days. In that context, for the purposes of marriage, the key causal entities are humans, and distinguishing them from nonhumans is crucial. Knowing how many humans are involved makes the difference between monogamy and an inadvertent, surly threesome. Biology is full of situations where crucial interactants are identifiable. So if genes really are the key directors of development and heredity, they ought to be just as recognizable. That claim thus becomes incoherent: a true determinant would act independently, but genes demonstrably do not. If genes really were the sole causal agents of phenotype, then the link from genotype to phenotype should be direct and readable from the DNA sequence alone, which everyone knows it is not.

This leads to a striking inconsistency between the epistemic humility required to identify a gene in practice and the rhetorical confidence with which DNA is said to "cause" things. In reality, genes can't be identified or even meaningfully discussed apart from the regulatory, cellular, and environmental systems in which they participate (Portin and Wilkins, 2017). The same evidence that makes gene definitions elusive also supports the view that DNA is a vital but insufficient contributor to development. Before turning to some adaptationist evo-devo payoffs of this perspective, the next section examines two more lines of evidence that reinforce it.

3.5 Moving toward common ground: unresolved phylogenies and missing heritability

Here are two more examples of empirical observations that are often interpreted differently by "genes-are-everything" and "genes-aren't-everything" biologists. Both cases highlight situations in which, if genes as more or less discrete units of DNA were truly the fundamental causes of phenotypes, as biologists' common daily use of language suggests, we would reasonably expect a much stronger genetic signal in the data than what actually appears. "Genes-are-everything" biologists suspect that, as sequencing technologies improve and more genomes are analyzed, the missing signals will eventually emerge. "Genes-aren't-everything" biologists, in contrast, see these results as expected consequences of systems in which genes are necessary but not sufficient. If developmental and therefore evolutionary processes depend on non-genetic components as well, then cases like these are exactly what would be expected to arise from time to time.

3.5.1 Missing phylogenetic resolution

For most of the history of biology, systematics has been a fairly frustrating endeavor. Systematists are biologists who use organismal attributes to try to figure out how species are related to one another. Their goal is to infer evolutionary relationships by sorting patterns of trait similarity and divergence across species. The products of their work are phylogenies, and their work involves phylogeny reconstruction. If the proportion of organismal features that vary is small with respect to the number of species a systematist wishes to include in a phylogeny, then hopes of resolving many nodes in a phylogenetic tree are dim. For example, the wood of trees is made up of just a few cell types. Trees transport water under very unusual circumstances, under negative pressure. Water under negative pressure is metastable, which means that if it weren't in tiny membrane-bound compartments, it should, dead to rights, be boiling. There aren't a lot of options available to plants or even to artificial engineering systems for transporting water under these conditions. This means that the selection pressures experienced by the transport networks of trees don't change much across species. This similarity in selective condition and the relatively small amount of variation in cell types and material (cellulose, lignin) across species means that the wood in tree lineages separated by sometimes hundreds of millions of years can look virtually identical.

In the early 20th century, some plant systematists turned their attention to wood as a source of information for reconstructing plant phylogeny, and the effort seems to have driven them to their wits' end. By the 1960s, their literature was a delirious, often unintelligible parade of "irreversible unidirectional progress," "levels of phylogenetic specialization," "refugia of primitive characters" in different plant parts, and "ontogeny recapitulates phylogeny" type reasoning (Olson, 2012b). Even by the standards of their day, most of the content of their literature seems to reflect a shared madness, and an antiquated one at that. Pretty much all of the elements of their reasoning had already been rejected in biology at large. For example, in 1916, before the wood systematics folks even got going in 1918 (Bailey and Tupper, 1918), E. S. Russell, in his brilliant *Form and Function* (Russell, 1916) declared the notion that phylogeny can reliably be reconstructed from the study of ontogeny conclusively rejected. If their work did start driving them a little crazy, the wood systematists can be forgiven. There are 300 thousand or so flowering plants, so the juxtaposition between the paltry variation they had to work with on the one hand, and the number of phylogenetic nodes they set out to resolve on the other hand, could challenge anyone's composure.

Amid scenarios like this, which played out to one degree of severity or another across all biological groups, DNA sequencing has been a godsend to systematics. Where the poor wood anatomists would struggle with their four main cell types, and the three or four different subtypes that each type could assume, systematists can now comb entire genomes for nucleotide variation. Even the smallest flowering plant genome currently known, that of the tiny plant *Genlisea aurea*, charismatic with its dancing yellow flowers and exquisite spiral subterranean traps that digest errant rotifers and nematodes, has 64,000,000 base pairs (Greilhuber *et al.*, 2006; Zedek *et al.*, 2024). And that's just the nuclear genome; even in tiny *Genlisea*, additional nucleotide sequences, and thus information for reconstructing phylogeny, are available in the chloroplast and mitochondrial genomes. Imagine telling this to the wood folks in 1960. This abundance of data means that the number of characters is very often higher than the number of nodes to resolve, an ideal situation. It's no wonder that molecular phylogenies have revolutionized systematics and biology at large. Biologists finally have solid phylogenies that make sense, with species formerly hard to place finally nestled among neighbors that share their morphology or lifestyle, even though these shared traits were understated and eluded notice, given pre-DNA tools. Given this bounty of phylogenetically informative characters, polytomies (unresolved nodes in phylogenies) could reasonably be expected to be a thing of the past.

Even so, molecular phylogenies, even those including millions of nucleotides, routinely come up against nodes that are resistant to resolution. Whether the rangy New

Caledonian treelet *Amborella trichopoda* is the single sister species to the rest of the flowering plants, or whether instead it plus the waterlilies are the sister group, is debated despite very large datasets (Xi *et al.*, 2014). In a similar way, it's not clear whether sedate sponges, shimmering comb jellies, or blobby Placozoa are sister to the rest of the animals, even though very high-coverage data are available (Juravel *et al.*, 2023; Schultz *et al.*, 2023). Nor are unresolved nodes limited to deep divergences. Within the manakins, little neotropical birds known for courtship dances and leks, a study using full-genome data still resulted in unresolved relationships (Zhao *et al.*, 2023).

Such examples are legion, and they include surprising cases such as plant chloroplast genomes that seemingly migrate across species at will, with variation being predicted not at all by species limits but instead by geography. In oak trees (genus *Quercus*), plastome phylogenies tend to show clades made up of trees that live near one another and hybridize (selection seems to maintain identity across these readily recognized species). Within these clades, rather than all samples from the same species being more closely related to one another than to samples from other species, the samples from different species tend to be scattered throughout the phylogeny, closest to geographically close samples, whether of the same species or not (Pham *et al.*, 2017). So, even with what would seem like a surfeit of genetic data, some relationships remain hard to work out.

Many perfectly standard molecular processes can account for these hard-to-resolve nodes, but the possibility that phenotypes are caused by more than just genes has only begun to be explored. A common cause of unresolved phylogenetic nodes is incomplete lineage sorting. Ideally, for the purposes of phylogeny reconstruction, genetic variants within a species should be all more closely related to one another within the species than with other species. Given isolation between species and enough time, gene flow between individuals and genetic drift lead to loss of genetic variation within species. In these situations, the lineages, manifest as genetic variants, "sort" themselves by species: variant a in species A, and variant b in species B. But nature is not always obliging toward systematists, and speciation often occurs before such "sorting" occurs, when a given species is still polymorphic at one or more loci. When the ancestral species splits into two or more daughter species, some individuals in species A can inherit variant a, while others inherit variant b. The same thing can happen in the other species. This results in a situation in which some individuals in species A are more closely related to those in species B according to the locus in question. Because the genetic variants don't "sort" themselves neatly into their respective species, it's called incomplete lineage sorting (if it's helpful, the name in Spanish— "sorteo de linajes"—depicts the process better: "sorteo" means "lottery," which evokes the chance involved in which species inherits which variants. "Sorteo" is not Spanish for "sorting," though; "sorteo de linajes" just sounds similar to "lineage sorting").

Looking to DNA to reconstruct evolutionary history makes perfect sense; diverged species should accumulate distinctive variation sooner or later, and this variation is precisely that needed to infer phylogeny. But if genes are only one part of the developmental machinery, then it makes sense that some percentage of evolutionary divergences might elude detection through sequence data alone. If some developmental changes associated with species differences are largely invisible just by reference to the genome, then this could help explain at least some perplexing situations in which genetic divergence is much less than would be expected given morphological disparity. For example, if divergence in phenotype or reproductive isolation was triggered or accelerated by extragenomic factors in the presence of allelic polymorphism, the result could be an evolutionary split at the organismal level not reflected at the molecular level. Considerations of this sort lead Singh and Ahi (2022) to hypothesize that rapid evolution of high phenotypic diversity across species, as in the African Rift Lake cichlid species flocks, could plausibly involve extensive posttranslational modification with relatively little DNA sequence diversity. In at least some cases, lack of phylogenetic resolution may be a symptom of the fact that DNA, while essential, is not sufficient for explaining developmental divergence.

3.5.2 Missing heritability

Many organismal traits exhibit consistent heritability. Traditional methods for estimating heritability include regression of offspring traits on parental traits, twin studies (comparing monozygotic and dizygotic twins), sibling or family-based studies, and artificial selection experiments (Visscher *et al.*, 2008). Heritability estimates from such studies estimate that as much as 90% of human height variation can be attributed to genetic factors (Conery and Grant, 2023). If genes are straightforwardly the source of heritability, as language such as "the genetic basis" of traits, "genotype-phenotype maps," or "genes for" certain features suggests, then it should be possible to account for heritable phenotypic variation through corresponding genotypic variation.

However, studies designed to identify the genetic variants underlying heritable traits commonly fail to account for as much genetic variation as expected. In genome-wide association studies (GWAS), biologists gather DNA samples from a large group of individuals across a species. In humans, these studies often compare individuals with a specific trait or medical condition to those without it. Biologists then analyze millions of genetic markers, most commonly single-nucleotide polymorphisms (SNPs), spread across the entire genome. The aim is to see whether certain variants appear more frequently in the group with the trait compared to the group without it. If genetic variation explained phenotypic variation in ways that language, such as "genes for" traits and "the genetic basis for traits" often suggests, then GWAS would reasonably be expected to account for a large proportion of heritable variation.

GWAS often explain far less phenotypic variation than anticipated if genetic factors were straightforwardly the sole contributors; this discrepancy is termed "missing heritability" (Manolio *et al.*, 2009; Stringer *et al.*, 2011; Wray *et al.*, 2011; Gage *et al.*, 2016; Torkamani *et al.*, 2018; Baverstock, 2021). For example, GWAS with very large sample sizes explained surprisingly little of the variation in human height, which you will remember is renowned for its consistently very high heritability. The variance accounted for by SNPs was just 25% (Yengo *et al.*, 2018). Thousands of SNPs across hundreds of loci covering over 20% of the genome were implicated, which stands to reason, given how many bones, cartilages, and other tissues are involved in "height" (Conery and Grant, 2023). The missing heritability problem is even more pronounced for complex traits like schizophrenia, which has an estimated heritability of 60 to 80%, yet SNP-based GWAS explain less than 1% of the variance within studied populations (Visscher *et al.*, 2008; Schubert *et al.*, 2020).

Explaining missing heritability pushes both "genes-are-everything" and "genes-aren't-everything" biologists in generally similar directions. An idea that motivated GWAS in the first place was that genes cause phenotypes in a fairly straightforward way, as in talk of the search for "causative mutations" in early GWAS studies (Klein *et al.*, 2005). Even the most committed "genes-are-everything" biologists acknowledge that, if genes are the fundamental causes of phenotypes, the results of GWAS are puzzling if not outright disappointing. Understated consternation comes through in remarks such as "there is growing evidence that much more of the genome than previously thought is required to explain the heritability of complex phenotypes" (McKinney and Pajewski, 2012, p. 1).

In response, GWAS practitioners start looking beyond simple assumptions about the genotype-phenotype relationship. If your reaction is something like "we've always known that genetic architecture is complex!," don't look at me: it's widely recognized that early-stage GWAS were based on an additive model, in which each locus contributes independently to the phenotype (McKinney and Pajewski, 2012; Palmer *et al.*, 2023). In going beyond an additive model, some researchers examine the potential contribution of dominance, where instead of alleles contributing independently to trait variance, one allele can mask or modify the effect of another. Since standard GWAS assume additive effects, dominance interactions are not captured well (Palmer *et al.*, 2023). Others focus on epistasis, when the effect of one gene depends on the presence or absence of variants in another gene. Epistasis can obscure or enhance associations, leading to missing or underestimated genetic effects (McKinney and Pajewski, 2012). Beyond genetic interactions, some researchers look to epigenetic modifications, such as DNA methylation, as a factor in the explanatory deficit of GWAS (Flanagan, 2015; Yengo *et al.*, 2018). Even farther afield, others consider environmental influences, which can confound GWAS by mimicking genetic effects (Gage *et al.*, 2016), echoing classic insights from cultural inheritance theory, where learned behaviors can produce heritable patterns independent of DNA (Cavalli-Sforza and Feldman, 1973).

Notice the way that interpreting the results of GWAS moves biologists from initial assumptions of a purely additive model—just 20 years ago, not some relic of the 1940s—in which genes straightforwardly specify phenotypes toward explanations that extend ever farther from this starting point. Dominance and epistasis introduce allele-allele and gene-gene interactions, depicting genetic effects not simply as the sums of independent contributions but interactions that can mask, amplify, or depend on each other in unexpected ways. Epigenetic factors, ones beyond just the DNA sequence, are even farther removed from a strict genes-cause-phenotypes scenario. And including the environment goes even farther in including resources traditionally regarded as "external" to organisms inside the causal network necessary to understand the causes of phenotypes. All of these expansions of the set of factors that must be examined to explain GWAS results come, essentially, from traditional genes-are-everything researchers; after all, no one else would conduct a GWAS.

Yet the shift from additive genetic effects to allele interactions, to gene-gene interactions, to epigenetic regulation (a potentially very large and heterogeneous set of processes; see Jablonka and Lamb, 2005) to even "external," "environmental" factors, would warm the cockles of the hearts of the most committed "genes-aren't-everything" biologists. "Genes-aren't-everything" biologists reason that because DNA only contributes to the formation of new individuals within highly specific cellular and environmental conditions, these nongenetic factors must also play a role in evolution (Akbay, 2022). By following the causal chain wherever it leads, biologists of all perspectives head more or less in the same direction: understanding development, and therefore heredity, involves far more than just genes. Such reasoning leads some workers to the principle of "causal parity."

3.6 Developmental resources and causal parity

DNA obviously has a distinctive role in development and inheritance. The molecular structure of DNA, with its prodigiously long and stable chains of base pairs, gives it a unique capacity in supporting the production of molecular sequences across generations. Biologists who point to considerations like these are surely correct—DNA is both essential and unquestionably has a distinctive structure and distinctive ways of participating in developmental processes.

But, the "genes-aren't-everything" view goes, it is a mistake to infer from this uniqueness that DNA is the *only* important resource in development, understood as the construction of a new individual that resembles its parents. Mode of inheritance, or any notion of "information" content, is not the same as causal sufficiency in development (see the musings of Morgan (1917) along these lines). Rather than having a more essential role than other developmental factors, DNA is an utterly necessary but not sufficient member of a broader ensemble of developmental resources, the conditions and substances required for building a new individual or part of one.

Necessity in a system is not the same as primacy, point out the "genes-aren't-everything" workers. In practice, the removal of any essential nongenetic resource can halt development just as effectively as the removal of DNA. Eliminating all lipids from an embryo is as lethal as eliminating its genome. Likewise, depriving an embryo of water will bring development to a stop. If this line of reasoning, the basis of knockout experiments, is taken to imply that the targeted element contains development-directing information, then lipids and water must be regarded as carriers of essential developmental information, too (Maynard Smith, 2000). To be fair, part of the reason knockout experiments have focused so heavily on DNA is experimental tractability: DNA sequences can be precisely targeted and manipulated, whereas disrupting broad categories like "lipids" or "cytoplasmic texture" is chemically and technically more diffuse. But this practical asymmetry should not be mistaken for a difference in causal necessity. Even beyond issues of tractability, knockout reasoning is not applied symmetrically: DNA is routinely described as containing "instructions" or "information," while other indispensable resources are treated as mere background conditions.

The principle of *causal parity* of developmental resources (Oyama *et al.*, 2003) holds that when multiple resources are equally necessary for development, no single one should be treated as more fundamental than the others. "Causal parity" can be a somewhat misleading term if taken to imply that all causes contribute equally in strength, magnitude, or timing in every developmental moment. What it really refers to is that if developmental resources (genes, environment, cells, etc.) are necessary, then none should be privileged a priori as *the* cause of a trait or outcome (Griffiths and Knight, 1998; Griffiths and Gray, 2005). "Parity of necessariness" perhaps captures the intended meaning. The core idea is that many components, including but not limited to DNA, may be necessary for a particular outcome. The components can still vary in their relative influence at different moments or in different contexts. Some examples of causal parity include the following.

3.6.1 *Opn*, photons, and eye development

Eye development in mammals navigates the contradictory terrain of supplying the growing lens with blood and nutrients while producing a perfectly transparent structure. In mammalian fetuses, a vital pipeline spans the developing eye: the hyaloid artery, a delicate vascular cobweb that stretches through the vitreous body from the retina to the lens. It irrigates the lens with oxygen and nutrients, sustaining it in darkness. But an artery running straight through the visual field would clearly impair sight. In normal development, the hyaloid artery is resorbed before birth, leaving behind a clear visual path and functional vision (Vemaraju *et al.*, 2024).

Rao *et al.* (2013) showed that in mice with mutations in the *Opn4* gene (these individuals produce abnormal variants of melanopsin), or whose mothers are raised in total darkness during pregnancy, the eye does not undergo the necessary vascular withdrawal pattern during maturation to produce normal vision. Abnormal melanopsin is associated with a failure of resorption of the hyaloid artery. On the traditional view, one of the functions of melanopsin is clearly the correct resorption of the hyaloid artery. This is also clearly a selected function because non-withdrawal of the artery is a major impediment to vision.

But when mother mice with normal *Opn4* gestate in complete darkness, their offspring still develop the same phenotype—a persistent hyaloid artery. It turns out that photon flux through the mother's visceral cavity, and impinging on the fetal retinas, is just as essential for the resorption of the hyaloid artery as a functional copy of *Opn4*. This light-dependent process occurs around day 16 (late in gestation) and requires the presence of melanopsin in the fetus. In this example, photons have an utterly equivalent participation in leading to functional eyes as *Opn4*. Knock out either, and the same phenotype results. This equivalence underscores a central point: both a genetic component (*Opn4*) and an "environmental" one (photon flux) function as developmentally indispensable resources. Judging from the prevalence of functional eyes in mice, photon flux through the visceral cavity is evidently a reliable input, present ontogeny after ontogeny, generation after generation. A resource so important causally and so persistent surely requires taking into account in explaining how normal phenotypes are dependably reconstructed.

3.6.2 Vitamin C

Vitamin C is essential for a large number of metabolic reactions. Practically all organisms, including plants and animals, including almost all primates, can synthesize this acid. In humans, the gene involved in the production of the enzyme L-gulonolactone oxidase has undergone mutations that prevent the synthesis of the acid. This mutation was invisible to natural selection because vitamin C must have been abundant in the ancestral human diet. In other primates, vitamin C is produced involving a gene; in humans, who have lost the gene, it is acquired in the diet. There isn't any difference in the biological role of vitamin C in human physiology versus in other primates for which vitamin C is produced involving a gene.

The identical physiological role of endogenous and dietary vitamin C makes it an excellent example of causal parity. In both cases, vitamin C has the same metabolic role. What makes the example so useful is that biologists often treat "endogenous" resources as causally privileged, assumed to be more fundamental or biologically significant than externally acquired equivalents. If humans possessed L-gluconolactone oxidase, the enzyme that enables endogenous vitamin C synthesis, this bias would go unnoticed. From the traditional perspective that genes are fundamental causes, the presence of the enzyme would be attributed to the gene that "codes for" it, reinforcing notions that gene-directed synthesis is developmentally primary. But there are no good reasons to claim that vitamin C produced through gene expression is more metabolically or evolutionarily significant than the same molecule obtained through diet. To be anthropomorphic about it: natural selection doesn't care where a resource comes from, as long as it works. The next example will drive this point home even more vividly.

3.6.3 Carnivorous versus "protocarnivorous" plants

The distinction between carnivorous and protocarnivorous plants provides another example of causal parity. In many parts of the world, on nitrogen-poor soils, "carnivorous" plants grow. These plants trap animals, usually insects, with a variety of mechanisms (three are shown in Figure 3.5). Once the insect is trapped, the plant secretes enzymes that digest the animal. The leaf absorbs the released nutrients. The noncarnivorous relatives of these species do not have digestive enzymes. The conventional view is that these species are "carnivorous" because they both trap and digest their prey. In contrast to "carnivorous" plants, some species are considered "protocarnivorous" because they trap prey, but outsource the digestion of the prey to other organisms, such as insects (Figure 3.6) (Ellis and Midgley, 1996; Roberts *et al.*, 2018). One striking case comes from the South African genus *Roridula*, which traps insects on sticky leaves but lacks digestive enzymes. Instead, it relies on a symbiotic relationship with so-called "*Roridula* bugs," hemipterans that eat the trapped insects and excrete nutrient-rich waste onto the plant's leaves. The plant then absorbs nitrogen from this waste.

To say that "natural selection does not care" where a resource comes from means the following. It means that a variant associated with greater survivorship

Figure 3.5 Carnivorous plants. On nitrogen-poor soils in many parts of the world, carnivorous plants trap insects with a wide variety of trap types, secrete digestive enzymes, and absorb the nitrogen from their prey. (A) Butterworts (*Pinguicula* spp.), like these specimens growing on a moss-covered boulder in Santa María Huitepec, Oaxaca, trap small insects on their sticky leaves. (B) Sundews (*Drosera* spp.) trap insects in mucilage droplets borne on motile tentacles, as in *Drosera arcturi* in a Tasmanian bog. (C) The bright colors and nectar of tropical pitcher plants in the genus *Nepenthes*, as in this *N. viellardii* from New Caledonia, attract insects, which slip on the waxy inner surface of the pitchers into the digestive fluid below. Botanists traditionally regard plants that trap *and* digest their prey as "carnivorous."

or reproduction will come to predominate within a population whether or not the variation is due to "endogenous" causes, such as genes or whether they come from "environmental" resources. Let's assume that, in past populations, in the ancestors of the plants in Figure 3.5, individuals that had mutations in genes involved in the production of proteolytic enzymes on the leaf surfaces had more nitrogen available than conspecifics without such mutations. These digestive-enzyme-secreting variants came to predominate, and this sort of scenario would help explain why carnivorous plants trap and digest insects today. To use the metaphor of natural selection, "seeing" a given variant, in carnivorous plants, selection saw and favored individuals bearing mutations associated with the production of proteolytic enzymes on leaf surfaces (Givnish, 2015).

The same sort of scenarios apply in protocarnivorous plants, but instead of enzyme secretion, the relevant variation is the presence or absence of *Roridula* bugs. An adaptive explanation for proto-carnivory implies that in populations of *Roridula* ancestors, there must have been variation between individuals in their propensity to attract and host *Roridula* bug ancestors. Those with a greater propensity to attract the bugs had greater amounts of *Roridula* bug poop on their leaves, and therefore greater nitrogen inputs. To continue with the "seeing" metaphor, it makes absolutely no difference to natural selection whether the source of nitrogen comes from endogenously produced enzymes or from digestion by an insect. Selection is utterly blind to where a resource comes from. Endogenous digestive enzymes are causally equivalent to those produced by *Roridula* bugs. As long as it is associated with differences in survivorship and reproduction, a resource can be recruited into developmental processes. The next example will illustrate that developmental resources can come not just from the "environment," from the mother, or from other organisms, but from the developing organism itself.

Figure 3.6 "Proto-carnivorous" plants. The plant family Roridulaceae is endemic to South Africa, where there are two species in the only genus, *Roridula.* (A) Mucilage on the tentacles of *Roridula gorgonias* glistens on a moist hillside in the Kleinrivier Mountains. (B) *Roridula dentata* on an open sandy flat in the rugged Cederberg Mountains. Like *Drosera* and *Pinguicula, Roridula* also traps insects with its sticky leaves. (C) Unlike *Drosera* and *Pinguicula, Roridula* does not secrete digestive enzymes. Instead, two species of hemipteran insects, *Roridula* bugs, feed on the insects trapped in the plant's mucilage. The plant, in turn, absorbs nitrogen from the poop of the *Roridula* bugs. The photo shows *Pameridea roridulae* on *R. gorgonias.* These insects are only found on the two species of *Roridula* and walk without getting stuck on the leaves as they forage for insects.

3.6.4 Duckling auditory development and gaping cichlids

Buried deep in a long series of descriptions of detailed experiments, in a section entitled "Effect of prenatal auditory deprivation on the postnatal auditory discrimination of the mallard maternal call...," Gilbert Gottlieb (1971) provides a remarkable example of causal parity, in the form of what can be called self-construction in development. Previous experiments of his had shown that mallard ducklings raised in isolation, so that they had never heard the calls of their own species, could nevertheless distinguish mallard calls from hen (female chicken) calls when finally exposed to them. Duck and hen maternal calls are, apparently, very similar in both frequency and rate, meaning that baby birds need fine auditory perception to distinguish them. This is the sort of "innate" behavior often assumed to have a "genetic basis." Chicks (baby chickens) and ducklings can distinguish the calls of their own species and run to the appropriate mothers, presumably because selection has favored this discrimination. Within a population of chickens, chicks that run to mother ducks would have lower survivorship, say, following the mother and her brood into the water. Selection would therefore favor genetic mutations that allow chicks to identify the calls of hens and ducklings to identify the calls of ducks. Whether or not you think that this sort of scenario applies in this particular case, recognize that scenarios that say things about a trait having a "genetic basis" are very common in biology.

A series of trial-and-error experiments led Gottlieb to develop a technique to mute ducklings and then see how muteness affected their discrimination between species calls. After trying many different techniques, immobilizing membranes on the syrinx, the vocal organ at the base of a bird's trachea, turned out to mute the ducklings completely. Amazingly, this muting procedure can be done by opening the egg and performing a quick operation on the embryo. The egg is then closed up and the embryo goes about its normal development.

Gottlieb then incubated these muted ducklings in isolation, so that they couldn't hear the vocalizations of other ducklings or adults, nor, of course, their own. What he observed was a striking contrast: ducklings that had been isolated but *not* muted could still distinguish between duck and hen calls. But those that were both isolated *and* muted could not. Gottlieb's results, along with much subsequent data (Colombelli-Négrel *et al.*, 2021), mean that hearing *their own* vocalizations during development, not just external duck quacks, is critical for species recognition. Developing ducklings appear to require feedback from their own calls in constructing discrimination of species-specific sounds. This suggests that vocal activity *in ovo* modulates neural and sensory development in ways that shape behavior, potentially even influencing gene expression patterns like *slowpoke.* Here we have another example of causal parity. In Gottlieb's experiments, knocking out fetal duckling vocalizations showed that the duckling's own calls were just as essential in the development of species-appropriate sorting behavior as any other resource, including genes. This is a particularly striking example of a feature of development across organisms, which is self-construction, the participation of a developing individual itself in the conditions that reconstruct the favored phenotype.

Another example of an essential nongenetic, self-constructed resource in development comes from the Rift Lake cichlids, which span over 1,000 species and a famously varied range of morphologies, from massive algae- or snail-eaters to swift predators and stealthy scale- or, appallingly, eye-eaters (Meier *et al.*, 2017). Genetic divergences between many of these species are remarkably small (Won *et al.*, 2005). The study of Hu and Albertson (2017) illustrates how nongenetic resources are essential links in chains of developmental causation. One of the main axes of divergence across Rift Lake cichlids is diet, and diet diversity is expressed to a large degree in fish jaw morphology. As with auditory development in Gottlieb's study of ducklings, Hu and Albertson showed that craniofacial development in cichlid larvae is influenced by the behavior of the fish themselves. Not long after hatching, and before their jaws ossify, tiny cichlid larvae begin a gaping behavior, vigorously opening and closing their mouths. How fast or slow the gaping is not only varies among species but also turns out to correlate with differences in bone deposition in ways that are associated with differences in adult jaw morphology and movement. So, the behavior of the baby fish is a key participant in producing their adult morphology.

Hu and Albertson's experiments, performed on single broods to minimize genetic variation, showed that slowing gaping in fast-gaping species led to morphologies resembling those of the slow-gaping species. At the same time, accelerating gaping in slow-gaping species led to morphologies resembling those of the fast-gaping species. Suppose, for argument's sake, that variation in gaping behavior is caused by genetic differences, fast-gaping alleles versus slow-gaping alleles. Hu and Albertson show that different gaping behaviors are associated with different patterns of expression in *ptch1* expression, which plausibly has a role in generating the different morphologies. So, even if we assume the initial cause of behavioral variation is genetic, and that morphology results from differential gene expression, the developing individual's behavior is nevertheless an essential, nongenetic link in the causal chain, sandwiched between genes and necessary for turning upstream genetic variation into downstream morphological outcomes. Altering behavior alone, without any genetic change, leads to the phenotype of another species. These examples are important because they illustrate that nongenetic resources can be essential, and that they can come from many different sources. Organisms participate in multiple ways in their own development, a theme we'll encounter again shortly. For now, though, these examples bring us back to the central idea of causal parity.

3.6.5 Parity of necessariness

Photons in eye development, vitamin C, protocarnivorous plants, and duckling auditory development all illustrate that developmental resources come in many forms: DNA, water, lipids, air, diet, photons, including one's own actions. Not only do these resources differ in type, but no category ranks itself above the others in essentialness. Removing one deranges development just as surely as removing another. Rather than genes being more fundamental than other resources, both genetic and nongenetic contributors are equal in their necessariness, exemplifying the principle of causal parity. Parity of necessariness isn't the claim that every resource contributes equally, constantly, or in the same way at all times: this relative contribution is substrate for rich empirical work. Reflecting this empirical reality, much of biology and biomedicine are shifting away from seeing genes as commanding authorities that direct development from a privileged position, and toward viewing them as integral but not sufficient components of a broader developmental system. A few analogies can help bring this shift into focus.

3.7 Distributed causation and three analogies of the genome

In discussing development, evolutionary biologists and ecologists almost always refer to genes as though they were making things happen on their own. They talk about genes causing, coding for, regulating, or underlying traits. They less commonly use language about certain loci being simply "involved" in the production of or merely "associated" with certain traits. However, a look at how cellular processes actually work shows that it is impossible to extricate genes or any other key resource as elevated above the others as more necessary (Noble, 2013). This is the principle of causal parity treated in the previous section. If not exclusively in the genes, then where does the causation lie? The notion of "distributed causation" highlights that it is in the characteristics of the network of interactants. Here are three analogies that help illustrate the notion of distributed causation in biological development.

3.7.1 The automobile engine analogy

McKenna *et al.* (2022) use the analogy of a car's engine in thinking about the "genetic basis" of biological traits. No analogy is perfect, so what we want to do with these analogies is use them to help examine notions of genes as the fundamental directors of ontogeny and cellular processes. McKenna *et al.* use the analogy to lead us beautifully into thinking in terms of a system. Cars provide a useful analogy because, like an organism, an engine is made up of many parts that interact with each other in ways that produce a global outcome, moving the crankshaft, as well as homeostasis-like behaviors such as maintaining a stable temperature. These engine-level functionings are not dictated by any one component but arise from the system as a whole.

Given the engine-organism analogy, if it makes sense to look for a single type of resource as the "cause" of an organismal phenotype, then it should make just as much sense to look for the singular cause or director of an engine's running. Knockout logic applies here, too: remove any number of components, a spark plug, the fuel pump, or the battery, and the engine stops. But stopping the engine doesn't reveal which part "directs" it. McKenna *et al.* invite us to imagine disconnecting the battery, observing that the engine fails to start, and then concluding that the battery must be the central controller. But then we notice that the battery is charged by the alternator, so perhaps the alternator is in charge? Yet the alternator, in turn, only functions once the engine is already running, because it is powered by the spinning of the engine's own shaft. The analogy points to the same problem in biology: knockout results can tell us what's necessary, but not what's primary or directing in isolation. DNA, like a car battery, is essential, but not alone in its necessity.

3.7.2 The analogy of the army general

Another strategy is to look for an example of an autonomous causal agent, something that sends instructions from the top, and see how DNA stacks up in the analogy. Army generals wield this kind of authority: they can set entire systems in motion simply by issuing orders. They don't have to be in the trenches; they can command from behind a mahogany desk or by texting from a nightclub. Their authority moves with them. If genes are the generals of development, autonomous agents issuing instructions, then we should be able to recognize them on sight, just as we can recognize a general by uniform and insignia, whether they are giving orders or standing in line at McDonald's. But if a gene's causal role only emerges through its interaction with cellular processes, then it isn't a general with inherent authority, it's a participant whose function depends entirely on context.

Someone might reply that even a general's uniform and insignia are context-dependent, that without the military structure of a particular society to interpret them, they're just a person in fancy clothes. But this only reinforces the point. If even a general's identity depends on institutional context, how much more so does a gene's causal role depend on the biochemical and developmental systems in which it operates? The analogy is meant to underscore the point that a general isn't a general without an army, and DNA isn't a developmental resource without a cell, an organism, and an environment.

3.7.3 The analogy of digital music files

Biologists routinely discuss DNA as though it were the master program of life, specifying everything an organism will become, just as a digital music file encodes a song. But does encoded information alone really dictate outcomes? To examine this issue, consider another analogy. Imagine some extraterrestrials from Planet Klia making a pit stop in Mexico. They observe biology students at parties connecting their phones to speakers, and as the music plays, students dance in various ways. Some patterns of sound lead to cumbia moves, others merengue, others salsa, still others bachata. Intrigued, the Kliites examine the phones and discover memory chips, where the songs are stored as long sequences of 0s and 1s, tiny differences in voltage potential. Just as DNA is made of nucleotides, these music files are made of binary code.

The Kliites suspect that the binary code compels the students to dance. To test their hypothesis, the aliens transfer these files to other student groups, first in Mexico, then in Central America, then in Colombia. Everywhere they go, similar dance patterns emerge: bachata rhythms predict bachata moves, cumbia rhythms predict cumbia moves. They shuffle the song order and see dance sequences shift accordingly. As in DNA knockout experiments, they corrupt the files, and dancing stops. From these observations, they conclude that the binary code is what compels the students to move in certain ways. Like biologists performing gene knockouts, the Kliites attribute causal primacy to the code alone, overlooking the social, cultural, emotional, and cognitive conditions that actually make dancing possible.

The analogy we are trying to make is between the 0s and 1s of the digital music files and the GATC of DNA. The digital codes are essential, but they do not cause the students to dance in the sense that they compel them to do so. The students are not helplessly embodying directives encoded in the digital file. Ask them why they dance, and you'll hear about joy, play, tradition, and flirtation. They understand, intuitively, that the music file is just one part of a much larger system. The same is true of DNA. Even though music "knockout" experiments produce results parallel to genetic ones, no one imagines that the music files alone inexorably determine behavior.

The invitation in this analogy is to consider that the role of DNA in developmental processes is like that of digital music files in students' parties: crucial but not sufficient, structured by selection but not autonomous, and embedded in a broader system. Yes, DNA can be regarded as coded "information," in a form shaped by selection in its functional role. Yes, it is utterly essential, but it is a manifest empirical fact that it is just one essential part of a system with multiple resources. And since "information" so often crops up in discussion of DNA, it's worth turning to this notion for a moment.

3.8 DNA contains information: yes and no

A key motivation behind the "genes-are-everything" perspective might be summarized as follows: with its chains of nucleotides, DNA appears to contain coded information, whereas factors like lipids, water, temperature, or pH do not. Therefore, the argument goes, DNA must be the fundamental director of ontogeny and the true locus of inheritance. As mentioned above, a conspicuous asymmetry arises: when knocking out a gene leads to developmental failure, this is often taken as evidence that the gene provides information for the trait. Yet when the removal of nongenetic factors, such as water, oxygen, or photons, produces the same outcome, those factors are rarely described as containing information (Maynard Smith, 2000).

This interpretive asymmetry highlights how the assumption that information resides in DNA reflects metaphor more than mechanism, a metaphor imposed by scientists, not revealed by nature (Section 5.7.10 explains how to recognize a metaphor). The notion of "information" in DNA, when examined closely, turns out to be very different from information as we experience it in human communication. The following considerations help reveal why the "information" attributed to DNA is metaphorical, not literal.

The first consideration is the relationship between the medium of information transmission and the message (Eco, 2008; McLuhan, 2017; Strate, 2017). In human communication, the various media we use, such as printed text, handwritten letters, digital screens, or spoken words, serve as vehicles for conveying meaning. For instance, when biologists discuss the relationship between zooxanthellae and corals, the content remains unchanged regardless of whether it appears in a printed book, is scribbled on a page, broadcast digitally, or spoken aloud. The medium itself does not physically embody the biological characteristics of corals or zooxanthellae, and it certainly is not physically necessary for the coral-zooxanthellae relationship. So, in human experience, the medium of information transmission often has no material connection to its message.

Nothing could be more different in the case of DNA. In the case of DNA, the medium is the message. Instead of the "message" being independent of its medium, DNA is an essential material participant in cellular processes. The message isn't independent because DNA directly interacts with ribosomes, proteins, messenger RNAs, histones, DNA polymerase, and all of the other components essential in gene expression. This is entirely unlike the relationship between print, digital media, handwriting, speech, and corals and zooxanthellae. Rather than DNA transmitting "information" that is independent of its characteristics, instead, DNA is a physically necessary part of biochemical processes within cells (e.g. Mahner and Bunge, 1997). This leads us to the second point.

The second is that metaphors are dispensable. We will discuss biological metaphors and how to use them in more detail in Section 5.7.10, but for now, it's enough to point out that a key characteristic of metaphors in biology is that they are not actually necessary to describe a given biological process. So, while it's indisputable that DNA contains chains of nucleotides, and that the sequences of these chains are related to amino acid sequences, the term "information" is not actually necessary for describing these reactions. A science of heredity and development can function perfectly well by referencing these processes without ever using the term (Sarkar, 1996). For example, we can say "gene expression is the process in which RNA polymerase and various transcription factors transcribe DNA into a pre-mRNA, which undergoes splicing, capping, and polyadenylation to become a mature mRNA exported to the cytoplasm, where ribosomes, transfer RNAs, and translation factors work together in assembling amino acids into a functional protein." Notice that there was no need at all to use the term "information," and in fact informativeness actually increased when paraphrasing away "information." This dispensability is a diagnostic of metaphors, and points clearly to the "information" in DNA being metaphorical.

The meaning of "information" in biology is an ongoing field of discussion (Sarkar, 1996; Griffiths and Stotz, 2013; Griffiths, 2017; Baker, 2021; Thorvaldsen *et al.*, 2024), but for the purposes of explaining organismal form, it is most helpful to regard DNA as containing "information" only in a metaphorical sense (Mahner and Bunge, 1997; Robert, 2004; Levy, 2011). While DNA is undeniably unique, its chains of nucleotides participating in intricate interactions during gene expression, this special status does not make it more essential than the many other components involved in building a new organism. More importantly, it should not be taken to mean that development proceeds according to a predefined plan stored in the genome.

3.9 No plan in development

Many metaphors of development evoke the unfolding of a plan. We often talk about developmental "pathways" or ontogenetic "trajectories." "Paths" and similar metaphors cast ontogeny as something like a human adult, confidently running toward a clearly visible finish line at the end of a track. Instead, ontogeny is more like a stumbling toddler, whose wobbling attempts to walk are supported at one moment by a parent, then steadied briefly by a nearby sofa. On the next step, the toddler nearly falls before being caught again, then staggers against a wall. At each moment, it might fall in any direction, grab onto a nearby object, drop to all fours and crawl, or simply remain swaying in place. In the same way, at each step, a developing organism has the potential to go in any number of directions (Snell-Rood and Ehlman, 2023). One clear source of evidence for this spectrum of latent potential comes from teratologies, abnormal developmental outcomes (Minelli, 2009; Held, Jr., 2021).

Temperature-shock-induced teratologies are an excellent illustration of the constellation of developmental potential latent at each stage of development. Guinea pigs are one of the best-studied mammals with regard to high-temperature-induced teratologies, thanks to a series of studies conducted after an Australian researcher noticed a high number of abortions and birth defects in a guinea pig colony housed in a metal shed during a hot summer near Sydney. His observation led to a series of experiments in which researchers briefly and slightly elevated mother guinea pig temperatures at points throughout gestation (Edwards *et al.*, 1995). The researchers recorded the effects on offspring of temperature shock applied at different times. In doing so, they found two things. First, sensitivity to temperature differs at different moments in development, with some moments being highly sensitive to high temperatures and others unsusceptible. The second was that exposure to high temperatures at different moments in development was associated with different developmental outcomes (see Table 3.1).

Table 3.1 Sensitivity of guinea pig prenatal development to high temperatures.[a]

Defect produced	Mouse (days)	Rat (days)	Guinea pig (days)	Equivalent stage in humans (weeks)
Exencephaly	7–8	9	13–14	3–4
Encephalocele	8	9	13–14	3–4
Microphthalmia	7–9	9	13–14	3–4
Vertebral kyphosis/scoliosis	8		12–14	3–4
Cataract			11–25, 39–46, 53–60	3–18
Micrencephaly	12–15	13–14	20–30, 39–46, 53–60	3–18
Maxillary/mandibular hypoplasia		9–10		3–4
Facial clefts	8	9		3–4
Talipes			18–25	4–6
Hypodactyly	9	12–14	19–25	4–6
Dental hypoplasia			15–32	4–9
Exomphalos		9		3–4
Arthrogryposis			30–46	~20

[a]Table from Edwards, M., *et al.* 1995. Hyperthermia and birth defects. *Reproductive Toxicology* 9: 411–425, reproduced with permission.

The researchers identified three moments in development when short exposures to high temperature were particularly associated with teratologies. On days 13 and 14 of the ~70-day guinea pig gestation, corresponding to the initial period of formation of the neural plate (a nervous system precursor), raising the temperature of guinea pig mothers just 3–4°C for an hour was associated with exencephaly (the brain outside of the skull), encephaloceles (pseudopodia-like extensions of the brain and associated membranes outside of the skull), and microphthalmia (abnormally small eyes). Temperature shocks on days 18–25, 39–46, and 53–60 were associated with micrencephaly (abnormally small brains). The days in between these sensitive periods showed little incidence of micrencephaly, even though the mothers were exposed to the same temperature shocks. Finally, they found that high temperature on days 53–60 were associated with less myelin production and lower numbers of the oligodendrocytes associated with myelin. No amount of combing the genome would ever predict these potentialities of development, but had these same malformations been associated with genetic mutations, they would almost certainly be interpreted as evidence that the genes in question contain information controlling brain and eye development. But because the same outcomes arise from changes in maternal temperature, we rarely claim that temperature contains "instructions" for development, despite its consistent and specific causal role.

The developmental machinery required to produce exencephaly, encephaloceles, microphthalmia, micrencephaly, reduced myelin production, and other teratological outcomes clearly exists in guinea pigs. For all their poor survivorships, they are consistent, repeatable morphological outcomes that appear under specific conditions. Isidore Geoffroy de Saint-Hilaire commented on this repeatability even of very unusual ontogenetic outcomes in his 1832 *Treatise on Teratology*: "The same monstrosity is ordinarily reproduced in several individuals, such that there exists among them an analogy, and often even a striking resemblance, so that they may be considered as formed on the same type" (Geoffroy de Saint-Hilaire, 1832, p. 104). If, for any reason, selection were to favor one of these configurations, it could in principle be stabilized; guinea pigs can reliably "make" these forms under specific circumstances. Yet no one would suggest that these outcomes are encoded in the genome in the form of instructions or blueprints. These outcomes emerge not from changes to a genetic program, but from changes to the parental bodies, behaviors, and environmental structures that guide the wobbling toddler of ontogeny.

By the same token, normal development, the reliable reappearance of the ancestral phenotype, is not spelled out in the genome either. It simply reflects a developmental system that staggered through a familiar set of supports, more or less retracing the ancestral stagger. What makes that form repeat is not that it was depicted in the genome, but that the necessary developmental resources—the supports—were present and functional. It is not the identity of any one component, but the reliability of the whole system that explains the reappearance of form.

3.10 Reliability of recruitment is what makes a resource

A common objection to thinking about extragenomic factors as key resources in development is that they are somehow not as durable as genes. The argument is that, yes, things like photons, water molecules, or a certain environmental temperature might be important *now*, but these are so changeable that they cannot possibly be bases for heredity. This is reflected in statements such as Richard Dawkins's that "It is its potential immortality that makes a gene a good candidate as the basic unit of natural selection" (Dawkins, 1976, p. 38). From this viewpoint, the only factors of heredity are the genes, and it must be this way because nothing else survives transmission across many generations.

However, there are many examples that suggest that a resource is recruited by virtue of its dependability and not by virtue of its identity (gene versus anything else) (Moore and Lickliter, 2023). "Dependability" means that it is reliably present generation after generation. This reliable presence means that it can be recruited (incorporated by natural selection) as a component of the normal development of a given species.

The fact that dependability is the main requisite for recruitment as a developmental resource is why selection is "blind" to the genetic/nongenetic status of a resource. This was illustrated in Section 3.6.3 with the example of carnivorous versus protocarnivorous plants. Selection could recruit *Roridula* bugs as digesters into the *Roridula* lifestyle because the bugs were present generation after generation. Likewise, selection was blind to humans losing endogenous vitamin C production precisely because the presence of vitamin C in the diet was so dependable, and regardless of origin, the physiological role of vitamin C is identical. Selection would have quickly weeded out variants incapable of producing if vitamin C had not been present dependably in our diet generation after generation. Presence generation after generation is what allows a resource to be recruited as a part of inheritance, not the genetic/nongenetic distinction.

Resources are not merely found in the environment; they are summoned, shaped, and sustained by the developing individual itself, in moments reliably reenacted with each unfolding of ontogeny. To think that for sighted young to emerge, a mother mouse must, on day 16 of gestation, leave the shelter of her lint-lined den and scurry into the sun, risking the gaze of predators in exposing her unborn to photons, is as humbling as it is astonishing. And mother mice do this not once, but pregnancy after pregnancy, generation after generation. The delicate dependency, in which light must pass through the mother's body and into the eye of the embryo, discussed in Section 3.6.1, has been documented in lab mice and probably occurs in humans as well (Rao *et al.*, 2013; Nguyen *et al.*, 2019). If this dependency on light does stretch back to the common ancestor of rodents and humans or before, then the number of mothers of so many species reliably exposing themselves to light at the right moment is truly mind-boggling. But even if it is limited to mice, whether the species is a few hundred thousand or a few million years old (Patnaik *et al.* 1996, Afzali 2024), it still implies millions upon millions of pregnancies, each with its own daring, photon-gathering mission. The repeated, reliable incorporation of such a nongenetic resource over evolutionary time should be enough to convince us that these resources can be every bit as enduring, as present, and as vital as genes. Reliable presence, not genetic status, is what characterizes the elements of inheritance. This leads to the principle that natural selection shapes not genotypes but development.

3.11 Natural selection as the biasing of development

Now is a good time to pause on the term "normal" development. A normal developmental outcome is simply the most common one, not the most fundamental, not the most "intended," and certainly not the only possibility embedded in an organism's developmental repertoire. As Mark Blumberg writes in his must-read *Freaks of Nature* (2009, p. 95), "The embryo's potential to produce two heads is no less ancient, and no less fundamental, than its potential to produce just one. So there it sits, like so many other embryonic potentials, waiting in the wings should evolution ever see fit to call it forward for active duty." As examples such as seven cervical vertebrae in mammals illustrate, the field of developmental potential available to any given individual is not necessarily continuous, in the sense intended in Section 1.10. But even so, every single individual has the potential to express a multiplicity of developmental outcomes. Given the right conditions, any one of us might have developed diprosopia, exencephaly, encephalocele, microphthalmia, or micrencephaly. These outcomes are just as much part of the human developmental repertoire as the more familiar ones. Most of these possible outcomes are rare because they are associated with low survivorship or reproduction. Over time, natural selection filters out variants that tend to produce them, generation after generation. The toddler of development keeps staggering along the same well-worn path, not because there are no other paths, but because the world is arranged, not through the necessity of a blueprint or a "map," but through history, to nudge it that way. Normal ontogeny, then, is simply the most common outcome.

This leads to one of the central points of adaptationist evo-devo: *natural selection is the biasing of ontogeny such that the favored outcomes tend to be produced* (Waddington, 1942; Lala *et al.*, 2024). Look no further than your own ontogeny: you were protected in your mother's body for months, in a controlled environment different in almost every important variable from conditions just outside, in temperature, humidity, salinity, light and sound levels, and the buoyancy of a liquid medium. Gestation is a particularly vivid example of the general principle that selection favors developmental conditions that increase the likelihood of producing fit outcomes. This principle is visible throughout the natural world (Snell-Rood and Ehlman, 2023). Island plants lack the wings and fluff that make the seeds of their mainland relatives disperse so widely on the wind. Loss of dispersibility increases the chance that island plant seeds will fall on soil, where they have a propitious place to develop, rather than the inhospitable ocean (Carlquist, 1974). Seed plants enclose their developing seeds deep within fruits, which provide a moist environment buffered from excessive ultraviolet radiation, abrasion, and herbivory. Birds and reptiles build nests, which offer the right temperature, humidity, and microbial communities for proper development (Campos-Cerda and Bohannan, 2020). At each developmental step, the right combination of resources is present, from genes to proteins to ligands, transcription factors, membranes, temperature, pH, humidity, light, and so many others, such that, out of the universe of developmental possibilities available to every individual, the individual tends to produce the favored variant, just as its parents did. Natural selection is the biasing of development itself. This leads us to recast heritability as emerging from developmental construction.

3.12 Heritability through the reconstruction of the parental phenotype in development

It now makes sense to regard heritability as emerging through the reconstruction of the ancestral or parental phenotype. As explored in Section 1.1.2, natural selection can occur when offspring tend to resemble their parents in traits associated with differential survival and reproduction. Each organism develops from an interplay of influences that permit a range of possible outcomes. For progeny to resemble their parents, they must follow a comparable sequence of developmental steps. Natural selection acts on this process by shaping the availability and reliability of developmental resources. As guinea pig thermal shocks and photon flux in mice demonstrate, if an individual encounters different resources at each stage compared to its parents, it will trace a different developmental path. The degree to which that path aligns with or diverges from the parental one determines the resemblance or dissimilarity of the offspring. Heritability, then, arises from the recurrence of developmental processes that reconstruct similar outcomes. This raises an important question: if not genes alone, what is the unit of inheritance relevant to evolution? To get to that question, it's helpful to flag a couple of terms that often come up in the discussion, "extended inheritance" and "niche construction."

3.13 Inheritance involving more than genes

In recognition of the expansive set of resources involved in any ontogeny, many biologists use the term "extended inheritance," to highlight that the reconstruction of parental phenotypes involves much more than genes. Extended inheritance comes in many forms (Jablonka and Lamb, 2005; Chakrabortee *et al.*, 2016; Bonduriansky and Day, 2018; Samhita, 2021; Skinner and Nilsson, 2021; Tando and Matsui, 2023; Tello-Ramos *et al.*, 2024). Some are very close to the genome. Types of extragenomic inheritance closely linked to the genome include chromatin marking and histone methylation. Others are very far removed from the genome. Spending countless hours on ladders in the Kalahari measuring white-browed sparrow weaver nests, María Cristina Tello-Ramos and her colleagues (2024) discovered that groups of individuals build nests with distinct architectural styles. These styles, varying in size, entry and exit tube lengths, and the angles between them, are consistent within groups but differ between them. Genetic similarity does not explain these patterns. Instead, the differences appear to result from cultural transmission, with individuals copying and perpetuating group-specific building behaviors, yet another mechanism of reliable reconstruction across generations.

Another kind of inheritance, known as structural inheritance, leads to reliable outcomes without involving changes to the DNA sequence itself, instead involving cellular structures as templates. Many cellular division systems replicate existing structures through templating mechanisms, in which preexisting cellular forms serve as guides for the formation of new ones. These lead to structural features being transmitted without genetic variation. A classic example is structural inheritance in the orientation of cilia in ciliates like *Paramecium* and *Tetrahymena* (Ng and Frankel, 1977). These protozoa move through water by waving rows of multitudinous cilia. These cilia are slightly curved, and any change in their orientation, whether natural or artificially induced, is faithfully reproduced during cellular replication (Figure 3.7). This is because the existing structures are templates or scaffolds on which the new structures are constructed (see Caporael *et al.* (2014) on the notion of "scaffolds" in development and evolution). Differences in cilia orientation perfectly meet the three necessary conditions for natural selection. First, alternate cilia configurations represent variation. Second, these variants are heritable, given that existing orientations are "templates" built on during replication. Third, variants are associated with differences in fitness. The orientation of cilia directly impacts locomotion.

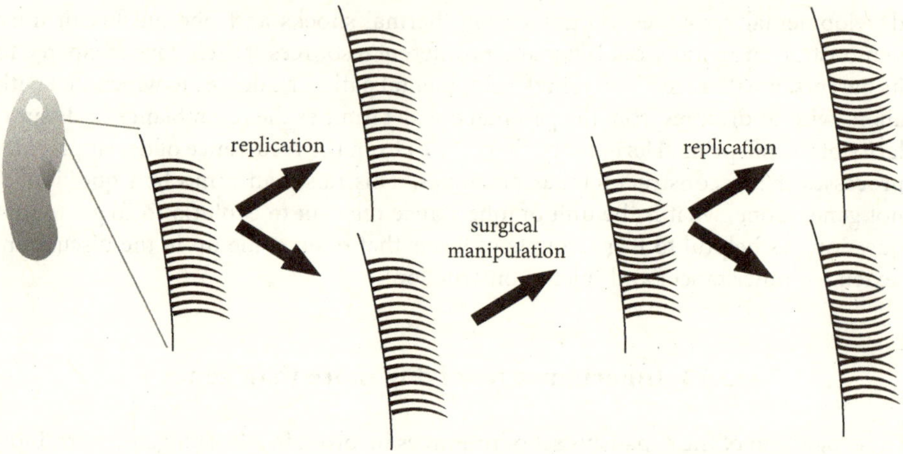

Figure 3.7 Scaffolded inheritance in ciliates. Certain cell division systems involve the construction of new cells following the structure present in the original cell. Any difference in configuration is inherited by the cells produced, without any genetic differences. Such inheritance is referred to by terms such as "structural" inheritance (Jablonka and Lamb, 2005) or "scaffolded" development (Caporael *et al.*, 2014). A classic example is the structural inheritance in the orientation of cilia in ciliates like *Paramecium* and *Tetrahymena*. Surgical alterations to cilia orientation are reliably inherited across cell division and impact performance and fitness but are neither specified nor inherited genetically.

A protozoan's movement is more efficient in the direction opposite to the curvature of the cilia, and because it would impede the organism's movement, an individual with a mixed pattern of ciliary orientations (like the surgically altered paramecia in Figure 3.7) could be fatal. Therefore, we have yet another example where natural selection could lead to adaptation in a heritable trait without a genetic basis.

Such scaffolding is not limited to relatively large structures like cilia, but also operates at the molecular level, in self-templating proteins (Chakrabortee *et al.*, 2016). Self-templating proteins can adopt alternative conformations that serve as structural templates. When these altered proteins encounter normally folded versions of the same protein, they catalyze a conformational change in those proteins, inducing them to adopt the novel folding pattern. The self-templating attribute of some macromolecules is exploited in biomimetics, where a variety of structures are achieved on a genetically uniform background (Chung *et al.*, 2011). In living systems, self-templating in proteins leads to the new conformation being passed on during cell division. Many proteins involved in these processes contain intrinsically disordered segments (Holehouse and Kragelund, 2024). These regions allow proteins to adopt multiple conformations, which makes them versatile in their interactions with other proteins. The flexible regions help establish a self-sustaining change from one cell generation to the next, so the protein state is maintained even as the amount of the protein becomes very dilute. Molecular chaperones like Hsp104, Hsp70, and Hsp90 play a key role in these protein interactions, either helping in maintaining the altered states or, when inhibited, reversing them (Taipale *et al.*, 2010).

By either enabling a conformational switch or blocking it, these interactions help determine which protein state persists, and then making that state heritable across cell generations. Changed proteins are stored in the cytoplasm, and are passed on to new cells during cell division or even transferred between cells through processes such as cytoduction, which involves cytoplasmic exchange between cells without nuclear fusion. This direct passing of altered proteins, plus scaffolded replication independent of nuclear genetic material, leads to altered phenotypic states persisting in daughter cells. These heritable protein states, apparently often, lead to increased fit between cell performance and selective context, such as increased growth or resistance to novel environmental conditions (Chakrabortee *et al.*, 2016). This epigenetic process, involving often large spaces of possible protein conformations, offers yet another mechanism by which stable, heritable traits can arise and be passed on, and one that likely plays a major role in how organisms respond to changing conditions. Extended inheritance is well discussed in friendly treatments (Jablonka and Lamb, 2005), so it's not necessary to catalogue them here beyond flagging that as long as a mechanism can be reliably reproduced generation after generation, it can contribute to the recreation of the ancestral phenotype. Moreover, even environments can be inherited, the subject of the next section.

3.14 Niche construction

"Niche construction" is another term that causes resistance among some biologists, who say that it's just a new term for a well-known phenomenon, while other biologists highlight that, even if it were, it points to important issues that are worth paying more attention to (Keller, 2003; Odling-Smee *et al.*, 2003; Brodie III, 2005). Niche construction is a fairly recent term used to designate the alteration of the conditions of selection experienced by organisms by the activity of the individuals themselves. Biologists have discussed interactions between organisms and the environment for a long time (Woodger, 1930; Haskell, 1940; Lindeman, 1942; Hutchinson, 1948; Odum, 1953; Simpson, 1953). But the niche construction literature has brought home a key point in a way that had not quite been done before (Wade and Sultan, 2023). The "niche construction" term focuses attention on the ways that organisms guide their own development and therefore evolution. These interactions become a focus of research in their own right, and the term now brings together workers from different biological systems who traditionally interacted little (Odling-Smee *et al.*, 2003; Walsh, 2015; Stotz, 2017; Deffner, 2023). Biologists often speak as though the "environment" were an external entity that acts inexorably on passive living matter, for example, "an insect living in a humid forest can evolve to tolerate the ambient temperature and humidity, but the ambient temperature and humidity won't be changed by that adaptation." In this way, it is common to think of the biotic-abiotic variable distinction as a non-arbitrary one, and "the environment" as always separable from organisms.

One reason to reject the notion of "the environment" as separable from organisms is what philosopher of biology Denis Walsh has called the "experienced environment" (Walsh, 2015; see also Facoetti, 2021). This is the observation that the way an organism reacts to an environmental variable depends fundamentally on its own traits. Walsh uses the Reynolds number as an example. The Reynolds number (*Re*) is a dimensionless quantity that expresses the ratio of inertial to viscous forces in a fluid. At low *Re*, typical for very small organisms, viscous forces dominate, meaning movement through the fluid feels more like swimming through honey than water. At high *Re*, inertia dominates and organisms can coast or glide through the fluid. Thus, small organisms do not experience familiar fluid dynamics like turbulent eddies or gliding after a stroke; instead, they live in a world of purely laminar flow, where motion stops the instant effort stops (Vogel, 1994). Biologists often consider the viscosity of water to be an "abiotic" variable, but given that *Re* depends on the size of the organism experiencing it, the selection pressure exerted by the same water differs depending on organism size (Ryerson and Deban, 2010; Allan *et al.*, 2021).

Ryerson and Deban (2010) provide a dramatic example of the evolutionary importance of the organism-specific experience of fluid resistance in the growth of *Xenopus* frog tadpoles, which pump water through their mouths as part of breathing and feeding. *Xenopus* tadpoles begin life very small—4 mm snout-vent length—and end up nearly 20 mm long when metamorphosis sets in. Assuming a rough geometric approximation in which body volume (and mouth volume) scale as length^3, then *Xenopus* tadpoles increase in volume by over two orders of magnitude (64 to 8000 mm^3). This enormous increase in size means they traverse a wide range of *Re*. Ryerson and Deban show that the scaling of buccal pumping volume and velocity does not increase isometrically with body size (all parts maintain the same linear proportions across body sizes), as would be expected if the experienced properties of water were the same across the entire size range traversed by *Xenopus* tadpoles. Instead, the allometric scaling of buccal pumping volume and velocity showed a slope less than 1 on a log-log scale, meaning that smaller tadpoles pump proportionally more water than larger ones. This indicates that they must work harder to move fluid, consistent with the idea that, at small sizes, water feels more viscous and thus demands more effort to displace. The point, then, is that the same water does not constitute the same environment for all organisms. It is experienced differently depending on the characteristics of the organism moving through it, and therefore, the selection pressures it imposes are inseparable from those organismal traits.

Walsh's experienced environment shows that "the environment" is not a fixed backdrop, that not even water can be meaningfully characterized apart from the organisms that experience it. A porpoise and a nauplius larva might swim past each other in the exact same patch of seawater, but their experiences of that water, and the evolved morphologies they bear in response to their experiences, could hardly be more different. The sleek, streamlined porpoise, 1–2 meters in length, experiences water as effectively inviscid, where inertial forces dominate. In contrast, the <1 mm long nauplius lives in a world governed by stickiness. Its squat body and extravagant headdress of bristly limbs (the illustrations in the key of Dahms *et al.* (2006) are very much worth marveling over) would induce catastrophic drag at porpoise size but are perfect for rowing through the goo at very low Reynolds numbers. These examples illustrate that selection is not simply imposed by an external environment, but inevitably shaped through the interaction between organism and world, even in a situation as seemingly straightforward as swimming through the same column of water.

Moreover, organisms do not merely experience environments, but they select and modify them too (Baedke and Fábregas-Tejeda, 2023). Anyone who has stood in the midday sun of a tropical pasture knows the feeling: humans and cows alike head for the cool shade beneath a tree. That same tree is not just a provider of shade but a participant in the experience. It, too, benefits from the lowered temperatures and higher humidity beneath its crown and moisture around its roots. These conditions are not merely "abiotic"; they are constructed, at least in part, by the tree itself. Many tree species can only establish in closed-canopy forests, where the cool, moist shade they require is created by trees, including their own parents. Forest trees, in this sense, create the very conditions required for their own reproduction. Their roots also show just how active this engagement can be. Roots do not simply encounter soil; they burrow into it. They coat their tips in mucilage, wedge themselves into cracks in rock, and once so established, build up cellular pressure until they split the rock or push deeper into the soil. These are not simply responses to some autonomous "environment" experienced the same by all of its occupants, but adaptations born of the root's own behavior. Its downward growth trajectory exposes it to a very different selective context than those faced by stems or by roots of epiphytes (plants that grow perched on other plants), whose roots cling to tree trunks (Figure 2.1D), form aerial masses, or even grow upward, as in some orchids and aroids. Some mangroves, too, provide a striking inversion, with snorkel-like pneumatophores poking above the mud into the air (Figure 3.8). Burrowing animals likewise illustrate how organisms shape their own selective regimes. Shovel-like forelimbs, reduced eyes, and elevated hemoglobin affinity in low-oxygen environments are adaptive responses to the animal's own habit of burrowing. The subterranean world is not a fixed background that exerts selection. It is a niche constructed, in part, by the organism's own "choices," its consistent tendency to go down into the ground rather than up a tree or into the water.

Even though biologists have always recognized organism-environment interplay, focusing attention on it is helpful in at least two ways. For one, if there is no non-arbitrary distinction between biotic and abiotic variables, for any so-called "abiotic" factor can be shown to be shaped by living things in some way, then nothing important should rest on that dichotomy. The distinction might arguably serve occasionally as a convenient shorthand, but certainly nothing of importance should be built on it, for example, an analysis attempting to partition the relative contributions of "biotic" and "abiotic" factors on a given pattern. Since the division is arbitrary, so will be any conclusions that rely on it. Even more useful for our purposes is that if organisms shape their own development by altering the conditions they experience, then those self-altered conditions must count among the core resources required for building a new organism (Bonduriansky and Day, 2018). If individual organisms are involved in creating the conditions for their own ontogeny, then development cannot simply be an irrelevant black-box unfolding of the phenotype prefigured in some omniscient genetic instruction (Depew and Weber, 1997; Oyama *et al.*, 2003; Jablonka and Lamb, 2005). Recognizing this invites a foundational question: if not the gene, what, then, is the unit of inheritance?

Figure 3.8 Organismal characteristics and the experienced environment. In most plants, roots grow downward into soil rich in oxygen and fresh water. But in mangroves, such as *Avicennia germinans*, some roots grow upward, breaking the surface and growing into the air. Spongy bark brings air to the roots that traverse the saline, anaerobic mud below. Roots that grow in the soil and into the air highlight that "the environment" is not a fixed backdrop but a context partly shaped by the organism itself, and what counts as a selective pressure depends on the root's own orientation and location.

3.15 The unit of inheritance as the developmental system and "evolution" as change in the developmental system

The traditional unit of inheritance in evolutionary biology is the gene, but examining how organisms reconstruct themselves anew each generation points to a different conclusion: the entire *developmental system* as the unit of inheritance (Oyama *et al.*, 2003; cf. the "cellular phenotype" of Baverstock, 2021). The "unit of inheritance" is the minimum necessary for a new organism to reconstruct the parental phenotype.

And by this criterion, genes alone cannot possibly qualify. One reason is the observation, which I don't want to belabor, that DNA left to its own devices outside of a cell only degrades, and does not cause, specify, or direct things to happen. If DNA really were sufficient for specifying phenotypes, then it would, by definition, be able to do so without the cellular machinery that transcribes, translates, modifies, and regulates it. It might sound trivial, but again, remember how often terms such as "genetic basis," "genetic cause," "genetic underpinning," and so many others are routinely deployed in everyday biological discourse. As discussed in Section 3.4, even identifying what counts as a gene depends on observing the participation of DNA within cellular processes. The very boundaries of genes, where they begin and end, which variants they produce, can only be determined by reference to how they function in a system. Meanwhile, the examples of photons, vitamin C, *Roridula* bugs, cichlid gaping, and duckling auditory development in Section 3.6 illustrate the variety of nongenetic resources that organisms depend on in reconstructing the parental phenotype.

These are not irrelevant backgrounds; they are necessary developmental resources. The entire set of resources necessary for constructing a new individual is the developmental system. It is necessarily the system that is passed on between generations. Simply passing on DNA will not result in a new individual. Instead, new individuals build themselves in the image of their parents because they have at their disposal a set of resources sufficient to trace the ancestral ontogenetic pathway anew.

If the unit of inheritance is the developmental system and not the gene, then this implies a definition of "evolution" distinct from the traditional one, and that evolutionary change is a single-step and not a two-step process (Figure 3.9). The traditional account of selection is that the information necessary for building a new individual is inherited in the genetic material. This information is then read out in ontogeny, maybe with adaptive modulation of "the" environment, and then the resulting readout is exposed to natural selection. The bearers of some genetic variants survive and reproduce better than others, and so some genetic variants become disproportionately represented relative to others. This view is what leads biologists to say things like "evolution is by definition a change in allele frequencies" (Ghalambor *et al.*, 2007, p. 394; Urry *et al.*, 2021). These variants are passed on in the next generation and expressed again during ontogeny. So, from the traditional point of view, evolution consists of repeated rounds of two steps per generation: 1. transmission of genetic information, and 2. exposure of genetically encoded variation to selection (Lickliter and Harshaw, 2010).

From the developmental systems view, by contrast, there is no alternation between discrete phases of transmission and selection. Instead, evolution is a single, continuous process, the succession of changes within developmental systems across generations. These systems consist of all the resources, genetic, environmental, epigenetic, social, and otherwise, required to recreate favored phenotypes. As such, they are not reassembled anew each generation by decoding transmitted genetic instructions. Instead, the developmental system as a whole is maintained and reproduced. This includes, of course, the genetic material but also the context, scaffolds, and interactions that reliably persist or are summoned from one generation to the next. As a result, selection does not operate only after variation is generated, as a separate phase. Rather, as Badyaev (2011, p. 1924) says that "natural selection can be viewed not as an

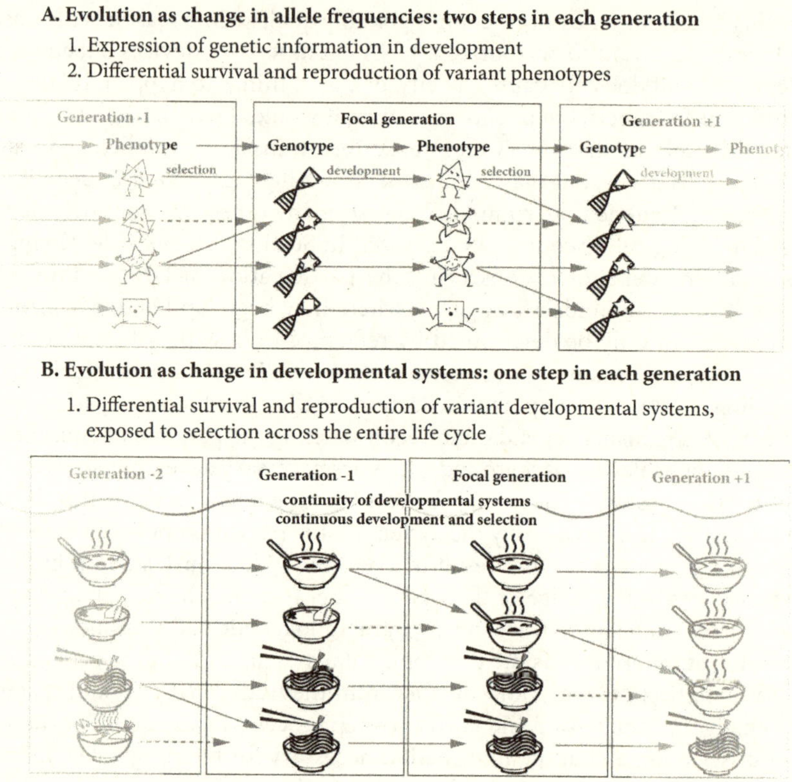

Figure 3.9 Two views of evolution: allele frequency change versus change in developmental systems. (A) If evolution is defined as "change in allele frequency," then each generation unfolds in two steps: genes produce phenotypes through development, and those phenotypes are subjected to selection, leading to changes in allele frequencies. Genetic variation is symbolized here by different shapes (stars, triangles, circles), which produce phenotypic variation. At least implicitly, under this view, development is just the reading out of the genetic information, not a key locus of evolution. (B) In contrast, the developmental systems view treats the entire inherited system, genes, cellular context, ecological conditions, and more, as the unit that evolves. Each developmental system is subject to selection across its full life cycle. All necessary non-DNA changes must also be passed on, and the continuity of developmental systems across generations reflects evolution as change in development itself.

additional external force in charge of filtering, sorting and eliminating developmental outcomes. . . but as a process that continuously and directly arises from the generation of organismal forms within a population." What vary and reproduce themselves, varying between generations as a function of their heritable features, are developmental systems. Related perspectives can be found across biology in approaches that emphasize networks, feedback, and context. In the next section, we turn to a few such windows into systems thinking, starting with the friendly inroads of ant biology.

3.16 Order without bosses: Deborah Gordon's work on ants

Stanford biologist Deborah Gordon's research reveals the ways that ant colonies can show such remarkable order (Gordon, 2010, 2023). "Order" in this context means that the collective behavior of a colony shows predictable responses to recurring situations that the colony faces. One example is foraging: if there is a dense concentration of food near a nest, ants preponderantly tend to head in that direction. Another is colony repair: shoving the nest mound of an ant colony immediately calls hordes of ants to action in repairing the disruption. How this remarkable coordination occurs is of importance way beyond entomology, reaching into the very heart of the debate over the cause of development and heredity.

Ant colony behavior is relevant to thinking about development and heredity because it provides a very clear example of a situation in which there is order without a central director, no one sending instructions to each ant. While there is indeed a queen in an ant colony, her role is limited to producing eggs, not overseeing or directing the actions of the ants making up the colony. In a series of studies on red harvester ants (*Pogonomyrmex barbatus*), Gordon dissected the ways that regularities in behavior in ant colonies emerge from simple "rules" that each ant follows. For instance, in the morning, "patroller" ants make initial forays into the landscape around the colony. If they return, "forager" ants then head out and gather food; if the patrollers do not return, the foragers remain in the nest (Gordon, 2002). Presumably, this behavior is adaptive, avoiding the loss of too many colony members if weather or predators make it unsafe to venture aboveground.

In addition, each colony has a distinctive profile of cuticular hydrocarbons, greasy substances that ants produce and spread over themselves. Within a colony, slight variants in cuticular hydrocarbon profile are associated with different tasks, such as nest maintenance, foraging, or patrolling. In a series of carefully designed experiments, Greene and Gordon (2003, 2007) captured patrollers before they could return to the nest in the morning. They extracted hydrocarbons from patrollers and coated small glass beads with these patroller hydrocarbons. They then dropped the patroller-smelling beads into nests at different rates. They found that not only were patroller hydrocarbons on the beads necessary in accelerating foraging, but also that the rate of patroller-bead entry into the nests was crucial. Too slow and foraging was not stimulated, but too fast as well. A rate of 1 bead per 10 seconds significantly boosted foraging as compared to faster or slower rates.

What makes this example so broadly relevant is that, instead of there being a central authority directing the colony, each ant can be thought of as having an algorithm or set of rules in its tiny ant mind. Gordon (2010, p. 48) distills such rules as "If I meet another ant with odor A about three times in the next 30 seconds, the probability that I will go out to forage will increase by about 10%; if not, it will go down by about 20%." Instead of control from the queen, each ant applies its decision algorithm independently based on the local stimuli that it experiences. From these local interactions, the higher-level order of colony behavior emerges.

If such emergence of higher-level order from simple rules for decision-making based on local interactions can occur in ant colonies, then it can underlie order in development as well (Kirschner *et al.*, 2000; Detrain and Deneubourg, 2006).

In fact, it would be very implausible if order emerged through the interactions of decentralized units at all other levels of biological organization—ants interacting in producing foraging behavior, neurons interacting in forming thoughts, cells forming tissues, ecological communities stabilizing over time—but somehow, uniquely, ontogeny required a boss, genes sitting in the gene boss chair. Even if we assume that each ant's algorithm is somehow genetically programmed in its brain, it's still clear that much of the activity of the colony—the tendency to go where food is, to emerge to forage or remain underground—is to a large degree the result of the interactions between individual ants. If biologists can agree that decentralized control does indeed exist in one system, such as ant colony behavior, then this is fertile common ground for entertaining the possibility of such control in other contexts, such as development.

3.17 The physical properties of biological materials in development: more common ground

Developing organisms are, in their ways, ant colonies. What is meant by this is that the amount of order, understood as behavior precisely tuned to variation in the environment, in the ant colonies that Deborah Gordon studied, exemplifies what some scientists refer to as "order for free." "Order" refers to reliable outcomes in development. This order being "for free" refers to the way that, because many details of development emerge as the result of the interaction of simple rules between lower-level interactants, not all details are specified in the genome. If their being specified in full detail in the genome is regarded as involving metabolic cost, their not needing to be fully specified can be thought of as emerging "for free."

"Order for free" points the way toward convergence in thinking across biology. Genes-aren't-everything biologists often referred to as "structuralists" highlight the importance of "order for free" type behavior of biological tissues in development (Dwyer, 1984; Goodwin, 1990, 2001; Griffiths, 1996a; Kirschner *et al.*, 2000; Newman, 2003, 2014, 2015; Steinberg, 2003; Baverstock, 2021; see also Held, 2021; Brown, 2023). For example, when pushed or pulled, biological tissues tend to bend, fold, or stretch in characteristic ways. These properties transcend biological matter; any material with the same mechanical properties as a given tissue will behave the same way when subjected to the same perturbation. These material properties mean that some configurations, such as folds, tubes, or sheets, even quite complex ones, can emerge not because every detail is specified in the genome but because the right tissue material conditions are present, subject to the right mechanical forces.

As an example of such outsourced order, Santos-Durán et al. (2024) describe ways that variation in tissue mechanical properties is involved in differences in crocodile scale shapes and textures on different parts of the head. Stiffer tissues, characterized by highly aligned collagen fibers (like those on the top of the jaw) resist deformation and produce long, aligned scales. Softer, more flexible tissues with more randomly distributed fiber orientations (like on the sides of the jaw) fold more freely under pressure, creating smaller, more polygonal or labyrinthine scales. Differences in growth rates between skin layers also contribute. When the outer layer grows faster than the inner ones, compressive stress builds up, causing the skin to buckle in distinct ways.

Their experiments and simulations confirmed that these simple physical properties are involved in reliably generating the varied scale patterns that give crocodiles much of their curmudgeonly charm.

To pinpoint the order for free in the crocodile example, let's assume that genes specify the collagen fiber alignment, the rates of keratinization, growth rates of tissue layers, the positioning of scale types, and any other factors that affect tissue mechanical properties and the stresses those tissues experience. Even so, the folding of the tissues into different patterns—labyrinthine, brainy patterns, bark-like polygonal plates, or long, parallel striations—occurs because of the material properties under specific stresses. The genes aren't specifying these basic biomechanical behaviors. They emerge because any material with the same mechanical properties subjected to the same stresses would pock, fold, or wrinkle in the same way. Even if we regard the genes as providing all the conditions necessary, the actual folding is still outsourced to basic material properties.

The same is true for the examples of duckling auditory development, cichlid jaw development, and protocarnivorous plants in Section 3.6. In the duckling case, *in ovo* vocalizations were necessary for discrimination of maternal calls from those of other species. Even if we assume that the propensity for *in ovo* quacking or peeping is genetically programmed, we can still see that there is a nongenetic step that is essential, one outsourced to the organism's own vocalizations and not all spelled out in the genes. In the cichlid case, faster or slower gaping speeds in larvae were necessary for different adult jaw morphologies. Again, even if we assume that different gaping speeds must be genetically programmed, and that gaping speed signals cascades of gene expression that produce jaw morphology, we can still recognize that there is a crucial nongenetic participant in the process, and that the signal to produce a given jaw morphology is outsourced to behavior and not fully depicted in the genes. Likewise, even if we assume that genetic differences among *Roridula* plants cause them to attract more or fewer bugs, it's clear that digestion is outsourced to the bugs, not to endogenous, "genetically encoded" enzymes.

These outsourced sources of order in development are not limited to tissue-level interactions but are found at all levels, including the molecular level. Biologists know, as second nature, that the folding of amino acid chains into proteins follows processes that are not entirely specified in the genome. The final three-dimensional structure and overall form of a protein are not prefigured in any direct way in DNA. Instead, folding arises from molecular interactions that play out in specific physical contexts, such as pH, the presence of ligands, cofactors, chaperones, temperature, pressure, and the crowding of other macromolecules around the emerging protein (Dill and MacCallum, 2012). The attractions and repulsions between residues, whether hydrophobic or hydrophilic, charged or neutral, are not themselves encoded by the genome, but are physical properties of the constituent atoms and molecules. The genome is involved in generating the protein sequence; the folding emerges from molecular biophysics. The formation of favored protein configurations is "order for free," a structure that arises through molecular properties and interactions, not because all aspects of final form are represented in the genome.

If all biologists agree that at least some of development is "outsourced" to nongenomic aspects such as biomechanics or molecular physics, then the real question is simply one of proportions. On one end of the continuum are those who believe that relatively little is outsourced and that nearly everything is specified in the genes: these biologists are toward the "genes-are-everything" end of the continuum. On the other end are those who see a large share of developmental outcomes as emerging from material or systemic interactions not reducible to genetic instruction; this is the "genes-aren't-everything" view. Finding where the real-world sweet spot on the continuum lies, therefore, simply becomes a project for empirical research— common ground and a shared mission across evolutionary biology if there ever were any (Baedke and Fábregas-Tejeda, 2023; Brown, 2023).

3.18 A systems exercise

A nonbiological example can help illustrate the way that "order" can emerge from systems of parts that interact via simple "rules." Again, "order" for our purposes means the salient regularities of ontogeny and evolution. Examples include the ways that organisms compartmentalize themselves into cells of different types, tissues of different types, and the ways that different species are distinguishable from one another. It includes regularities such as the reliable outcomes of development, and what we have called the mystery of synapomorphy, the observation that some structures are reliably produced and recognizable as "the same" even over millions of years of divergence and extreme disparity in function (Section 2.1.3). Alongside these features that apparently resist change despite variation in function, "order" also includes the fit between form and function that is so widespread in organisms. Order includes the discontinuities between taxa (e.g. the absence of morphological intermediates between humans and chimpanzees) as well as the tendency for some teratologies to be produced and for others never to arrive (Wilder, 1908; Alberch, 1989; Blumberg, 2009; Minelli, 2009). All of this "order" emerges in ontogeny.

What sends biologists looking for sources of "order" is that the genome can't possibly specify all by itself all of organismal complexity in the ways suggested by the metaphors that biologists so often use. There are perhaps 40,000 genes in the human genome (Amaral *et al.*, 2023), but the human brain alone contains around 100 billion neurons. Clearly, it's impossible for 40 thousand genes to specify the construction and positioning of 100 billion neurons in any direct sort of way, certainly not any way that resembles a blueprint or any of the other ways that might license talking about a "genotype-phenotype map," at least not a kind of map any mortal has ever seen. If the genome can't provide the "order" that the simplistic metaphors that biologists consistently use suggest that it does, then there must be some additional source of order harnessed during the developmental process of self-construction.

Much of the research on processes of self-construction, such as ontogeny, focuses on the properties of systems (Kauffman, 1996; Barabási, 2002). Systems in the sense intended by these authors are characterized by having many interacting parts. Each part has potential states that it can adopt, and there are connections between some of the parts, such that the state of one part influences the state of another. As in Deborah Gordon's ant colonies, no global overseer, no single type of entity, directs all the others. For our purposes, the most important characteristic of a system is that the parts can interact following even very simple rules, yet the behavior of the entire system can exhibit a high degree of order.

An example could be Friday evening traffic on the southern exit of Mexico City. If we were observing from above, we might conclude that some commander ordered thousands of drivers to leave the city. However, this isn't the case. Each driver has independently decided to spend the weekend at the beach or in the warm lowlands, and straight from work, they leave the city on Friday night, causing a massive congestion in the south. They obey other rules such as "pay tolls at the highway tollbooth" and "don't hit other cars." The kilometers-long line of stopped cars isn't the result of central command or a "blueprint" but the response of each car's occupants to simple rules, such as their work schedules, available budget, and the dream of Acapulco sand between the toes by sunset.

The states that each part can adopt define a space of possible states for the whole system, and the rules determine how the system moves through this possible state space. Alternative states include, for example, whether a car stops or goes, or the metabolic state of a cell. The "rules" govern interactions between parts. An example rule is: "if a cell's two neighbors aren't producing a crucial enzyme, it should produce it; if any neighbor starts producing it, the cell stops." The total number of possible configurations of the elements defines the system's state space. In a system of a thousand cells, each of which can either produce or not produce a substance—two states per cell—the system can be in any of 2^{1000} or 1.07×10^{301}, an enormous state space.

A staggering aspect of a system's state space is that if the interaction rules and the number of possible states for each element are just right, the system can settle into stable patterns called attractors. Attractors are regions in state space where the system tends to stay or return to after small disturbances. To make this concrete (following Kauffman, 1996; see also Solé *et al.*, 2000), consider a system of three thermostats controlling different rooms in a building. Each thermostat can be either ON (heating) or OFF. The total number of possible states for the system is $2^3 = 8$, since each thermostat has two possible settings (Table 3.2). We can define a simple rule set (Table 3.3): for example, if thermostat 2 and 3 are OFF, thermostat 1 turns OFF. In fact, only if both 2 and 3 are ON will thermostat 3 turn ON.

Now we can pick any initial combination of ON and OFF and observe the path the system takes. For example, if all thermostats are ON, according to our rules, they will stay ON in the next step. We've found a loop (Figure 3.10). What if we start with thermostats 1 and 3 ON, and thermostat 2 OFF? According to our rules, in the next step, the first thermostat turns OFF, the second turns ON, and the third stays ON. In the next iteration, we reach the same loop again: all the thermostats will turn ON, and the system will stay in that state. The same happens if we start with ON-OFF-OFF or ON-ON-OFF for thermostats 1, 2, and 3. These "loops" correspond to attractors, and all states of the system leading to the same attractor form its "basin of attraction." In this case, the basin of the all-ON attractor has four states. In contrast, if we start with OFF-OFF-ON or OFF-ON-OFF, we reach a different attractor. The same goes for starting with OFF-OFF-OFF, which is an attractor with only one state. With this simple example, we see the truly remarkable property of a system to exhibit attractors as a consequence of the interaction rules between the parts. The rules and the initial conditions (the state where we start) determine the attractor the system will reach.

Table 3.2 The full set of possible combinations of states is called the "state space" of a system. A system with three thermostats, each of which can be either ON (heating) or OFF, has $2^3 = 8$ possible system states. Each row shows one possible state of the system, a different combination of ON and OFF.

Thermostat 1	Thermostat 2	Thermostat 3
OFF	OFF	OFF
OFF	OFF	ON
OFF	ON	OFF
ON	OFF	OFF
ON	ON	OFF
OFF	ON	ON
ON	OFF	ON
ON	ON	ON

Table 3.3 Each thermostat updates based on the state of its two neighbors at the previous step. To use the table, choose a focal thermostat and locate the row matching its neighbors' states. The final column shows the focal thermostat's next state. Repeat for the other two thermostats.

Focal Thermostat	Neighbor A	State A	Neighbor B	State B	Effect on Thermostat
Thermostat 1	T2	OFF	T3	OFF	OFF
Thermostat 1	T2	OFF	T3	ON	OFF
Thermostat 1	T2	ON	T3	OFF	OFF
Thermostat 1	T2	ON	T3	ON	ON
Thermostat 2	T1	OFF	T3	OFF	OFF
Thermostat 2	T1	OFF	T3	ON	ON
Thermostat 2	T1	ON	T3	OFF	ON
Thermostat 2	T1	ON	T3	ON	ON
Thermostat 3	T1	OFF	T2	OFF	OFF
Thermostat 3	T1	OFF	T2	ON	ON
Thermostat 3	T1	ON	T2	OFF	ON
Thermostat 3	T1	ON	T2	ON	ON

It's striking that even a tiny, non-living system with simple rules can converge on and remain in a small portion of its state space. In larger systems, there may be just a few attractors with huge basins, or many attractors each with narrow ones. The biological significance of this is profound. Attractors offer a compelling possibility to account for the orderly outcomes of development without resorting to a gene-as-blueprint model. If we take the number of protein-coding genes in the human

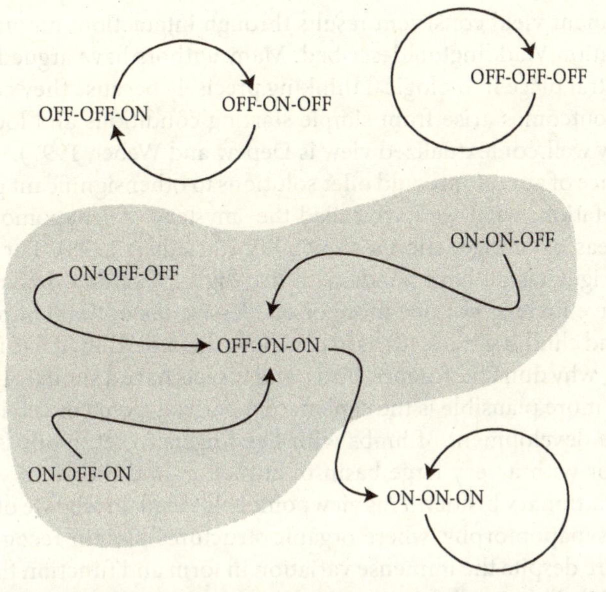

Figure 3.10 "Order for free": attractors in a simple network. In this simple example with three thermostats, it's remarkable that three attractors emerge purely from simple local rules of interaction. The "state space" consists of all possible ON/OFF combinations across the three thermostats. "Attractors" are the subsets of this space where the system tends to settle and remain. The rules of interaction are "local," meaning that each thermostat updates its state based only on the state of its neighbors, without any global knowledge or external control. The "basin of attraction" of an attractor includes all initial states that eventually lead to it. Two of the attractors, at the top, are very small. The third has a basin of four different starting states, each of which eventually leads to the same attractor. If we imagine these thermostats as standing in for resources in developmental systems, it's easy to see how "order," like the specification of cell types, tissue patterns, or morphological structures, can emerge from local rules, without needing to be fully encoded in the genome.

genome to be about 20,000, and assume that there is an equal number of noncoding RNA genes (Amaral *et al.*, 2023), this number is still very small when juxtaposed against the complexity of the body, with its trillions of cells organized into hundreds of types. This disparity poses a fundamental question: how can so much complexity arise from so little information?

One possibility is to view development not as the execution of a detailed genetic script, but as the behavior of an interacting system composed of proteins, membranes, nucleic acids, cells, tissues, oxygen, water, photons, and all the other elements whose states influence each other. In such systems, stable outcomes like cell or tissue types or bodily sectors may correspond to attractors. C. H. Waddington noted that "while it may be possible to steer a mass of developing tissue into one of a number of possible paths, it is difficult to persuade it to differentiate into something intermediate between two of the normal possibilities" (1942, p. 563). Just as the thermostat system settles into stable configurations based only on local rules and initial states, so too does development yield consistent results through interactions among components, as in the situation Waddington described. Many authors have argued that attractors deserve a central place in biological thinking precisely because they can help explain how reliable outcomes arise from simple starting conditions and local interactions (a particularly well-contextualized view is Depew and Weber, 1997).

The existence of attractors could offer solutions to other significant problems in the study of adaptation: what we have called the "mystery of synapomorphy" (Section 2.1.3) and areas of ontogenetic inaccessibility (Striedter, 1998). For instance, bats, humans, and iguanas all have hands with five digits. A standard adaptationist view implies that five fingers, not one more or one less, is *the* optimal number for flying, swimming, and climbing trees. It's hard to believe this assertion. If five fingers are optimal for flying, why don't pterosaurs, birds, and insects have a similar structure in their wings? Much more plausible is the explanation derived from conceiving ontogeny as a system. The development of limbs with five fingers in tetrapods is characterized by an attractor with a very large basin of attraction and therefore very difficult to change in evolutionary history. This viewpoint helps explain why we observe the phenomenon of synapomorphy, where organic structures remain recognizable as "the same" structure despite the immense variation in form and function they can present (Striedter, 1998). This perspective also explains why some structures are *not* observed. If they represent a part of the state space that is inaccessible or outside an attractor, it's practically impossible to produce those morphologies.

If we think of species themselves as occupants of large attractors within a space of possibilities, we find a possible explanation for many species' characteristics that have so far been hard to explain. First, we often find that species are very distinct in their morphology, and there are no species or individuals with intermediate morphologies. If species occupy adjacent attractors, any individual starting its ontogeny at a point between the two species will follow the basin of attraction until reaching the morphology of one or the other species. This would help account for developmental potential that has inaccessible or difficult-to-access regions. The leap from one attractor to another could explain why evolutionary change can be very rapid and apparently without a graded series of evolutionary intermediates (Bak and Boettcher, 1997).

The view of development as traversing spaces of possibility and impossibility from attractor to attractor does not at all obviate the importance of selection. In any adaptationist evo-devo scenario, some variants out of the possible have higher fitness than others, and these are preferentially produced. If the space of the possible is understood as corresponding to the location of attractors, then some attractors correspond to higher survivorship and reproduction and selection consists of biasing ontogeny toward their production. By the same token, the structure of the network itself is almost certainly subject to selection such that certain positions and numbers of attractors are produced over others. This tuning could occur at fine scales, for example in the repeated production of one morphological variant over closely related alternatives.

Systems theorist Stuart Kauffman also envisions selection occurring at a higher level (see Section 5.8 for more on selection involving entities above the species level) keeping biological systems poised "at the edge of chaos." In this context, a "chaotic" system is one in which basins of attraction and attractors are so large that with respect to the time frames of relevance for organismal evolution, for all practical purposes, there are no attractors. No attractors means that systems cannot reliably reconstruct ancestral configurations over and over across generations. In a system without heritability, selection cannot occur and the lineage will go extinct. On the other hand, systems with a great deal of structure, such as those with a single attractor, are so rigid that evolution cannot occur. Selection, therefore, favors systems in which networks are sensitive to changes in patterns or numbers of connections between parts such that a moderate number of attractors occur, and that some change can happen in the numbers and positions of these attractors. Systems that are excessively sensitive are in the region of chaotic behavior. Because selection favors systems that exhibit some instability but not too much, Kauffman describes selection as pushing to "the edge of chaos." This attractors-based perspective thus helps explain how systems with so few components, such as so few genes, can build structures in which the numbers of cells and connections between them can so exceed the numbers of genes available to specify them directly. It also helps explain morphological discontinuities between taxa, how speciation can occur quickly and relatively discontinuously, why some features are so stable over the eons, as well as how lineages can retain their identities over time and yet maintain developmental flexibility (Morrison and Badyaev, 2017; Potticary *et al.*, 2020). It does not by any means exclude the importance of selection, but what such thinking has done is open the door to the flowering field of systems biology.

3.19 Systems biology

Increasing computing power and the availability of large biological datasets means that the vision charted by Kauffman and others has firmly taken off in the form of the field of systems biology. The field is burgeoning and has many expressions, such as journals like *Systems Biology*, *Molecular Systems Biology*, and *BMC Systems Biology*, and field-specific software and approaches (Karp *et al.*, 2021). Institutions such as Harvard Medical School, Columbia, UMass Chan Medical School, and the City of Hope Cancer Center all host dedicated programs in systems biology. Biologists and biomedical scientists from countless starting points are drawn by the sheer structures of their study systems to examine networks of interactions. As Portin and Wilkins note, "genetics is becoming systems biology. . ." (2017, p. 1361; see also Qi and Ge, 2006).

Bearing out this assessment, studies of what are called "gene regulatory networks" (GRNs) actually include much more than genes. GRNs, a key focus of current systems biology, model the way specific DNA loci, from entire genes to individual binding sites, transcription factors and other proteins like enzyme or structural proteins, and RNAs (like micro-RNAs or long noncoding RNAs), interact, including concentrations, reaction rates, binding kinetics, and other interactants and considerations (Ingram *et al.*, 2006; Karlebach and Shamir, 2008; Lai *et al.*, 2016). Even in the simplest GRN, where genes are the only nodes, arrows of interaction between them are presumed to represent transcription factors (Badia-i-Mompel *et al.*, 2023). By illustrating the necessity of interactions involving non-DNA components, these models bear out the point that genes alone do not direct development.

The potential of GRNs to represent biological interaction is, in principle, limited only by computing power and the patience of biologists. A simple starting point is a GRN in which a given transcription factor invariably produces the same effect on a particular gene, regardless of context (Levine and Davidson, 2005). Most proteins, including transcription factors, include intrinsically disordered regions, and these can profoundly influence network structure (Dunker *et al.*, 2005; Holehouse and Kragelund, 2024). Intrinsically disordered regions are portions of a protein that, instead of folding into a tidy tertiary structure, flutter among a sometimes very large array of different conformations. This structural lability means that proteins with intrinsically disordered regions can exhibit one binding behavior in one context, and a different behavior in another (Downes *et al.*, 2014; Tuttle *et al.*, 2018; Moses *et al.*, 2023). So, a protein can behave as a transcription factor in one context, and a cofactor in another. Nature readily seizes on economy of resource use and precision of regulation. So, it's no surprise that multi-use, finely context-sensitive proteins should be yet another layer of regulation in cellular processes. This context-sensitivity of protein behavior means that predicting what a given protein will do requires considering a wide array of factors, such as pH, osmotic pressure, solute concentration, solute identity, availability of free water, the presence of other proteins, all the way to temperature and pressure (Kirschner *et al.*, 2000; Bondos *et al.*, 2022). On top of this, chromatin and a host of proteins, including transcription factors, wander through the cell largely by Brownian motion. Their collisions and interactions, sometimes occurring at just the right concentrations, can trigger gene expression as a probabilistic outcome of stochastic processes. Stochastic gene expression, far from being a biological glitch, is likely ubiquitous and appears to be actively harnessed in many cellular processes (Kupiec, 2014). These examples illustrate that the very nature of cellular processes inevitably leads biologists to wider and wider sets of interactants, far beyond genes alone.

The aim of including all relevant participants into GRN models, broadly construed, is a tall order. Even "simple" GRNs can involve dozens of interactants (Levine and Davidson, 2005). Intrinsic disorder in protein structure means that cellular context and protein-protein interactions must be taken into account to determine whether a given protein even behaves as a transcription factor at all, markedly affecting network structure. Section 3.4 explored striking examples of alternative splicing, such as the thousands of *Dscam* variants. Such flexibility in protein sequence is not only likely to play a key role in rapid adaptation, but also represents yet another axis of expansion of GRNs (Shin and Manley, 2004; Singh and Ahi, 2022), given that each splice variant can alter network architecture and regulatory outcomes involving the same genetic locus. Posttranslational modification, as when the same protein is subjected to different patterns of function-altering acetylation or phosphorylation, is another set of evolutionarily important processes that can affect network structure (Guan *et al.*, 2014). The protein that a given DNA region is involved in producing might be phosphorylated, acetylated, cleaved, or otherwise modified in becoming active, or it might be quickly degraded depending on cellular conditions. These modifications depend on enzymes, cellular energy levels, and redox states. A given protein might only function in certain compartments, such as the nucleus, membrane, cytoplasm, or chloroplast (if you have them), and if product A doesn't get to the right place, it won't affect its target, regardless of how much is produced. These, again, are considerations that can be included in GRNs. Similarly, the arrangement of molecules within cells can affect processes within the cell. Scaffold proteins, membrane domains, cytoskeletal elements, and local concentrations of reactants can all influence whether a given protein actually finds and interacts with a given gene's promoter or its regulatory machinery. And even if that protein reaches the DNA near a given gene, that gene's chromatin might be condensed or modified, as via methylation, making it inaccessible to transcriptional activation or repression. At higher organization levels, tension in tissues, matrix stiffness, cell shape, and other mechanical cues can influence nuclear architecture and gene expression (Discher *et al.*, 2009; Albergante *et al.*, 2016). Mechanisms such as these highlight the ways that manifold components, not just genes, lie at the heart of development and metabolism, components that some biologists already explicitly incorporate into GRNs, and others seek to accommodate as part of broader regulatory logic (Niklas *et al.*, 2015; Snell-Rood and Ehlman, 2023).

In the face of the manifold factors that intervene in development, Gjuvsland *et al.* (2013, p. 2056) summarize both the aim and its formidable challenges: "The causality flows from the system state through a change in use of DNA (as an inert system component) that results in a change in the production of RNA and protein, which in turn perturbs the system's dynamics. In those cases where variations in DNA cause changes in the perturbation regime it may lead to different system dynamics and thus physiological variation. . .without the massive, combined use of existing and new mathematics, high-dimensional data analysis, computer science and advanced engineering methodology, biological research will not be able to probe very deeply into the genotype-phenotype relation." Even with this goal still out of sight over the horizon, systems biology has already opened up important avenues. To underscore the vitality and fertility of a systems perspective, a few of these avenues are summarized here.

3.19.1 Systems biology insights: network architecture

There are many ways to connect a given number of nodes (genes, proteins, neurons, etc.), so the question of why certain architectures and not others are observed becomes central (Keller, 2005; Khanin and Wit, 2006; Broido and Clauset, 2019). Studies of biological and other networks suggest that some patterns of connection may be more common than others. Often in biological networks, such as large gene regulatory circuits, protein-protein interaction maps, or metabolic pathways, most nodes (e.g. genes or proteins) tend to have few connections, while a small number of "hub" nodes have many. This is a very different configuration from networks in which each node is connected to every other, say, or where all nodes have exactly the same number of connections. As the study of networks was gaining momentum, there was a great deal of excitement over the possibility that many real-world networks, biological, social, and even the World Wide Web, might all be "scale-free" (Keller, 2005).

"Scale-free" is a technically correct but totally unhelpful term for describing network structure. It's confusing in part because biologists are attuned to the ways that different patterns and processes emerge as they zoom in and out on different scales of time or space. Circadian rhythms appear on the scale of days, while the periodicity of mass extinctions only becomes visible on the scales of hundreds of millions of years (Gould, 1985; Jablonski, 2007). Biologists also use "scale" casually to refer to hierarchical levels of organization, as in "the tissue scale" versus "the organ scale." But none of this is what's meant by "scale-free" network structure. In network parlance, "scale" refers to the typical number of connections that nodes have. If at a party everyone knows ten people, then the social network has a characteristic scale, ten. But if most people only know one or two, and a few know dozens or hundreds, such that the distribution of connections follows a power law, then there isn't a characteristic scale (Barabási, 2002). That's what "scale-free" means. A term like "life-of-the-party network" might be more evocative. Explorations of real biological networks suggest that most don't strictly follow a power law (Keller, 2005; Khanin and Wit, 2006; Broido and Clauset, 2019; Gamermann *et al.*, 2019). Still, while they may not be strictly scale-free, biological networks do often show strong heterogeneity in connectivity: most nodes are sparsely connected, and a few are highly connected.

It seems likely that selection shapes such architectures because of the properties they give rise to. For example, certain network structures may lead to patterns of variation in gene expression that are favored under different selective conditions in different species (Chen and Wu, 2007). Under some conditions, selection might favor networks that minimize distances between any two nodes (Wagner and Fell, 2001). Minimizing distances would maximize the speed of propagation of interactions across a network. Maximal speed would mean that network responses, say those leading to important metabolic shifts like from aerobic to anaerobic respiration, are swift in response to changing conditions. Perhaps more commonly, selection favors networks that have distances that are longer than the theoretical minimum, but have the virtue of modularity (Zhang and Zhang, 2009).

3.19.2 Systems biology insights: modularity

Biologists have always been intrigued that some sectors of organismal bodies develop and evolve to some extent independently from others. Modularity is the property of an organism to be made up of subunits or parts, with interactions being stronger or more frequent within parts than with other parts (Wagner, 1996; Schlosser and Wagner, 2004; Serban, 2020). Lewontin (2000a) calls this "quasi-independence." This organization is manifest in relative independence in development and, therefore, evolution. For example, the plant genus *Euphorbia* is perhaps the most varied of all plant groups in its vegetative body, which ranges from tiny herbs to lianas to trees to some of the weirdest succulents on earth. Across this exceptional diversity of vegetative bodies, the flowers remain relatively uniform. Only if development of the vegetative body were relatively independent of the flowers could there be so much variation in the vegetative body without concomitant variation in the flowers. In human development, the eye emerges as an outgrowth of the forebrain and develops along a largely autonomous trajectory (Edward and Kaufman, 2003), allowing its distinctive features to evolve without dragging the surrounding orbital mesenchyme along with it.

Networks in which there are relatively few hubs (nodes that interact with many other nodes) and a much larger number of nodes that have relatively few connections are inherently modular. The "parts" or modules of a network are made up of the hubs and their net of closely connected nodes, with relatively few connections between hubs. These networks manifest the key modularity property of interactions being stronger or more frequent within parts than with other parts of the system (Qi and Ge, 2006; Serban, 2020; Hernández *et al.*, 2022). Modularity themes include whether modularity at higher levels requires modularity at lower levels, and why modularity might arise in the first place (Wagner *et al.*, 2007; Klingenberg, 2014; Serban, 2020).

Hypotheses on the evolutionary origin of modularity vary wildly. On the one extreme are scenarios that depict selection ruthlessly favoring variants with lower metabolic costs (Clune *et al.*, 2013). These hypotheses start by noting that the number of nodes intervening between any other two nodes should at least approximately be related to metabolic cost. Given that interactions in a network involve proteins, genes, or other metabolically costly elements, this assumption is entirely plausible. If this is true, then individuals with similar performance but shorter networks would have greater metabolic surpluses directed to crucial aspects like growth, immune responses, or reproduction (Section 1.8 treated energy efficiency and fitness). Networks in which all nodes are connected to every other node, maximizing the number of connections, would be among the most metabolically costly in connection maintenance. At the other extreme, long linear networks with few connections may be metabolically cheap but functionally costly, since any signal must traverse the entire chain. Modular networks, by contrast, with their dense connections within modules and limited but strategic links between them, allow for rapid coordination across the whole system while keeping overall connection costs low. They could therefore result from competing selection pressures maximizing connectivity while minimizing metabolic cost or organ volume (Clune *et al.*, 2013). In neural systems, this logic appears in what has been called the "save wire hypothesis," under which selection should minimize the amount of neural tissue needed for a specific level of connectivity (Mitchison, 1991; Cherniak, 1995). In this view, modular network structure emerges as the product of natural selection balancing connectivity with energetic efficiency.

At the other extreme are hypotheses that depict energetically profligate developmental systems free to proliferate in their components (Wagner *et al.*, 2007). One version holds that modularity arises when a gene, along with its interactions and products, is duplicated. The resulting set is tightly connected internally but loosely connected to the rest of the system, forming a quasi-autonomous module. The original remains subject to selection, while the duplicate, insulated by its limited external connections, is free to diverge in function.

Yet another set of hypotheses has to do with selection favoring alignment of developmentally quasi-independent subunits with functional domains. These often involve selection at the above-species level, with different patterns of modularity associated with differential persistence between lineages, as well as standard population-level selection. The argument goes that given variation within a population, individuals in which developmentally quasi-independent units coincide better with functional domains are more likely to have higher form-function fits. If functional subunits are quasi-independent developmentally, then they can diverge mutually to a greater extent than in individuals with different functional domains that are tightly linked (Breuker *et al.*, 2006; Hintze and Adami, 2008). On this view, lineages with modular development, and thus a greater likelihood of fine adaptation, are prone to diverge into morphologically disparate species and persist longer than those with less developmental independence between parts. The cause of modularity remains a vigorous area of research, with hypotheses ranging from traditional population-level selection favoring energetic efficiency to scenarios of highly permissive selection, to those invoking differential proliferation or persistence among lineages (see Section 5.8). Modularity is thought to contribute to robustness, another central systems biology theme.

3.19.3 Systems biology insights: robustness

Robustness is the persistence of function in a system despite perturbation, and is manifest in organismal attributes like homeostasis, in which a metabolic state is maintained despite environmental buffeting (Wagner, 2008; Silva-Rocha and De Lorenzo, 2010; Bateson and Gluckman, 2012; Félix and Barkoulas, 2015; Bertolaso *et al.*, 2018; Sappington and Mohanty, 2025). Another manifestation of robustness appears in organismal development, which involves buffering or working around perturbations even in normal development, a theme explored in the next chapter. Perturbations take many forms, from mutations that degrade function in a given protein, environmental insults, such as toxins or temperature shocks, or disruptions to the architecture of regulatory networks.

An example of a perturbation to network structure—it's hard to think of bigger ones that happen so frequently—is polyploidy and hybridization. Polyploidy, commonly associated with speciation in plants, results from whole-genome duplication (Alix *et al.*, 2017). If anything qualifies as an alteration in network structure, surely doubling the entire genetic complement is a good candidate (Lachowiec *et al.*, 2016). One might reasonably expect that a descendant individual with a doubled genome would show twice as much mRNA as its ancestor with just a diploid genetic complement. Instead, transcriptional output in polyploids generally falls well short of what would be expected if transcript levels scaled proportionally with genome size, and is often similar to that of diploids (Doyle and Coate, 2019). This is an example of robustness in the extreme: even in the face of a major alteration to network structure, a clear buffering potential results in entirely functional outcomes. Hybridization is another example of drastic network alteration, which likewise results in viable offspring, especially in plants. Results of hybridization-reworked networks are so viable, in fact, that, like polyploidy, hybridization is commonly involved in speciation in plants (Alix *et al.*, 2017). All of these are manifestations of robustness even in cases of drastic alteration in network structure.

Systems biology provides unprecedented access to the potential bases of robustness, with one line of evidence provided by network motifs. "Motifs" are recurring patterns of interaction among components of biological networks, such as genes, proteins, RNAs, or signaling molecules, that show up in regulatory networks across different organisms (Milo *et al.*, 2002; Qi and Ge, 2006; Stone *et al.*, 2019). A well-studied example is provided by feed-forward loops, in which a regulator X controls a second regulator Y, and both X and Y regulate a downstream target (Alon, 2007). One kind of feed-forward loop (Mangan *et al.*, 2003) reflects the sort of economy of resource use that natural selection so often leads to. In this loop, Z is expressed only if conditions for its expression are sustained, even if the relevant stimuli fluctuate quickly, sometimes present, sometimes absent. Such fluctuating signals are a very important aspect of selection shaping organismal responses, avoiding, insofar as is possible, costly energetic expenditures resulting from misread cues. Most cactus seeds, for example, never germinate at the first sign of water, but only after repeated soaking and drying, which correlates better with the sustained soil moisture conditions necessary for seedling establishment as compared to a single splash of water. Feed-forward loops represent this sort of bet-hedging behavior at a regulatory level. Both X and Y are needed to activate Z. Y takes a while to accumulate, though, and so it is only present when X receives sufficiently sustained stimuli. So, the feedback loop only leads to activation of Z if the conditions are sustained. As soon as conditions abate, X disappears and Z is quickly deactivated.

Many other motif structures are known, and their spare logic of feedback and control will be instantly recognizable to any biologist, accustomed as we are to encountering such mechanisms across countless biological systems. For decades, biologists worked out these feedbacks with arduous studies of single steps in metabolic or gene regulation pathways (Jacob and Monod, 1961). Systems biology has rapidly expanded the recognition of these patterns across many different systems, proposed tools for their discovery in network data, quantified the prevalence of different motifs in different contexts, and analyzed their differing behavior in differing contexts. Some workers have gone so far as to frame motifs, rather than genes, as basic "building blocks" of regulatory networks (Milo *et al.*, 2002). But just as a gene's function depends on its context, the behavior of a motif depends on how that network structure is instantiated in real biological systems (Ingram *et al.*, 2006), leading us back to causal parity.

3.19.4 Systems biology insights: causal parity, again

Consider a gene regulatory network (GRN) that includes not just DNA and transcription factors, but also RNAs, cofactors, cells, tissue-level mechanical properties, water, oxygen, photons, pH, vocalizations, molecular crowding, and *Roridula* bugs.

Picture the GRN as a diagram: each interactant is represented by a point (a node), and each interaction by a line (an edge) connecting them. As in all complex GRNs, the result is a dense tracery of interdependence (Levine and Davidson, 2005). Now imagine removing all the labels that identify the components—"gene X," "transcription factor Y," "O_2," and so on. What remains is a web of interactions: some nodes are highly connected, others sparsely connected, and many interactions occur only under specific cellular conditions. In this unlabeled network, it's no longer possible to tell which nodes are genes and which are not, based solely on position, connectivity, or influence. Genes have no universal topological signature. They are not necessarily central hubs or initiating elements; many are activated only in particular contexts and participate more as endpoints than regulators. Likewise, nongenetic elements such as oxygen concentration, local pH, or cell geometry can serve as necessary conditions for regulatory activity, and thus have just as much causal traction in the system as any "gene." The role of DNA as a vital participant but not a master director emerges unbidden in systems biology, study after study.

3.19.5 Systems: an attractor for all of biology

The point of this section is not to give anything like a comprehensive overview of systems biology, but to underscore the message of this chapter. What systems biology finally makes visible is the possibility of regarding organisms as systems of interacting components, in which morphology and behavior arise from networks of relationships. This perspective is among the most inspiring and important developments in modern biology, and it fully deserves the dedicated attention it has received, Ball (2023) being an excellent example. Here, however, this brief mention of insights from systems biology is meant simply to illustrate the flourishing alternative to thinking of DNA alone as containing the directions or instructions for development. As we saw in the survey of the journal *Evolution* in Section 3.3, biologists still routinely invoke genes as if they were, by themselves, the causes of organismal phenotypes, the units of inheritance, and the directors of development. Yet a detailed study of genes and their roles in real developmental processes paints a very different empirical picture.

It might be argued that the difference between the "genes-are-everything" camp and the "genes-aren't-everything" perspective is largely moot. In a way, it is, because regardless of ideological position, the nature of DNA as an inert participant in regulatory networks becomes clear whenever anyone tries to explain how phenotypes actually emerge from biological processes. Whether or not one holds to gene-centric models, empirical inquiry leads biologists and biomedical scientists (Tsongalis, 2018) inexorably away from the view of genes as autonomous directors of phenotype. It leads away, too, from formulations that invoke an arbitrary "genes plus environment" model. We have seen in this chapter that the dichotomy between genes and everything else is itself artificial, as is the split between biotic and abiotic variables (Section 3.14). Instead, actual examination of the way form emerges in development leads to systems biology. Systems biology can, however, awaken some lingering attachment to seeing DNA as a vital but not sufficient participant; some forms of this reluctance, to keep an eye out for, follow.

3.20 Genetic "eppur si muove"

An almost certainly apocryphal story says that Galileo Galilei, after being forced to renounce his position that the earth revolves around the sun, grumpily muttered "eppur si muove," something like "it [the earth] really does move, though." In evolutionary biology class, students often reach their own "eppur si muove" moment with respect to genetic causation. After exploring examples involving ducklings, guinea pigs, *Roridula* bugs, and attractors, they see that genes can't possibly act in isolation. They become convinced, or at least see, that they are unable to provide anything like an alternative interpretation, and that development depends on a system of interacting resources, not on DNA alone. And yet, despite this recognition, biologists, students and professionals alike, often relapse into gene-as-agent language (Noble, 2015). These are the modern "eppur si muove" moments, when, even after everything, we still find ourselves acting as though genes move the world. Here are three examples of genetic "eppur si muove."

3.20.1 Genetic causation residing in unseen antecedent steps

The main "eppur si muove" move is to invent genetic causation in a step previous to the observed one (Lala *et al.*, 2024 call this "genes standing in for development"). For example, a biologist might state that the propensity for mother rats to head into the light on day 16 of gestation is genetically encoded, and so photon flux as a participant in normal eye development reduces to the genetic cause that sets events in motion in a previous developmental moment. They might argue that the physiological role of Vitamin C is genetically encoded, and thus the participation of Vitamin C in metabolic processes can ultimately be ascribed to a genetic instruction embedded further back in the sequence. They might argue that interindividual variation in *Roridula* bug abundance between the members of ancestral *Roridula* populations must have been due to some genetically caused differences between individuals, such as volatile attractants or navigability of the sticky tentacles on the leaves. From this point of view, variation between *Roridula* plants is all reduced to genes in an unseen step previous to habitation by *Roridula* bugs and the causal role of *Roridula* bugs in digestion is all due to the genes in the plants. They might argue that the propensity for *in ovo* ducklings to vocalize is genetically determined, and so the role of vocalizations in development is ultimately just a question of genetic causation at a step previous to the vocalizations. They might argue that the algorithms in the ants studied by Deborah Gordon are the genes telling the ants what to do, so colony behavior is ultimately ascribed to genes. They might point out that at *this* moment, the behavior of a GRN arises from regulatory interactions that aren't reducible to DNA sequence alone, but network structure was specified by the genome in antecedent steps. The net effect of these maneuvers is to reassert genetic primacy: nongenetic elements may participate in development at step B, but only as downstream executors of genetic instructions implemented at step A.

However, it is always step B. At every step, DNA is a necessary but never sufficient participant in developmental systems. This "eppur si muove" move has two steps.

First, it acknowledges that at any given focal moment in development, DNA cannot be shown to act alone as a causal agent, directing ontogeny or triggering events independently. Second, in response, it shifts the genetic causality one or more steps back: yes, extragenomic factors are involved right now, in step B, but the preceding step, step A, must have been caused by DNA bosses sitting in DNA boss chairs. But there is no previous step in which DNA was, in fact, making everything happen all on its own. DNA's necessary-but-not-sufficient role holds not just in a few quirky cases but across all moments in development.

3.20.2 Genetic backfill

Another maneuver is to discount the "order for free" properties of biological systems. This maneuver begins by acknowledging that developmental outcomes emerge from component interactions, even basic ones like electrostatic charges between amino acid residues. The argument concedes that physical propensities, not directly specified by the genome, do indeed govern such interactions and thus account for much of protein folding. However, the maneuver then pivots: because the amino acid sequence is encoded in the genome, the genome must still be the most important part of the explanation.

This rhetorical move can be understood as a kind of genetic backfilling. In construction, backfilling refers to filling a hole, such as one dug to lay a foundation, with earth or gravel, packing material up against the structure. In genetic backfilling, the "hole" is the explanatory gap left by physical laws or system-level properties that are not and cannot be specified in the genome in the way that biologists' language would make it seem. That gap is then filled in with DNA, as if the genome, by participating in the system, also fully explains it.

By ascribing all of these effects to genes, genetic backfilling glosses over the crucial role of non-genomic properties such as the folding of proteins, which depends on factors like pH, temperature, molecular crowding, and the inherently stochastic behavior of molecules in cells. It similarly glosses over the mechanical properties of tissues, as in crocodile scales (Section 3.17), or the way order can emerge "for free" as the result of interactions between parts, as in the way that high-level ant colony behavior emerges from ant-ant interaction (Section 3.16). These are real and functionally decisive, yet ascribing development to DNA and are treated as "givens" or background rather than as causes in their own right. DNA is, of course, involved in every biological process. But to say it specifies "order for free" outcomes like protein folding or tissue architecture is to obscure the nature of causation in development.

3.20.3 Environmental cues versus developmental resources

A final "eppur si muove" is to dismiss resources such as photons as "environmental cues." To use the example of photons in retinal development from Section 3.6.1, an "eppur si muove" move would be to say that photons are only important in development to the extent that the genetically determined response is "resorb the hyaloid artery if light is present" versus "do not resorb the hyaloid artery if light is not present." The problem with this position is that it confuses developmental resources and environmental cues.

"Environmental cue" implies detection-and-response systems favored by selection in the face of fluctuating conditions, systems that produce alternative adaptive developmental responses based on the cue. Trees in tropical dry forests, for instance, are highly sensitive to soil moisture. When water is abundant, they retain their leaves and photosynthesize; when soil moisture drops below a threshold, they shed their leaves, conserving water. Soil moisture, then, functions as an environmental cue: it's detected, and a response follows. The cue "abundant water" is associated with continued photosynthesis and growth, while the cue "water scarcity" triggers leaf drop, preventing hydraulic failure. Both responses are functionally appropriate, selected outcomes of the same detection-and-response system, and together they illustrate how environmental cues guide adaptive developmental decisions.

Developmental resources are not environmental cues, as the photon example makes clear. Cues are variables that trigger alternative, adaptive outcomes depending on their state. In tropical dry forest trees, high or low soil moisture leads to different leaf responses, retain or shed. But in eye development, the presence or absence of photons does not result in a pair of viable alternatives, it results in either a functional eye or a developmental failure. If photon flux were a cue, we would expect animals to develop one type of eye in the light and another equally functional alternative in darkness. But this isn't the case. Mice inhabit a wide range of environments, yet none live in total darkness, and none develop non-visual morphs when light is absent. Like genes, photons are indispensable for normal development, and when they're absent, the result is not a viable alternative but a pathological outcome, unless as sometimes happens, compensatory mechanisms are triggered.

Regarding photons as an environmental cue obligates us to regard genes themselves, or vitamin C, as cues. In the same way as photon presence in retinal development is associated either with a functional outcome or one with drastically reduced function, silencing or knocking out a gene is also often associated with drastically reduced function. Knocking out *Opn* complex genes results in phenotypes identical to those produced by rearing in the dark. In the same way, the absence of dietary vitamin C in humans does produce an alternate phenotype, known as scurvy. The absence of developmental resources, such as *Opn*, vitamin C, or photons, results in a pathological outcome.

It doesn't matter that there are liminal cases—there always are in biology. The point is that it is incorrect to dismiss any crucial developmental resource as a mere stimulus simply because it isn't DNA. Any number of examples could plausibly be classified as cues, resources, or both. Variation in oxygen concentration, for example, can lead to acclimatory responses, cue-like, but oxygen is also utterly essential for development, and concentrations that are too low result in malformation or even death, clearly a resource. The existence of such cases doesn't undermine the distinction; it underscores the mistake of treating all nongenetic inputs as if they were the same kind of thing. The point is not to enforce tidy dichotomies, but to understand the mechanistic details that shape the range of possible variation.

While a systems perspective is certainly not obligatory for adaptationist evo-devo, gently allowing "eppur si muove" moments to pass, should they arise, does allow new research opportunities to present themselves. With this in mind, we turn to what a developmental systems perspective potentially contributes to adaptationist evo-devo.

3.21 What a developmental systems perspective means for adaptationist evo-devo

A developmental systems perspective reshapes day-to-day evolutionary biology, primarily by opening up significant fields of scientific inquiry. There is even a current of thought in the philosophy of biology known as Developmental Systems Theory (Griffiths and Gray, 1994; Oyama *et al.*, 2003; Griffiths and Stotz, 2018) that focuses attention on these implications. Here are some examples.

3.21.1 Relaxing and expanding the notion of inheritance and the unit of inheritance

As emphasized in Section 1.1.2, heritability is one of the three conditions necessary for observing natural selection. At a minimum, recognizing the breadth of the set of resources necessary for constructing a normal individual can relax some of the stringency around expectations regarding natural selection (Edelaar *et al.*, 2023). Specifically, many formulations of natural selection stipulate that the inter-individual variation with fitness consequences must be heritable due to a "genetic basis." However, heritability depends on developmental processes that involve a large number of necessary resources, including but not limited to DNA. Therefore, specifying that variation must be "genetic" not only inaccurately depicts biological processes but also imposes an unnecessary stipulation. For most hypotheses of natural selection, all that is necessary is that the relevant variation be "heritable." How that heritability occurs, through the reconstruction of parental phenotypes, is a matter for empirical investigation.

3.21.2 Where the limits of developmental systems lie

Identifying how the full range of developmental resources interact is perhaps the most obvious avenue that a developmental systems perspective opens for empirical researchers. Identifying the resources that comprise the developmental system and delineating the boundaries of an organism's developmental system are key issues for empirical research. Biologists have long recognized that air and water, and certain temperatures or pH levels are necessary for normal development. Recently, the power and flexibility of quantitative genetics is being leveraged in such efforts as quantitative epigenetics to reflect these considerations, illustrating the fields opened by a developmental systems perspective (Banta and Richards, 2018; see also Munar-Delgado *et al.*, 2023).

If biologists want to understand how a new organism builds itself, we need to map the causal relationships between the necessary factors, an idea so fertile it drives an entire subfield, ecological developmental biology (Gilbert and Epel, 2015). Delimiting the minimal set of resources necessary for constructing a new individual promises to present challenges as complex as any in the entire field of biology. It is likely that mapping the minimal set of resources necessary for building an individual will find no non-arbitrary boundary (Stotz, 2017; Johnson, 2020). For instance, if oxygen is crucial for the development of a species, then all the species that produce oxygen, the entire atmosphere, Earth's gravitational field, and many other factors are implicated in the presence of oxygen during the development of an individual.

Biologists are no strangers to such questions. Determining where to draw these lines and why presents challenges analogous to those found in understanding organismal part-whole relationships, delimiting species, ongoing debates about what constitutes a biological individual, and not least the definition of "gene." Therefore, the question of what constitutes the developmental system fits comfortably in empirical biology, where grappling with boundaries that are not only inherently gradual or partial but also shift depending on conceptual focus is simply part of the terrain.

3.21.3 Expanding the causes of developmental potential and "mutation"

From a traditional genetic perspective, a "mutation" is a change in DNA sequence. These changes are faithfully replicated generation after generation, and those that impact function, such as affecting protein structure, can lead to reliable, heritable differences in fitness among individuals. A developmental systems view entails a different view, which is that any reliably reconstituted difference across generations, genetic or not, can count as a mutation in the sense of being a difference in what is inherited. Alteration to the sequence of a protein-coding gene associated with micrencephaly in guinea pigs represents a traditional mutation, the absence of a key instruction, which leads to an aberrant outcome. From this "genes-are-everything" standpoint, micrencephaly resulting from temperature shock (Section 3.9) is just an artefact of environmental perturbation, not a fundamental alteration of the instructions in the genes. But from a developmental systems perspective, the genetic mutation and the temperature shock exemplify causal parity. The genetic mutation is, to be sure, a significant change in a necessary resource. But at the same time, so is the temperature shock. They are both simply alterations to the necessary resources in the developmental system, and both lead to the same outcome. If resources beyond DNA are reliably present generation after generation and are essential for the reconstruction of the parental phenotype, and their alteration leads to alterations in developmental outcomes, then "mutation" can involve more than genes.

In addition to this empirical expansion, it is essential to explore the ways that a wider field of interactants affects the modeling of development and evolution. It seems likely that for many modeling purposes, the effects of expanding the set of resources necessary for development in evolution are fairly negligible. For example, most of quantitative "genetics" focuses on variances and covariances between traits, their heritabilities, and other statistical relationships, without studying the molecular basis of these traits (De Jong, 2007). Instead, this approach often involves making assumptions about the underlying "genetic architecture" without really identifying specific genes or molecular mechanisms. Quantitative genetics can be stated as "if the traits I am studying are caused exclusively by a genotype by environment interaction," then the "genetic architecture" must have such-and-such a structure. Any number of what-if invented models could account for the variances, covariances, and heritabilities measured in a given study.

To depict the relationship between the actual data in quantitative genetics studies and their causes, Pigliucci and Kaplan (2006), drawing on Shipley (2000), use the analogy of shadow puppets. The analogy highlights that any number of differently constructed puppets, or indeed objects, behind the screen could produce the same set of shadows. The shadow on the screen, therefore, is not a sufficient basis on which to infer the entire structure of the objects behind the screen. In this analogy, the shadow represents the statistical patterns of trait relationships documented in a quantitative genetics study. It is true that, if genes interact with the environment in the way depicted by quantitative genetics, then this would account for the statistical "shadows" of trait variances, covariances, heritabilities, responses to selection, and other attributes (Falconer and Mackay, 2009).

But many of the conclusions of quantitative genetics studies apply perfectly well to differing underlying models of inheritance and evolution. Trait variation and covariation, inheritance, and association with differing fitness can still occur even if one assumes a differing underlying model. It seems very likely, therefore, that substituting a wider array of resources for the "genes" in quantitative genetics would make very little difference to notions regarding the ways that traits are inherited, how and why they covary with one another, the ways that these covariations limit or facilitate adaptive evolution, and most other key quantitative genetics themes. The same goes for "genotype-phenotype maps," "genetic networks," "genetic bases," and a host of other such shadow puppets that biologists posit behind the metaphorical screens of organisms (Bourrat *et al.*, 2017). Such an expansion seems promising in explaining why quantitative genetics studies correspond only approximately to the findings of QTL and other studies that locate areas of the genome associated with traits of interest. Blossoming interest in "quantitative epigenetics" (Wang *et al.*, 2012; Banta and Richards, 2018) is evidence of this expansion.

These same realizations about the causes of inheritance, trait variation, and mutation also apply to the mapping of developmental potential. Recall that developmental potential is the sum total of all the configurations (developmental outcomes) that an individual, species, or higher grouping can produce. From this perspective, a "mutation" is a change in developmental potential. There is no separation between the genotype on one side and the phenotypic outcomes specified on the other. There is always just the developmental system. Some developmental doors are open to it at one time, and other doors at other times. If the developmental system changes due to "mutations" (genetic or otherwise), then this is simply a change in the developmental system and therefore in developmental potential. If "mutation" encompasses changes not just in genes but in any part of the developmental system, then the set of factors that shape developmental outcomes also expands. A wider range of features can make some outcomes likely, others rare, and still others unreachable. A benefit of this view is that the asymmetry mentioned in Sections 3.6 and 3.8 is alleviated. The asymmetry was that methods that we have for empirically inferring that DNA contains "information" also point to all other necessary resources, such as water, lipids, or photons, as also having information (Maynard Smith, 2000). A systems biology view points to the information metaphor as dispensable. A developmental systems view therefore simultaneously encompasses the key adaptationist evo-devo concerns of inheritance, selection, the causes of evolutionary change, and the causes of resistance to evolutionary change, manifested in developmental potential.

3.21.4 The term "phenotype"

At this point, it's time to confront one of the elephants that has hulked in the room throughout nearly all of the preceding discussions. It is a problem that I have been awkwardly writing around because there is, as of yet, no elegant way to address it. There might not even be a need to write around it. It is the possibility that the term "phenotype" has implications that are so undesirable that it is best avoided when trying to formulate precise notions of evolutionary causation (Pontarotti *et al.*, 2022). From the developmental systems outlook, there is just the developmental system and its potentialities that are expressed and those that are not at any given moment. Change in the developmental system and thus the potentialities and outcomes of development is all that there is. In other words, at all times there is only the phenotype. Only if genes really were autonomous agents, on their own causing developmental and metabolic events to occur all, and not as parts of systems, could we justifiably make a genotype-phenotype distinction non-arbitrarily.

Recall the analogy of the car engine, in which the battery is a necessary component, but for the engine to run, all of the parts must work together. In this analogy, we see that the relevant property that we are interested in—running—requires a functioning system. In the same way, in organisms, the relevant properties that we are interested in—development and therefore evolution—require a fully functioning developmental system. It is as arbitrary to separate genotype and phenotype in evolution as it is to separate the battery from the rest of the engine in running. With respect to a running car engine, the relevant unit is the engine, battery, and all. With regard to a developing individual reconstructing the parental configuration, the relevant unit is the developmental system, DNA and all. The state of the developmental system at any one moment is the phenotype of the organism. Because the fundamental unit of development is the developmental system, and because at any given moment there's just the state of the developmental system, all there ever is at any given point is "phenotype."

On this basis, it remains to be explored whether the term "phenotype" is useful, or whether it is unhelpful and requires replacement. On the unhelpful side, its use implies that the genotype-phenotype dichotomy is a valid one (Pontarotti *et al.*, 2022). It implies that something beyond the phenotype exists at some point in the evolutionary process. This is why, throughout, I have relied so often on terms such as "configurations" or "variants." The aim of such terms is to depict variation between individual developmental systems without implying a genotype-phenotype distinction and an evolutionary process that consists of inheritance of genetic material on the one hand, and then expression of genetic information and selection on the other. But the terms "configuration" or "variant" are not as convenient as "phenotype" in flagging the intent of what is being discussed. One thing that biology does not need is yet more exotic terminology. So, with some misgivings, I still use "phenotype" here and there, and it is likely that its use will continue. Exploring this use, though, whether it is beneficial or harmful to evolutionary biology, and what alternatives might exist, is yet another important direction for inquiry that the developmental systems view points to.

3.22 A developmental systems view is not required for adaptationist evo-devo, and more common ground

If you're sympathetic to the view that "genes aren't everything," great—there are fascinating questions to pursue from that perspective. But sympathy isn't required. Even if you are not at all convinced that nongenetic resources can make up reliable parts of development generation after generation, and even if you think that nongenetic aspects can be safely backgrounded in understanding development, you can still find value in the manifold approaches to developmental potential explored in Chapter 2, and the empirical studies in this chapter, for that matter.

Looking at biological discourse might make it seem like there is broad disagreement. A surprising number of researchers, busy ones with labs to run and grants and books to write, somehow find time and motivation to publish rebuttals and critiques of "genes-aren't-everything" perspectives (Brodie III, 2005; Coyne, 2009; Gross, 2012; Wray *et al.*, 2014; Futuyma, 2017; Svensson, 2023). On this basis, a reader could be forgiven for thinking that the "genes-are-everything" folks disqualify not just the discourse of the "aren't-everything" biologists but also their empirical work.

But here too, the common ground is just below the surface: even the most skeptical genes-are-everything biologists think the work coming out of "genes-aren't-everything" perspectives is valuable. A biologist builds a GRN that includes physical or cellular context? Excellent—those interactions matter. A molecular geneticist grimaces at how DNA is turning out to participate in cellular processes in increasingly indirect and protean ways? That's discovery in action. A researcher studies how nongenetic elements might be durably recruited across generations? Terrific—we need more of that. Temperature shocks at precise moments in development lead to predictable teratologies? Fantastic insights into key developmental stages. Someone investigates how tissue-level biomechanics shape the generation of form? That's not heresy, it's spectacular biology. The relationship is mutual—how could any biologist object to studies of variation, heritability, and fitness differences, or very real comparative patterns across species, no matter how "traditional" the motivation?

Real biological practice suggests that no matter what conceptual orientation a scientist adopts, they are likely to produce results that others, regardless of stance, will recognize as valuable. Our intellectual commitments shape the studies we carry out and how we interpret the evidence, but the evidence itself does have a reach that transcends subdisciplines. Moreover, independently of our particular doctrinal postures, as scientists we generally would dearly love for our theories to account for all of the available data. So, more and richer data are taken into account, sometimes with open arms, sometimes a bit grudgingly, across the board.

In none of the commentaries cited above critiquing "genes-aren't-everything" reasoning do the authors in any way discredit the empirical studies, just the framing stances of the authors. And so, even amid disagreement about what genes are or how development works, over what is novel and revolutionary or already well-known, scientific work continues to generate insights that move easily across conceptual boundaries. In this chapter, we saw evidence of common ground in the materials properties of biological materials (Section 3.17), order emerging from interaction among parts in systems (Sections 3.16–3.18), and the ongoing transformation of molecular genetics into systems biology (Section 3.19). The search for common ground forms the core of Chapter 5. Before we can get there, though, we turn to another concern central to adaptationist evo-devo, and one that also draws divergent interpretations depending on one's view of genetic centrality: the role of phenotypic plasticity in evolution.

4
Understanding phenotypic plasticity and its role in evolution

In constructing a well-supported explanation of organismal form, it's important to consider the role of phenotypic plasticity as well as its flip side, phenotypic accommodation. This chapter will clarify what phenotypic plasticity is and illustrate some of the very important roles that it has in evolution. All biologists have heard the term "phenotypic plasticity," and most feel some unease or uncertainty about it. This is understandable, given that the evolutionary biology and ecology literatures are filled with varied and often confusing uses of the term. In one of the most widespread confusions, it is common for biologists to wonder whether a given pattern of trait variation between environments is due to "genetic factors or mere plasticity." The genetic/plastic dichotomy often brings the implication that genetic causes are evolutionarily significant, whereas plastic ones are somehow less important. Another widespread and unhelpful notion is that plasticity "covers all types of environmentally induced phenotypic variation" (Stearns, 1989, p. 436). Such definitions lead authors to distinguish between adaptive plasticity and non-adaptive or maladaptive plasticity, which as we will see leads the notion of "plasticity" down some untenable roads. The present chapter clarifies these and other distinctions, building on a correct understanding of what plasticity is. Using this understanding, it is then possible to explore key roles that plasticity plays in adaptationist evo-devo. We start with the definition of plasticity.

4.1 What phenotypic plasticity is

Recognizing what plasticity is provides the key for effectively wielding plasticity in thinking about evolutionary processes and cutting through the confusion surrounding the term—first the definition, and then we can explore its different elements.

Phenotypic plasticity is the adaptive propensity for a given developmental system to produce the phenotype associated with the highest fitness in the environment that an individual finds itself in, relative to the other phenotypes that it can produce in that environment.

Now, here is some more detail on key elements of the definition. After that, we can introduce phenotypic accommodation and examine genes-as-followers evolution.

Adaptationist Evo-Devo. Mark E. Olson, Oxford University Press. © Oxford University Press (2026).
DOI: 10.1093/9780197831533.003.0005

4.2 Phenotypic plasticity is exclusively adaptive

The key to understanding phenotypic plasticity is that it is an adaptation. While it is often defined as "all types of environmentally induced phenotypic variation" (Stearns, 1989, p. 436; De Jong, 2005; Van Kleunen and Fischer, 2005; Ghalambor *et al.*, 2007; Visser, 2008; Morris and Rogers, 2013; Arnold *et al.*, 2019; Scheiner and Levis, 2021), these notions obscure the evolutionary distinctiveness of the phenomenon. If plasticity simply refers to any phenotypic difference triggered by the environment, then poisonings, injuries, and any alterations to developmental systems, such as guinea pig heat shock outcomes, all count, diluting the term until it becomes evolutionarily meaningless. But plasticity, in any biologically useful sense, refers to a mechanism favored by natural selection in producing functionally appropriate phenotypes under varying conditions. This is what makes plasticity an adaptation, and not just a passive consequence of any sort of environmental exposure.

Classically adaptive examples of plasticity include both morphological changes and physiological acclimation. For the first category, an obvious example all around us is tree height. Within a species, even among cuttings from the same individual, trees in dry places tend to remain short, while those in moist environments grow taller. This tuning of height to moisture availability is a plastic response. Trees in any given spot could stay very small and minimize drought risk, but they would also have lower fecundity, since tree size is linked to reproductive output. They could also grow larger, but taller trees face a higher risk of dying from drought. Adaptive plasticity allows trees to adjust their height in response to local moisture conditions, tall enough to capture abundant light and produce many propagules, but not so tall as to become excessively vulnerable to drought.

Another expression of plasticity is made up of classic physiological acclimation responses. Lowlander visitors to Mexico City, well above 2,000 meters above sea level, usually notice fatigue when climbing stairs immediately on arrival. After a few days, climbing stairs is much easier, a welcome result of well-studied acclimation responses to hypoxia, including increased hemoglobin concentration, increased oxygen-hemoglobin affinity, and elevated partial pressure of oxygen in arterial blood (MacNutt *et al.*, 2012). Being able to perform physical tasks with less fatigue at higher elevation is thought to be an adaptive plastic response. But so is the reverse: when returning to lower elevations, those same individuals typically down-regulate hemoglobin production and reverse other hypoxia-related adjustments. This reversal is also adaptive, since maintaining high red blood cell counts in oxygen-rich conditions comes with metabolic costs and increased blood viscosity, reducing cardiovascular efficiency. These phenomena involve clear phenotypic differences associated with environmental differences. These sorts of adaptive responses fit under any definition of plasticity, such as the "all types of environmentally induced phenotypic variation" variety.

But the same loose "all types of variation" definitions force us to include examples that clearly don't represent evolved plasticity. If you are exposed to very low levels of strychnine, you will continue to present your normal phenotype. Under moderate levels you would show restless agitation, muscle spasms, and breathing difficulties. Under high levels, you would die. These are marked phenotypic differences associated with different environments. If phenotypic plasticity is defined as *any* phenotypic difference associated with any environmental variation, then we are forced to accept strychnine poisoning as phenotypic plasticity. Likewise, if you expose yourself to an environment in which your diet is very low in saturated fatty acids, you will tend to have lower levels of low-density lipoprotein cholesterol, and therefore lower risk of atherosclerotic cardiovascular disease. If you expose yourself to an environment in which your diet is very high in saturated fatty acids, you will tend to have higher levels of low-density lipoprotein cholesterol, and therefore higher risk of atherosclerotic cardiovascular disease (Maki *et al.*, 2021). Heavy alcohol consumption during pregnancy can lead to babies being born with, among other features, a smooth philtrum rather than the normal groove below the nose and upper lip, and most commonly, alcohol-related neurodevelopmental disorder (Kodituwakku and Kodituwakku, 2022). Guinea pigs exposed to an environment of elevated temperatures at the right time in development will exhibit exencephaly and microphthalmia (Section 3.9). Fish that live in the environment of hydropower turbines tend to show trauma to all parts of their bodies (Bevelhimer *et al.*, 2019). Pollution often gives sensitive people asthma. By the lights of a definition of plasticity as *any* environmentally associated difference for the same genotype, we are forced to accept strychnine poisoning, atherosclerotic cardiovascular disease, fetal alcohol syndrome, exencephaly, micropthalmia, asthma, even injuries, as plastic responses. None of these involve evolved detection-and-response systems shaped by selection that lead to a match between phenotype and environment.

There are two reasons for excluding phenomena such as those in the previous paragraph from "plasticity." The first is pragmatic. If phenotypic plasticity includes any phenotypic difference associated with any environmental one, then the term "plasticity" is so wide that all phenotypic differences under different conditions fall under the umbrella of plasticity, whether it be acclimation to low oxygen concentration or heart attack due to atherosclerosis. For pragmatic purposes, categories are useful tools for communication because they help narrow down the variation in nature. If "phenotypic plasticity" is simply equivalent to "environmental variation between individuals," then we gain nothing by adding the term to the scientific lexicon. If the term "chimpanzee" included both humans and chimpanzees, it would be impossible to know if the blind date that you have been set up with, hoping for marriage and a family, would be zygotically compatible. If "natural selection" included drift plus ontogenetic impossibility, then attributing a given pattern of organismal trait distribution to selection would provide little illumination regarding process. So, a pragmatic consideration is that terminology can be aligned with nature in ways that are more and less useful for communication. If the "plasticity" category refers to something precise rather than acclimation, injury, poisoning, deformity, or pathology, then it is simply more useful.

The second reason is more important. Scientific terminology makes up an inextricable part of the scientific effort to reflect the causal structure of the world when terms align, insofar as possible, with entities and phenomena that exist independently of human imagination. The terms "chimpanzee" and "human" do seem to align with two different, real entities in nature. "Natural selection" does seem to pick out a real process in nature, distinct from other phenomena, such as population-level genetic drift or species selection (species selection is covered in Section 5.8). Picking out putatively real entities and processes in nature means that scientists can build around them in constructing explanations that postulate the causes that underlie the patterns in the natural world.

Defining plasticity as an adaptive phenomenon does seem to pick out a distinct phenomenon in the natural world, in the form of selection acting in qualitatively similar ways across species. To see why, let's use our standard adaptationist account (as in Sections 1.11.1–1.11.3). We start, as always, with variation. The relevant variation consists (or at one time consisted) of differences between conspecifics in their plastic responses, for example whether responses are gradual or sudden along an environmental gradient, or whether the breadths of possible plastic responses are wide or narrow. If plasticity is adaptive, then intrapopulational, interindividual variation should be heritable, that is, offspring tend to have similar plastic responses to their parents. Finally, if plasticity is adaptive, then the heritable patterns of detection-and-response of some individuals must be associated with greater survivorship and/or reproduction than others in the population. If plasticity is adaptive, then a process qualitatively similar applies across traits and species, whether acclimation to low oxygen concentration in humans or variation in leaf size with water availability in plants.

Adding in pathologies, poisoning, teratologies, injuries, and the like to the notion of "plasticity" results in a set of phenomena that do not pick out a qualitatively distinct biological phenomenon. Given two meanings for the term plasticity—the adaptive one that does pick out selection acting in a qualitatively similar way across species versus a heterogeneous meaning that does not pick out a distinct process—there is no reason to opt for the heterogeneous one. Appealing to the definition of plasticity in Section 4.1, we see that the distinct type of selection that plasticity picks out is the tendency to produce the configuration or configurations that correspond to the highest fitness in a given environment, out of the range of outcomes possible in that environment.

This kind of selection operates in a qualitatively similar manner across organisms, whether in Caribbean corals building streamlined, flow-resilient colonies in high-current zones versus more delicate but more effective light-gathering forms in sheltered reef flats or in Antarctic grasses growing tall in rocky crevices versus hugging the ground in open, wind-exposed areas. In all cases, out of the possible, organisms produce the phenotype that maximizes performance and fitness out of the possible. So, the key to understanding phenotypic plasticity is to recognize that it is an adaptation. Before showing how this adaptive definition clarifies "maladaptive plasticity" and other topics, we'll turn to the next most common confusion regarding plasticity—the "genetic/plastic" distinction.

4.3 What is really meant by the plastic–genetic distinction

Thinking regarding the causes of variation between individuals often seem to have the flavor that "genetic" differences are adaptive and significant, whereas others are merely plastic and somehow less significant (Lardner, 1998; Pigliucci, 2005;

Palacios *et al.*, 2011). For example, tree biologists investigating why some trees succumb to climate change-induced drought notice relevant variation across the members of a species: some individuals resist drought better than other conspecifics. They then wonder whether this variation is "plastic" or "genetic." Or, they might notice that across species, some factor such as leaf size tracks water availability, and they ask whether this variation across the species is "plastic" or "genetic." Here, I want to work through what biologists do and don't seem to mean with these questions and the rather surprising real content that such questions contain.

Here is one plausible interpretation of the dichotomy. Under the "genetic" view, differences between individuals arise from allelic variation: each individual carries a particular allele that determines a fixed trait value, expressed consistently regardless of environmental context. In contrast, under the "plastic" view, all individuals share the same genotype at the relevant loci, and trait differences arise entirely from variation in environmental conditions. To continue with the example from trees: in the genetic scenario, each tree's level of drought resistance is fixed by its genotype, unaffected by the environment in which it grows. In the plastic scenario, all trees are genetically identical, and observed differences in drought resistance reflect different positions along the same gradient of plastic response, say with individuals in wetter microsites showing lower drought resistance due to reduced exposure.

While this interpretation takes the "genetic or plastic?" question very much at face value, it doesn't seem likely to be what biologists are really interested in, because scenarios like this are rarely found in nature. Biologists don't ask "genetic or plastic?" about invariant traits, such as four limbs or one head in humans. Instead, they wonder about traits such as leaf size in plants across moisture gradients or resistance of trees to drought. The traits involved in "genetic or plastic?" questions practically always have at least some functionally significant plastic variation (Gianoli and Valladares, 2012). So, whatever biologists are trying to get at with the "genetic" part of the framing, it likely doesn't involve a situation of no plasticity at all within individuals. So, we need to look to a different interpretation.

Another plausible way to interpret the "genetic-plastic" distinction, and another one unlikely to be what biologists actually mean, is that plastic differences occur without the involvement of genes. To the extent that phenotypic variation involves DNA, it is also "genetic," so the distinction is unlikely to refer to variation that has no genetic component at all. Plasticity is an adaptation, and insofar as adaptation involves the favoring of some genetic variants over others, plasticity is "genetic." Moreover, differences in phenotype usually involve differences in patterns of gene expression, and so can be regarded as "genetic" on that basis too. Whether genetic variation is completely uniform at some locus, or has low variation, or even some other level of variation, isn't usually the motivating point. For these reasons, situations in which DNA plays no role at all in generating phenotypic differences do not seem to capture what biologists intend when they invoke a "genetic versus plastic" distinction.

Instead, what biologists often seem to care most about, when invoking a "genetic-plastic" distinction, is whether the range of plastic responses is wide or narrow within individuals, relative to the variation observed across a species or group of species. To illustrate this point, Figure 4.1 depicts a range of leaf sizes in different species growing

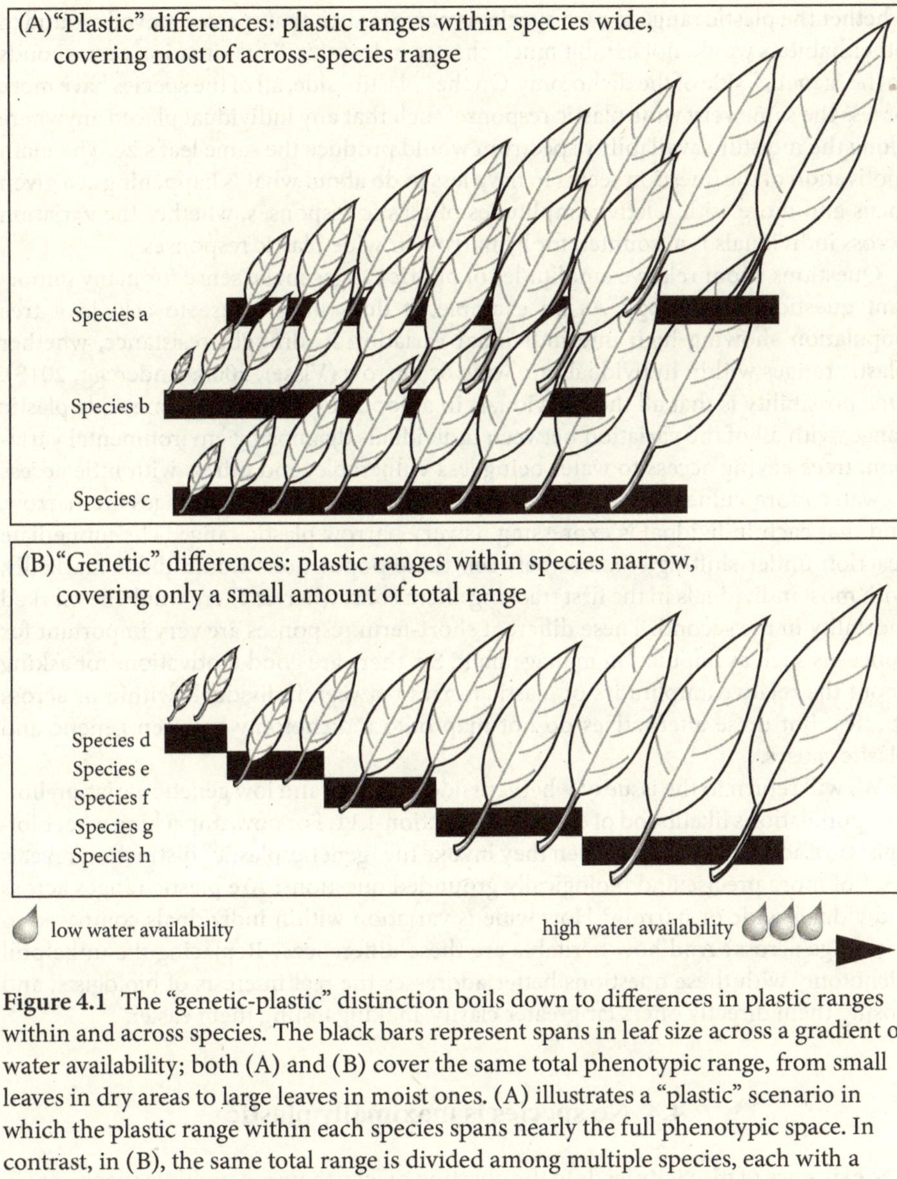

Figure 4.1 The "genetic-plastic" distinction boils down to differences in plastic ranges within and across species. The black bars represent spans in leaf size across a gradient of water availability; both (A) and (B) cover the same total phenotypic range, from small leaves in dry areas to large leaves in moist ones. (A) illustrates a "plastic" scenario in which the plastic range within each species spans nearly the full phenotypic space. In contrast, in (B), the same total range is divided among multiple species, each with a narrow range of leaf sizes. This is not so much a question of whether plasticity is absent or present (there almost always is some in the traits for which the genetic/plastic question arises), nor is it a simple question of "genetics"; if "genetic" means reliably heritable across generations, then wide or narrow plastic ranges are "genetic." Instead, the key issue is simply whether plastic ranges are wide or narrow within individuals or species relative to the breadth across them.

in different environments, with small leaves being found in species in dry habitats and larger ones in species in moister habitats. What seems to be of most frequent interest in such situations has to do with two alternatives. On the one hand, the question is whether the plastic range in each species is narrow, such that species moved around to other habitats would not exhibit much change in leaf size. This situation corresponds to the "genetic" side of the dichotomy. On the "plastic" side, all of the species have more or less the same very wide plastic response, such that any individual placed anywhere along the moisture availability spectrum would produce the same leaf size. The main motivation of the question seems to have less to do about what is happening at a given locus and more with relative amplitudes of plastic responses, whether the variation across individuals is accounted for by narrow or wide plastic responses.

Questions about relative amplitudes of plastic ranges make sense for many important questions in biology. As an example, it does make sense to ask, in a tree population showing high interindividual variation in drought resistance, whether plastic ranges within individuals are wide or narrow (Visser, 2008; Anderegg, 2015). One possibility is that all the individuals in a population span the same wide plastic range, with all of the variation between individuals diagnosing environmental variation, trees having access to water being less vulnerable, and others with little access to water more vulnerable. Another possibility is that the plastic ranges are narrow, and that each individual is expressing its very narrow plastic range. The immediate reaction under shifting climates between these populations would differ markedly, with most individuals in the first tracking alterations, whereas there would be marked mortality in the second. These different short-term responses are very important for concerns such as population management. So, there are good motivations for asking about the relative amplitudes of plastic ranges between individuals within or across species. But these alternatives do not map onto a dichotomy between genetic and plastic causes.

We will return to the issue of whether wide plasticity and low genetic variation limits a population's likelihood of evolving in Section 4.11. For now, unpacking what biologists are actually getting at when they invoke the "genetic/plastic" distinction reveals a set of more precise and biologically grounded questions: Are plastic ranges across individuals wide or narrow? How wide is variation within individuals compared to the range across? And how heritable are these differences? Replacing the unhelpful dichotomy with these questions better addresses the real interests of biologists, and posing them directly offers far greater clarity, making testing them easier.

4.4 No species is maximally plastic

The existence of plasticity leads to the question of why there are multiple species at all, and not just a single maximally plastic species spanning all habitats. In this maximally plastic species, wherever an individual finds itself, whether deep in the ocean, on a desert flat, or on a sea cliff poised to launch itself into the air, it produces the gills, water storage tissue, or wings appropriate to the conditions it experiences. This is not the situation in nature, however. In nature, innumerable species exist, and each species has its limited geographical and morphological ranges, and limited ranges of environmental conditions under which they can thrive. Various ideas exist about why plastic ranges are always limited.

A prominent idea regarding why plastic ranges are limited involves what are known as the costs of plasticity (Pigliucci, 2001, 2005; Relyea, 2002; Van Kleunen and Fischer,

2005; Auld *et al.*, 2010). Saying that plasticity has a "cost" implies that the ontogenetic apparatus necessary for producing a range of phenotypic responses requires allocation of finite resources. For plastic differences to emerge, each point along a phenotype gradient requires a set of developmental resources different from other points along the gradient. Maintaining the set of resources associated with a given environment involves investing in that set and not in something else. Given an individual with a wide plastic response, and an individual with a narrow one, in an environment in which both of these individuals are producing the same phenotype, the one with the wide plastic response has allocated more metabolic resources to unexpressed plastic responses. The one with the narrow plastic response has allocated these resources to growth, immune response, reproduction, and other factors that contribute to it having higher fitness than the individual with the wide plastic response in that environment, all else being equal. Under this "cost of plasticity" hypothesis, the plastic range of a given species represents the negotiation between one vector of selection favoring a wider plastic range because it allows for functional phenotypes in a wide range of environments, and another vector favoring a narrower plastic range because it deviates metabolic resources away from survival and reproduction given the frequency of experiencing different environments.

Another possibility regarding why there are no maximally plastic species has to do with the interaction between the frequency of experienced selection events and the rate at which deleterious mutations arise. It's a little involved, but entirely plausible. Again, picture variation within a population, with some variants having narrower plastic responses and others wider ones (Kawecki, 1994; Murren *et al.*, 2015). As in the scenario involving costs of plasticity, in this particular Goldilocks selection scenario, selection tunes the amplitude of plastic ranges such that they are neither too wide nor too narrow, given the range of conditions that a given species encounters. The variants with very narrow responses would have lower fitness in most conditions, though they would have higher fitness in the narrow range of conditions that their narrow plastic response is best suited to. Because the range of conditions that the species tends to experience surpasses the range in which the narrow variants do best, on average across environments, they will tend to have low fitness and not be over-represented in successive generations. On the other side, variants with very wide amplitudes of variation will skip around from environment to environment, and in all of them will do reasonably well. Recall, though, from Section 1.6 that the return interval of environmental factors helps determine whether they exert much of a selection pressure or not. By jumping around many different environments, the reasoning goes, variants with very wide plastic ranges will be exposed to the relevant selective pressures of each environment only rarely. Over these long trips around many environments, they can accumulate deleterious mutations that degrade their performance in a particular environment. When they finally arrive back in a given environment, thanks to the deleterious mutations that have not been exposed to selection while the variants transited through other environments, they will have very low fitness. The variants that will tend to have the highest fitness on average, then, are those that have the "just right" range of plastic responses given the range of environments that they typically experience, plus mutation rate. This mechanism could certainly act in concert with costs of plasticity in limiting plastic ranges, but it could also limit plastic ranges in the absence of costs.

Whatever the exact mechanism limiting plastic ranges, one vector of selection favors individuals with wide amplitudes, a wide range of phenotypes corresponding to a wide range of environments. The other vector of selection favors individuals with very narrow plastic responses. Under both scenarios, if the range of environmental conditions typically experienced relative to generation time is wide, then selection will tend to favor broader plastic responses. If the range of environmental conditions typically experienced relative to generation time is narrow, then selection will tend to favor narrower plastic responses. In this way, the benefits of plasticity, the costs of plasticity, opportunities for accumulation of deleterious mutations, and the frequency of variation in the experienced environment shape the breadths of plastic ranges. Before examining the key role plasticity can have in adaptive evolution, it's time to drive home one more time the exclusively adaptive nature of plasticity.

4.5 Clarifying "maladaptive plasticity"

Recognizing that plasticity is exclusively an adaptive phenomenon instantly clarifies confusing notions of "maladaptive plasticity" (see examples in Pfennig, 2021a). In the biology literature, maladaptive or non-adaptive plasticity are situations in which individuals produce different phenotypes across environments, but these differences result in lower performance or fitness, rather than the usual higher, as compared to some possibility in another environment or an imaginary optimum.

In these cases, the scenarios imply the wrong comparisons between individuals. Remember that our definition of plasticity is "the adaptive propensity for a given developmental system to produce the phenotype associated with the highest fitness in the environment that an individual finds itself in, relative to the other phenotypes that it can produce *in that environment*." Contravening this italicized stipulation, most cases of maladaptive plasticity are actually arbitrary comparisons between individuals in *differing* environments. We will explore this in a moment with some examples. We will also examine uses of "maladaptive plasticity" as a confusing reference to situations in which the selective environment has changed.

One sense of "maladaptive plasticity" involves comparing what is possible in one environment with what is possible in another. Morris and Rogers (see also Ghalambor *et al.*, 2007; 2013, p. 526–527; Pfennig, 2021b) write:

> maladaptive plasticity, also referred to in the literature as non-adaptive plasticity, can be defined as the plastic response to changes in the environment, such that there is increased divergence between the expressed trait and the phenotypic optimum. ... Maladaptive plasticity is typically considered for a set of environments with identical phenotypic optima, such that in the original environment the optimum is attained (the 'ancestral' environment), whereas in the novel environment the optimum is not attained due to plasticity. For instance, the optimal fruit mass for a plant may not differ between a high nutrient and low nutrient environment, yet the low nutrient environment may induce smaller fruit sizes due to energy constraints.

Notice that this framing compares what is possible in one environment with what is possible in another, rather than what is possible *within a given environment.* In the fruit size example, the appropriate claim is that selection favors the fruit size that maximizes fitness *among those producible in that environment.* This might be the largest fruit size possible even if it is smaller than that producible in another environment. Phenomena such as selection and drift emerge as consequences of the variation that actually is producible within a population, not on hypotheticals from other environments or even other species.

To understand this, it can be helpful to refer to selection scenarios that involve static traits and not plastic ranges. Picture for example heritable white and brown variants within a cryptic rodent species that finds itself living on a brown background. The relevant issue is the relative fitness of the white versus brown variants *in the context of the brown background.* Asking how the variants would perform on a white background is irrelevant. Now introduce plastic ranges as the intrapopulational variants. The situation is the same. It is still the case that the relevant question is the relative fitness of the possible variants in the environmental situation that the variants actually experience. Whether one or more variants with higher absolute performance value are developmentally possible in some other environment is likewise irrelevant.

If maladaptive plasticity is lower fitness in one environment relative to another one, then, once again, practically anything counts as maladaptive plasticity, such as a person dying in the vacuum of space when ejected from a spaceship, or crushed by water pressure on the bottom of the sea. So, while the phenomenon of lowered fitness under change in selective conditions can be of interest, such situations simply highlight the completely standard situation that some conditions of selection result in higher or lower absolute performance values as compared to other conditions. Naming these situations with terminology such as "maladaptive plasticity" is just confusing.

Decoding "maladaptive plasticity" into its real population level content reveals very important issues hiding beneath the vague term. Morris and Rogers continue: "Alternatively, the optimum in a novel environment may differ from the ancestral optimum, with maladaptive plasticity moving the organism away from, rather than toward, this new optimum." Such shifts in what organisms can produce between ecological moments are undeniably important. In this scenario, Morris and Rogers highlight a situation in which under changed climate conditions, the plastic range of a population (the summed ranges of plasticity of the individuals making up the population) scores lower on an index of productivity or fitness as compared to a previous environment. Given climates that are rapidly changing, such mismatches are obviously an important phenomenon to highlight, and one of clear concern in attempting to mitigate the effects of climate change. Again, though, they don't represent a special, "maladaptive" type of plasticity, just that different environmental conditions are associated with different organismal performance.

Some uses of "maladaptive" plasticity refer to how organisms track shifting conditions through bet-hedging mechanisms. In human practice, bet hedging means spreading investments to reduce the risk of total loss. In nature, it refers to the production of varied phenotypes or timing strategies in ways that result in at least some offspring succeeding under uncertain conditions. One form is intergenerational plasticity, in which the environment experienced by parents or grandparents influences descendant phenotypes. When environmental change is slow relative to generation time, such mechanisms can lead to a closer match between offspring traits and conditions. This is a form of bet hedging, but in any given generation it does not necessarily match conditions accurately. As in the example in Section 3.19, many desert cactus seeds germinate only after repeated rains, avoiding false starts from isolated storms, yet can still fail if early rains are followed by drought. Intergenerational plasticity works in the same way: it can produce mismatches, especially when climate shifts faster than populations can track (Zhang *et al.*, 2024). Such mismatches are sometimes called "maladaptive plasticity" and are definitely important for managing populations facing rapid change. However, labeling these mechanisms as truly maladaptive misrepresents their role. Like other forms of bet hedging, intergenerational plasticity is presumably maintained because, despite occasional mismatches, it yields higher long-term fitness than would be expected without it, given the range of traits that can be produced. It exemplifies the same kind of selection described in Section 4.1: among the possible outcomes, heritable variation in detection-and-response patterns is differentially represented over generations in association with higher fitness in given environments.

Another important concern hiding beneath "maladaptive plasticity" is the question of why organisms have one trait and not another. To borrow from Timothy Shanahan's memorable question "why don't zebras have machine guns?"(Shanahan, 2008), asking why machine gun-carrying isn't part of the developmental repertoire of zebras, or why fruit size isn't invariant across nutrient levels, are important questions. But declaring any of these to represent "maladaptive plasticity" is unilluminating because, again, it gets natural selection wrong. Phenotypic plasticity is the tendency to produce the phenotype associated with the highest fitness in the environment that an individual finds itself in *relative to the other phenotypes that it can produce in that environment*, not relative to those producible in a different environment. Translating "maladaptive plasticity" reveals the important and accessible questions about developmental potential and selection lurking beneath the confused term.

A main disadvantage of the term "maladaptive plasticity" is that it makes it seem as though there is a qualitatively distinct phenomenon, a special type of plasticity that intervenes in organismal evolution. But the term "maladaptive plasticity" is as obfuscating a bait-and-switch as can be found in scientific terminology. It is like pointing to "maladaptive morphology" and then presenting as an example the color morphs of *Biston betularia*. Before the industrial revolution, the gray color morph was favored by selection. On soot-blackened trees, gray moths suddenly represented "maladaptive coloration." Evolutionary biologists already have vocabulary for these situations: the conditions of selection have changed. As with "constraint" as a vague reference to natural selection in Section 1.4, there is nothing to be gained by replacing clear existing terminology with vague novel terms. No distinct phenomenon or type of plasticity corresponds to "maladaptive plasticity." The term is usually just vague shorthand for "the conditions of natural selection changed." In discussing the ranges of plasticity of species under climate change, Visser (2008) provides a good model of how to discuss changes in selective conditions without invoking "maladaptive plasticity." The definition of plasticity given above shows the way out of confusion by clarifying the correct comparisons between individuals.

4.6 Change in developmental system, change in outcome

Another confusing sense of plasticity, when made sense of, points us to evo-devo adaptationism's concern for developmental potential. Stearns (1989, p. 439–440) notes that

> nonadaptive responses also are an inevitable result of the physicochemical nature of organisms. Although much of the information required to construct an organism resides in DNA, development is constrained by the properties of the materials that make up the organism. All chemical systems, living or dead, are sensitive to temperature, pressure, pH, and substrate concentration. Any organism not buffered against those environmental changes to which chemical reactions are sensitive will inevitably be phenotypically plastic. There is no genetic control of adaptation involved in this component of plasticity, which is as much an encumbrance as an opportunity.

Such "inevitable results"—differing developmental outcomes as a result of differing developmental conditions—are simply part and parcel of developmental potential. Recall that the developmental potential of an individual is the entire range of configurations that a developing individual can produce. The developmental potential of a species or a clade is the sum of the developmental potential of the individuals that make up the grouping, across all ontogenetic stages, from common, favored configurations to rare and teratological ones. Natural selection can only favor developmentally possible variants. To the extent that factors such as "temperature, pressure, pH, and substrate concentration" make certain developmental outcomes more or less likely, or even impossible, as an "inevitable result of the physicochemical nature of organisms," study of these factors is an essential component of an explanation of organismal form in any given case. Heat-shocked guinea pigs producing micrencephaly or exencephaly, humans exposed to alcohol in utero lacking a philtrum or producing diprosopia, simply illustrate certain regions of the developmental potential of those species under certain configurations of their developmental systems. Stearns's notion of "nonadaptive plasticity" is thus most profitably seen through the lens of adaptationist evo-devo. Section 4.11.1 explores the evolutionary significance of such unusual phenotypes, but to get there are a few more issues to explore.

4.7 "Mere plasticity": when plasticity is important and when it isn't in a functional explanation

There some questions for which knowing whether standing plastic ranges are wide or narrow does make a difference, and there are also aspects where the distinction does not make a difference. Because notions surrounding plasticity can be so confusing, it is important to have clarity regarding these different situations. As mentioned in Section 4.3, whether a stand of trees has wide or narrow ranges of plasticity is of great importance for managing forests under climate change. However, with regard to the biophysical reason for why we observed one developmental variant over another in a given environment, the "genetic/plastic" distinction is not important. Take as an example a broad pattern, the gradient in leaf size from Figure 4.1, with small leaves typically found in desert environments and large leaves more common in tall tropical rainforests. Suppose we observe this pattern not within a species, but between species, where each species has a fixed leaf size, a "genetic" difference. How would we explain this gradient in size?

In explaining the case of each species having a narrow range in leaf size ("genetic" differences), a plant biologist might construct an account rooted in the physics of leaf function, focusing on energy and heat budgets, leaf boundary layers (the thin layer of still air hugging a leaf's surface), stomatal conductance (how easily water vapor exits and CO_2 enters), and the efficiency of photosynthesis relative to the cost of building and maintaining a leaf. In deserts, where water is scarce and sunlight is intense, small leaves are advantageous because they dissipate heat more effectively. Their smaller surface area and thinner boundary layer make it easier for them to cool off through radiation and avoid overheating, crucial for survival in hot, dry environments where evaporative cooling is limited. Large leaves, in such conditions, would trap heat and risk damage to photosynthetic tissues. In contrast, in moist rainforests where water is plentiful, plants can afford to grow larger leaves. The ready availability of water permits churning abundant water through the transpiration stream, lowering leaf temperature via evaporative cooling. Selection acting in similar ways across species thus leads each species, regardless of its ancestral leaf size, to strike a particular balance between the costs and benefits of different leaf sizes along the moisture gradient. This biophysical scenario helps explain why small leaves predominate in dry areas, large ones in moist ones, in "genetically fixed" differences across species.

Now imagine cuttings from the same tree planted along a desert-to-rainforest gradient, spanning the same range of leaf sizes seen across species. In the desert, the cuttings produce small leaves; in the rainforest, the genetically identical clones develop large ones. In this case, the same pattern of morphology across environments is chalked up to phenotypic plasticity. But to explain why these clones produce different leaf sizes, we would appeal to exactly the same factors as in the interspecific case: energy and heat budgets, boundary layer thickness, transpiration rates, photosynthetic efficiency, and carbon return on investment. At each point along the moisture availability gradient, the plant expresses the leaf size associated with the highest fitness, drawn from the range of phenotypes developmentally available to it. From the standpoint of biophysical explanation for why one leaf size is observed in a given environment, the distinction between "genetic" and "plastic" origins is irrelevant. The structure–function relation is the same. This doesn't mean that plasticity doesn't matter, only that the question of whether a trait is plastic is orthogonal to the question of why that trait has the highest performance in a given environment. Recognizing when plasticity matters, and when it doesn't, offers useful orientation for both theoretical and empirical work.

4.8 "Phenotype" and "environment" *sensu lato*

Discussions of high generality are necessarily coarse-grained, and mention of "the favored phenotype" in the context of "the environment" in discussing the evolution of plasticity is one place where this coarseness can be seen. It is therefore important to add some texture to what "phenotype" and "environment" can refer to and the important territory this texture leads to.

"Phenotype" in the preceding discussion is intended in a broad sense. The definition given in Section 4.1 states that phenotypic plasticity is the adaptive propensity for a given developmental system to produce the phenotype associated with the highest fitness out of those possible in that environment. This can be taken as a general case that includes situations in which a range of phenotypes are favored. "Phenotype" can be understood to include not just morphology but also metabolism, physiology, behavior, coloration, and so on.

Even more interestingly, "the environment" typically evokes images of ostensibly external variables (see Section 3.14) such as air or water temperature, light intensity, wind velocity, or relative humidity. These are for sure important factors shaping organismal form and function. But a relevant selection pressure acting on a given trait can just as well be a variable that is an attribute of the same organism. In the example of long bone scaling in Section 2.3.8, body mass is clearly a key attribute contributing to why certain bone length–diameter proportions are favored over other developmentally possible ones. So, for the purpose of the present discussion, "the environment" refers to the factors associated with differential survivorship and reproduction by virtue of their interaction with variation in the trait of interest. From this angle, "the environment" that stimulates ontogenetic buffering on the part of a developing organism can include not just temperature, humidity, or body mass but also a deleterious mutation or an insult such as a harmful molecule present during development. Recognizing this breadth of construals as "the environment" reveals the flip side of the plasticity coin, which is the phenomenon of phenotypic accommodation.

4.9 Phenotypic accommodation: the other side of the plasticity coin

Organisms display an extraordinary continuum of developmental adjustment and buffering in ways that result in functional outcomes; it is helpful to think about these phenomena as two extremes of this continuum, with phenotypic plasticity at one end, and phenotypic accommodation at the other. We have already defined phenotypic plasticity as the adaptive propensity to produce *different* phenotypes in different environments. Phenotypic accommodation lies at the other end of the continuum: it is the tendency to produce a normal, or as normal as possible, phenotype despite varying conditions. Where plasticity leads to divergent outcomes across conditions, accommodation yields similar or maximally functional outcomes in spite of them.

Phenotypic accommodation is part of every normal ontogeny (Gluckman and Hanson, 2005). There are always challenges in development, understood as factors that can lead to non-normal outcomes. These include temperature fluctuations, variation in oxygen availability, variation in light availability, pH, solute concentrations, chemicals in the environment such as alcohol or other toxins, mechanical pressures, and the presence of unfavorable mutations. As discussed in Section 3.9, there is no plan in development, no foreseen goal. Instead, even in normal development, a developing individual detects the conditions that it is experiencing, including its own developmental state, and from there takes the next step. Once at the next step, the process of sensing and taking the next step begins again. A crucial part of this sensing-and-reacting process is the continual adjustment of ontogenetic "decisions" given the condition of the developing organism at a given stage, in ways that maximize performance out of the possibilities available at that moment. In short, "normal" development involves phenotypic accommodation.

A highly consequential illustration of this process of sensing-and-reacting comes from development of the human heart (Ramirez *et al.*, 2024). Drenckhahn (2009) summarizes decades of research on phenotypic accommodation in heart development in Figure 4.2 (see also Gardiner, 2005; Hutter, 2018). Figure 4.2 depicts just a part of the pathways of accommodation documented, showing only cell proliferation or muscle growth. Even so, it provides a window onto the intricacy and sensitivity of the buffering aspect of ontogeny. For example, it shows that damage to heart tissue is met by a buffering reaction of tissue proliferation. If cardiac load becomes lower, cardiac muscle proliferation lowers, whereas if cardiac load increases, cardiac muscle proliferation also increases. In this process, the developing heart adjusts sensitively to the conditions it experiences and the blood transport demands it is subjected to, given the finite pool of metabolic resources that can be allocated to competing demands. Drenckhahn paints a picture of a mechanism that detects, moment after moment during development, the conditions that the heart is exposed to and responses that adjust heart muscle cell size and number in a way that leads to the most suitable performance given the available resources. That even normal development involves continuous buffering of developmental conditions (Waddington, 1942) illustrates a key principle at the nexus of evolution and development.

The principle is that there isn't any single, invariant ontogeny. Even when development results in the typical configuration observed in a species, phenotypic accommodation is always at work. Even while broadly tracing the same trajectory as its ancestors, each individual navigates a distinct sequence of challenges, buffering them through timing and sequence unique to that individual (Silva-Rocha and De Lorenzo, 2010). Another example from the human heart will illustrate the buffering property of phenotypic accommodation in normal development (Snell-Rood and Ehlman, 2023). Because of its obvious significance to human well-being, the heart has received centuries of attention, and a great deal is known about the variants in the configuration of the aortic arch. The aorta is the largest artery, the one leaving the left ventricle of the heart. It surges upward out of the heart into the chest, before bending downward to form the aortic arch. The arteries that branch off of the aortic arch transport oxygen-rich blood to the upper part of the body.

Artery diameters across branching points are strikingly well matched to the volumes of blood they transport, following a principle known as Murray's Law (Vogel, 1994). According to Murray's Law, the cube of the radius of a parent artery is equal to the sum of the cubes of the radii of its daughter arteries. This scaling relation holds

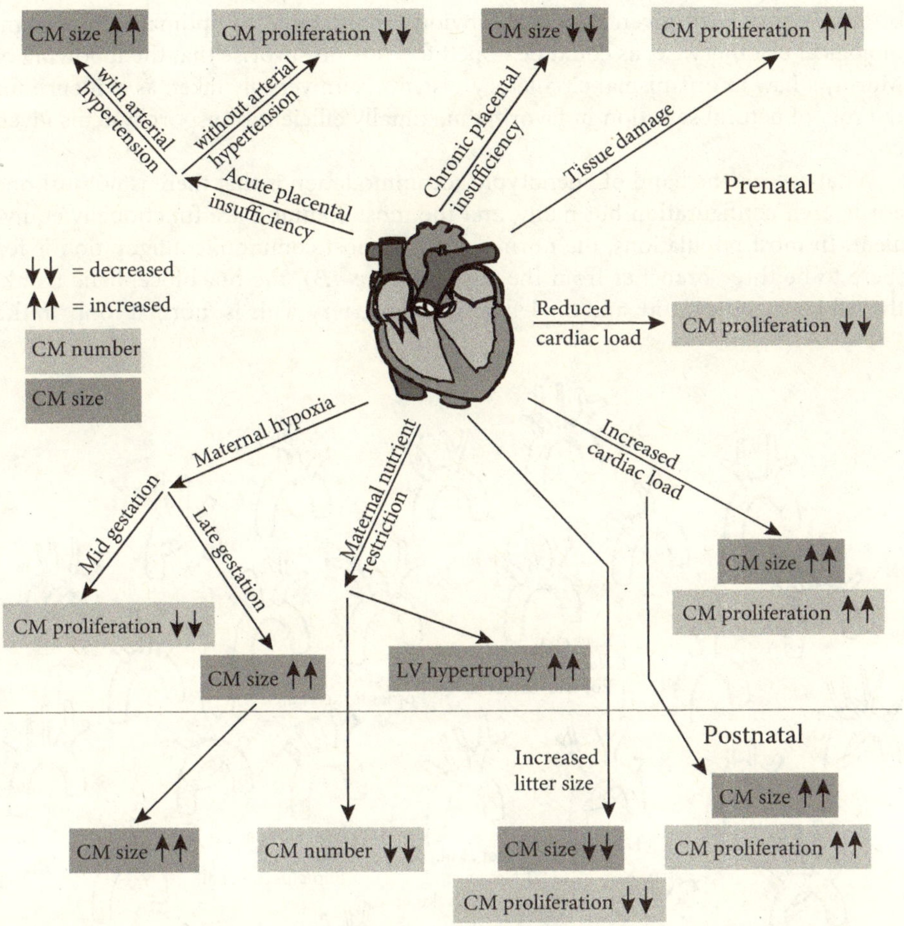

Figure 4.2 Phenotypic accommodation in the developing heart. Cardiac development in mammals illustrates the buffering aspect of ontogeny in producing maximally functional outcomes across a range of conditions. The arrows are labeled with developmental conditions or challenges, and the gray boxes represent two simplified responses: increases in cardiomyocyte (CM) size or number (LV = left ventricle). The continuous detection-and-response nature of development underscores that ontogeny is not the manifestation of a fixed adult form encoded in the genome, but a process shaped by ongoing, context-sensitive adjustments. Even "normal" development involves the accommodation of challenges within each individual trajectory. Adapted with permission from Drenckhahn, J. 2009. Growth plasticity of the embryonic and fetal heart. *BioEssays* 31: 1288–1298.

across successive branching orders, meaning that each split in the arterial tree preserves overall conductive volume while minimizing the combined cost of blood flow and vessel tissue. Each daughter artery's cross-sectional area is just right for the volume it carries, and each step away from the heart in the branching network narrows accordingly. A plumbing system in which the pipes are precisely tuned to the amount of fluid transported given cost is as obvious an example of optimality in design, biological or otherwise, as could be hoped for. It is no surprise that the following of Murray's Law in organismal circulatory systems is universally taken as evidence for the role of natural selection in favoring maximally efficient transport systems given cost.

What reveals the hand of phenotypic accommodation is that there is not just one aortic arch configuration but many, and the most common are functionally equivalent. In most populations, the normal, that is, most common, configuration is for there to be three branches from the aorta (Figure 4.3): the brachiocephalic trunk, the left common carotid, and the left subclavian artery. This is "normal" only in the

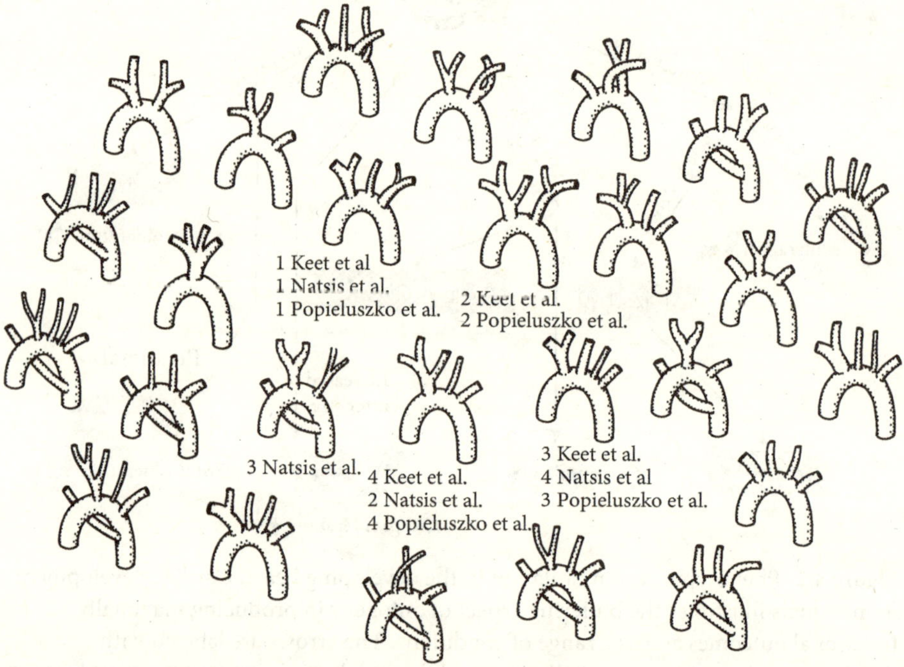

Figure 4.3 Alternative configurations of the human aortic arch. Rather than there being a single configuration for the aortic arch, there are numerous possible ones, some common, some less common, and some quite rare. The most common are clinically "functionally equivalent," which in this context means that they aren't predictably associated with health issues, and presumably provide similar blood flow at similar metabolic costs. Less common ones can be associated with difficulty breathing or swallowing, among other issues. Numbers 1–4 report the variants ranked in order of prevalence in three studies. Keet *et al.* (2019) studied a South African population; Natsis et al. (2009) 3 Indian, 4 US Caucasian, and 4 US African American populations; Popieluszko et al. (2018) included 51 studies, mostly from the US, China, and India, with most of the rest coming from other parts of Asia and Europe and included the data of Natsis *et al.* Even with the large number of variants depicted, this diagram is a simplification because some variants would be drawn the same but different branches correspond to different arteries.

sense of being the most common, being found perhaps 65–80% of the time, with variation in prevalence across human populations (Figure 4.3) (Natsis *et al.*, 2009; Müller *et al.*, 2011; Karacan *et al.*, 2014; Popieluszko *et al.*, 2018; Keet *et al.*, 2019; Tawfik *et al.*, 2019). In these variants, the brachiocephalic usually branches quickly into the right subclavian and the right common carotid, and so there are three branches from the arch. But in some relatively common cases, the first branch is known as the common trunk, with the left common carotid branching off from this, and then the brachiocephalic trunk, so there are just two branches from the arch. In other common variants, there are four branches from the aortic arch. Despite all this variation, performance is indistinguishable, with variation being of "no clinical significance" in the parlance of the doctors who deal day-to-day with cardiac performance and aortic arch variation (even Geoffroy de Saint-Hilaire, 1832 describes differences in artery conformation that don't affect function; Karacan *et al.*, 2014; Priya *et al.*, 2018). "No clinical significance" is the footprint of phenotypic accommodation, the buffering of variation during development in such a way that a phenotype that is maximally functional, even normal, is produced given the circumstances.

Just as in the development of fetal cardiac muscle, the formation of the aortic arch does not follow a rigid genetic blueprint, but rather unfolds through a flexible process of sensing and response. The walls of the aortic arch are richly supplied with baroreceptors, neurons sensitive to the stretching of the tissues they innervate (Elsaafien *et al.*, 2022). Through the baroreflex, the system continuously adjusts multiple variables, including arterial diameter. During development, vessel size is tuned to the flow it carries. Even when the genetic background is the same, a wide range of influences, from mechanical forces to vitamin deficiencies, can alter patterns of cell proliferation (Sugrue and Zohn, 2019). These influences can send ontogeny in different directions, leading to variation in the number of branches in the aortic arch. Yet in all such cases, the baroreceptive feedback process results in vessel diameters that maintain appropriate flow and configurations that still follow Murray's Law (Sather *et al.*, 2012). Rather than building to a fixed template, development continually responds to current conditions in achieving the most functional outcome possible given the circustances and the steps already taken, the doors already closed. There is no embedded image of the adult in the embryo; development proceeds step by step, with each step depending on the present state of the organism. Every individual confronts a particular sequence of conditions, so normal development involves buffering these unique combinations in ways that result, usually, in reconstruction of the parental phenotype. Normal development is possible thanks to phenotypic accommodation.

4.9.1 Phenotypic accommodation in trees and termite mounds

Phenotypic accommodation is not just a key participant in normal development, but it is also dramatically evident in situations ever more distant from the normal (Potticary *et al.*, 2020). With their continually ongoing ontogeny and long lives, trees always show examples of phenotypic accommodation. One example is branch cross-sectional shape. Just below the bark is a meristem called the vascular cambium, which produces wood to the inside of the plant in concentric layers. In most flowering plants, the cambium allocates wood where the stem experiences tension. If the stem is upright, then tensions are experienced more or less evenly all around and the concentric layers of wood are more or less circular. When the stem is strongly inclined, the upper part of the stem is subject to tension, and the cambium allocates more wood there. In stems that grow inclined for a very long time, especially in species with weak wood (which compensate weakness by allocating more wood in supporting their stems), the branches can come to be markedly elliptical. These elliptical branches have the long axis of their cross-section parallel to gravity. This orientation of material makes the branches stiffer than they would be if the same amount of material were invested in a cylindrical branch (Niklas, 1992). The mechanical demands placed on a tree change throughout its life, as the crown expands, as epiphytes load the branches, as branches are shed naturally, or when another tree falls through the crown, radically remodeling it. As new gaps of light open up, trees constantly adjust their canopies in ways that fill these gaps (Silvertown and Gordon, 1989). This continual adjustment of morphology and performance in ways that maintain function are easily seen examples of the ways that trees buffer constant challenges.

Of the challenges besetting trees, one of the most severe is the tree getting pushed over completely. But even here, trees detect mechanical stresses and allocate material accordingly, detect stem orientation relative to gravity and reorient growth, and sense the direction of light and grow in that direction, allowing trees to recover from being thrown to the ground. When a tree falls, the trunks that were previously oriented toward light and parallel with gravity now find themselves perpendicular to gravity. Some trees produce a special kind of wood fiber cell that has a sort of cellulose spring inside of it. The springs shorten after production, pulling the twigs a little closer to the vertical. The new shoots that are produced, often all along the now-horizontal trunk, grow erect, just as their parent trunk did for so long. The tree thus goes about recovering its position in the canopy.

This process of sensing gravity, light, and mechanical stresses reaches a particularly poignant apogee in the African tree *Moringa ovalifolia*. In most places where the tree lives through its range in Namibia and Angola, *M. ovalifolia* grows on rocky slopes protected from elephants. At the (now largely destroyed) stand known as the Sprokieswoud in Etosha National Park, though, *M. ovalifolia* trees grow on a silty plain where elephants roam at will. They push the *M. ovalifolia* trees over (Figure 4.4A) and gouge out the wet, pungent-smelling wood, which they apparently like to eat. The moringas resprout, sometimes all along their fallen trunks, thereby acquiring the appearance of a raft of stems, as the individual at right in Figure 4.4B. Once they recover sufficient size to attract the attention of idle elephants, the trees are usually pushed over again, and patiently start the process of recuperating an upright growth habit once again. This repeated process of upright growth, getting pushed over, and once again growing upright leads to very irregular shapes that are unusual for trees at large, but that are testimony to phenotypic accommodation, the aspect of ontogeny that produces functional outcomes even in the face of extreme perturbations. Trees pushed over can, with enough time, reconstruct phenotypes indistinguishable from individuals that have grown upright their entire lives. Such abilities are not limited to trees and can even be found in colonial animals such as corals and termites (Figure 4.5).

Figure 4.4 Phenotypic accommodation in trees. (A) *Moringa ovalifolia* trees in their native range in Namibia and Angola are often pushed over by elephants, who gouge out the fleshy, water-filled wood, as the individual in the foreground. The large trees standing are also *M. ovalifolia.* (B) After being pushed over, the trunks grow new wood and bark over the wounds. New trunks sprout and reestablish a vertical habit. The plants are eventually pushed over again, and the process repeats, often indefinitely over the life of a *M. ovalifolia.* The bloated base of the individual in the foreground of B represents the parts that have persisted these repeated topplings and resproutings. The individual on the right in B shows a raft of trunks that sprouted along the length of a tall trunk that was pushed over. Most *M. ovalifolia* trees in this area show how ontogeny can buffer challenges in ways that result in functional individuals, an expression of phenotypic accommodation.

4.9.2 Phenotypic accommodation and human stunting

While trees, corals, or termite colonies might, with time, reconstruct themselves entirely, there are many situations in which full buffering of challenges is impossible, and, as always, phenotypic accommodation leads to the most functional outcome *out of those possible in that situation.* The all-too-common case of stunting in human beings is an example. It is a sad fact that despite the wealth of some individuals and nations, the burning of fossil fuels to provide unprecedented ease to some, and the abundance and variety of foods available to many, that every year tens of millions of children under five experience life-altering, and often life-threatening, malnutrition (de Onís *et al.*, 1993; Semba, 2016). Protein malnutrition is particularly critical for children (UNICEF, 2019), because minimum threshold dietary protein levels are

Figure 4.5 Phenotypic accommodation in termite mounds. (A) The giant "tombstone" termite mounds of the northern Australian savannahs tower well over 2 meters tall, as in the intact mounds in the background, with Karola Stotz and Joe Olson providing scale. In the foreground is a mound that fell over and is well on its way to falling apart. On the left, though, can be seen the spires of a new tombstone mound, rising out of the fallen base of the old one. (B) Another mound farther along in the reconstruction process. Given time, the colonies will reconstruct their previous mound shapes and sizes, only to repeat the process if tipped over again.

required for proper brain development, with protein intakes below these thresholds leading to permanent cognitive impairment (Dewey and Begum, 2011, 2011; Ranade *et al.*, 2012). Malnutrition in children is a shameful feature of human society, but for our purposes here, the lesson that stunting provides is that the developmental system does its best to produce a maximally functional outcome despite scarcity.

Stunted children tend to be smaller, have weaker immune systems, and show greater learning disabilities than adequately nourished children, but there are phenotypic accommodations that lead to a maximally functional individual given the circumstances. This accommodation is Barker's "brain sparing model" (Gluckman and Hanson, 2005; Baker *et al.*, 2010). The hypothesis states that brain development is favored over musculoskeletal development *in utero*. Because tissue volumes are established more or less early, this *in utero* allocation then shapes lifetime morphology and physiology. In setting out why this preferential allocation to the brain over muscles is likely adaptive, Longman et al. (2017) postulate that "a well-fuelled brain may offer better survival odds than well-fuelled muscles when facing an environmental challenge."

From a public health and ethical standpoint, comparing environments, such as those that lead to stunting versus those that don't, is entirely appropriate and necessary. These comparisons help identify the conditions that support human flourishing and those associated with long-term degradation of opportunities. But from an evolutionary and developmental perspective, the relevant question is different. Evolutionary reasoning asks what is possible given the environment actually experienced, not what might have been possible under different conditions. In environments that lead to stunting, "brain sparing" guides resources preferentially to brain growth, buffering it from the worst effects of nutrient scarcity. The result is plausibly the developmental variant that yields the highest possible performance under those circumstances, in short, phenotypic accommodation.

4.10 Phenotypic plasticity and accommodation and novel phenotypes

One of the most striking and potentially evolutionarily important aspects of phenotypic plasticity and accommodation is that an organism can be taken well outside of its usual range of conditions and still produce a relatively functional outcome (West-Eberhard, 2003). An example is the etiolation response of trees and other self-supporting plants. Trees usually go their whole lives growing in light and supporting their own weight. Occasionally, though, a seedling will germinate in a deep crevice, will get buried, or a large tree will fall on it, blocking the light. A plastic response known as etiolation ensues. Etiolation involves drastic lengthening of stems relative to their diameters, accompanied by yellowing of tissues. Usually, trees support the weight of their own stems, and stems that are thick for their length help provide this support. Plants fuel most of their growth via photosynthesis, which involves capturing carbon from the atmosphere and converting it to sugars such as glucose, and then the cellulose their bodies are largely made of. So, a plant cut off from the light that drives photosynthesis is in a serious predicament. Plants do store a small amount of sugars in their living tissues, and can mobilize these sugars, fueling growth while reserves last. Etiolation maximizes the probability that the small amount of carbon that a light-deprived plant can draw upon will be able to reach the light. Etiolated stems are often so long and skinny that they can't support their own weight. In a plant digging its way out from under a landslide or a fallen tree, or slithering out of a deep crevice, the weight of the stem is supported by the substrate. Once the plant reaches the light, it can reactivate its photosynthetic apparatus, reinforce etiolated tissues, and resume self-supporting growth. Etiolation thus is a clear example of the plastic buffering characteristic making up a part of the adaptive ontogenetic arsenal of practically all plants.

Plastic buffering also results in functional outcomes even under entirely novel situations. Etiolation is an adaptive response to situations that present themselves generation after generation in the lives of plants, as when a seedling makes its way through the leaf litter in reaching the light. But organisms now and then find themselves in situations that are entirely novel. Sometimes this is because environmental conditions change drastically and suddenly. Or, individuals can be transported to novel areas that they have never experienced. Or, they can be born with novel teratologies or missing body parts or with additional body parts that are not part of the usual outcomes of a given species. The compensations that occur in the presence of skeletal deformities in humans are particularly well studied. Spinal deformities drastically affect the loading placed on the legs. In response, phenotypic accommodation responses include everything from altered stance, pelvic angles, angles of the knees, adjustments in muscle mass, and, perhaps most strikingly, remodeling of bone in ways that lead to stability in standing and walking despite the unusual loading caused by the spinal deformities (Shimizu *et al.*, 2021). These provide examples of how even conditions that are far outside the usual selective regime of a species are still met with phenotypic accommodations that result in functional phenotypes given the options available.

This characteristic of phenotypic plasticity plus accommodation to result in functional phenotypes that can be radically different from the normal ones almost certainly offers a crucial consideration for explanations of organismal form-function relations across species. What is so conspicuous about etiolated self-supporting plants is how much they resemble normal climbing plants. Many of the world's habitats, especially tropical ones, are host to an abundant and diverse array of plants whose normal stems do not support their own weight. These are known by names such as vines, climbers, scramblers, or lianas (Angyalossy *et al.*, 2015). Etiolated versions of self-supporting plants bear uncanny resemblances to normal lianas and other climbing plants. These sorts of similarities between plastic responses within species and normal phenotype differences across species leads the question of the role that plasticity might have in diversification across species, the subject of the next section.

4.11 Plasticity-first, genes-as followers evolution

The standard genes-causing-phenotypes account of adaptation makes sense, but there are reasons to think that it must be very slow in comparison with what plasticity can achieve. The standard account involves a population facing some selection pressure, say a change in environmental conditions. If mutation is the only source of adaptively relevant variation, and the variation isn't already present, then the population simply has to wait around until just the right mutation arises. Just to provide a ballpark, here is a "back of the envelope" illustration. Let's assume a plant species has a mutation rate per base pair per generation of around 1×10^{-8}, a genome size about three billion base pairs, and an effective population size (*Ne*) on the order of 10^4. Let's say a specific beneficial mutation could occur at one of 10 possible nucleotide positions. Then, the probability it arises in one individual in one generation $\approx 10 \times 10^{-8} = 1 \times 10^{-7}$. Total new mutations per generation = *Ne* × mutation probability = $10^4 \times 10^{-7} = 0.001$ new beneficial mutations per generation. So, on average, one copy of such a mutation might appear every 1,000 generations. For a plant with, say, a 5-year generation time, that's about 5,000 years just to get the mutation. Then, once the mutation arises, if strongly beneficial, with a selective advantage of 1%, it will still take on the order of hundreds to thousands of generations to spread through the population and reach fixation, depending on population size. Even under optimistic assumptions, this adds another few thousand years at 5 years per generation. So in total, we're looking at a rough estimate of 7,000 years or more for a beneficial mutation to arise and sweep through a population. However you parameterize such calculations, they serve to illustrate that the standard account, often called "genes-as-leaders" evolution, implies many generations going by before a favored mutation arises, and then more generations for it to spread throughout the population if it is to have anything like an effect on mean trait values.

Picture instead the same population exposed to the same environmental shift, but this time consider what plasticity and accommodation achieve. Practically all of the individuals will exhibit some plastic response, meaning that much of the population will immediately track the environmental change. For example, a lowering of light intensity will immediately lead to increases in leaf size in ways that increase light capture across an entire population of plants. Rather than waiting thousands of years to see a shift in average trait values, plasticity allows whole populations to respond almost instantly. If there were some way for these plastic changes to be stabilized by selection, then plasticity would provide a much faster mechanism of response to selection than the traditional mutation-driven one.

There is such a mechanism, known as genetic assimilation. The name "genetic assimilation" is unfortunate because it sounds as though somehow the genome is taking in some sort of externally imposed change. In reality, genetic assimilation is just standard natural selection. The variation in question is between individuals within a population in their plastic response thresholds. "Threshold" refers to how much environmental stimulus is needed to produce a given plastic response. In the case of light and leaf size, it implies that a small decrease in light level would be associated with only a few individuals presenting a plastic response, a slightly larger decrease in light levels would be associated with more individuals presenting the plastic response, and a large decrease in light levels would be associated with the largest number of individuals presenting the plastic response. Those presenting the plastic response with only a small decrease in light have a low threshold of induction, whereas those presenting the response only when exposed to a large decrease in light have a high threshold of induction. Seen through the lens of a standard adaptationist explanation, this situation looks familiar. There are individuals in a population that differ with respect to the value of a given trait. In this case, the trait is the threshold of induction of a plastic response. So, genetic assimilation requires standing variation in the threshold of induction of a plastic response among the members of a population, and that this variation be heritable.

Given heritable variation in plastic response thresholds, for genetic assimilation to occur, the variation needs to be associated with fitness differences. Because plasticity always pushes individuals toward more functional outcomes, variation in plastic response thresholds is very likely to be associated with differences in fitness. One example of a change in selective conditions that should be associated with genetic assimilation is a situation in which ancestral populations experienced wide ranges in selective environmental conditions, and had correspondingly wide plastic response ranges. Under the novel regime, conditions correspond to just a narrow slice of the ancestral environmental range. Under these new, narrow conditions, individuals that most readily produce the phenotype that best corresponds to the prevailing environmental conditions will be favored. These favored individuals will be those with the lowest threshold of induction of the favored phenotype. Another situation is one in which the novel conditions are outside of the range typically experienced by the ancestral population. This means that the favored phenotype will be the one that has the closest fit to the novel conditions. Therefore, the individuals that will be favored will be those that produce the phenotype at that extreme. As in the first situation, the favored individuals will be those with the lowest threshold of induction of the favored phenotype.

It would be much more transparent to call genetic assimilation "natural selection acting on standing variation in thresholds of induction of plastic responses." It's easy to see how such selection could lead to rapid evolutionary response to changed conditions across a population. Given the novel conditions, some individuals will produce the favored response, while others with higher induction thresholds will not. The individuals being "favored" means that they will have higher survivorship or reproductive output than the others, and will be disproportionately represented in subsequent generations. Beyond assimilation, epigenetic inheritance can also provide a rapid basis for intergenerational change (Vogt, 2017; Skinner and Nilsson, 2021; Santiago *et al.*, 2022).

When adaptation is driven by plasticity followed by genetic assimilation, notice how quick the population-level response will be as compared to the standard gene-driven scenario. First, plastic responses are typically exhibited by numerous, often most, individuals in a population. This is a far cry from the wait-around-for-a-mutation scenario in which the novel variant appears in a single, fragile individual. Moreover, this plastic response is not in just any direction. Instead, it is precisely a step, often a very large one, in the direction toward the highest-fitness phenotype that the individuals can produce developmentally under those conditions. This means that not only is much or most of the population showing significant change, but this change is in the direction favored in the novel environment. Again, contrast this with the mutation driven scenario, under which a mutation in just the right direction will not only be rare but even when in the favored direction it will be only a small step. Because it begins with a plastic response and is followed by selection on standing variation, adaptation via plastic response followed by genetic assimilation is often called "genes-as-followers" evolution or "plasticity-first" evolution, in contrast to the traditional "genes-as-leaders" evolution.

Because it seems certain to provide much larger intergenerational steps than genes-as-followers evolution, many biologists can't help but come to the conclusion that genes-as-leaders, the standard scenario in traditional evolutionary biology, can't possibly be the most common mode of adaptive evolutionary change. Because genes-as-followers evolution is plausibly so much quicker, they reason that it must be the preponderant mechanism (Schlichting and Pigliucci, 1998; Pigliucci and Murren, 2003; West-Eberhard, 2003; Pigliucci *et al.*, 2006; Schwander and Leimar, 2011). Empirical footprints of genetic assimilation suggest that it is probably a very important process in adaptive evolution (Diggle and Miller, 2013; Badyaev *et al.*, 2017; Levis *et al.*, 2018; see also Kovaka, 2019), but as discussed in Section 5.7.8, most lab selection experiments are not properly designed to detect it, and appropriate empirical approaches remain to be developed and widely implemented.

4.11.1 Unusual developmental outcomes can also be assimilated

While the adaptive changes of phenotypic plasticity make up the most plausible first steps in genes-as-followers evolution, even non-adaptive changes can be stabilized by genetic assimilation. Because developmental potential is broad, filled with a vast array of possible outcomes, it stands to reason that there is standing variation not only in the tendency to produce functional phenotypes but also in the sensitivity of development to perturbation: how readily development can be pushed off course into abnormal or rarely expressed outcomes. This kind of sensitivity, when exposed by environmental challenges, can itself vary among individuals and be subject to selection.

The most celebrated cases come from laboratory experiments. Classic work by C. H. Waddington on *Drosophila melanogaster* showed that exposure to heat shock or ether during early development could induce phenotypic changes, much like temperature shocks alter guinea pig development (Section 3.9). In one set of experiments, flies developed a "crossveinless" configuration, in which key veins between the longitudinal veins of the wing failed to form. In another work, he induced a "bithorax" phenotype, in which the third thoracic segment was transformed to resemble the second, producing a second pair of wings in place of the halteres (balancing organs) that are typical not just in *Drosophila* but also in all members of the order Diptera to which it belongs. Waddington found that these abnormal traits, neither clearly adaptive in any way, could be selectively bred over multiple generations. Eventually, they began to appear even in the absence of the original applied trigger; these results are what led Waddington to coin the term "genetic assimilation" in the first place (Waddington, 1942, 1953, 1956).

Such findings suggest that any form of developmental sensitivity that varies heritably can, in principle, be shaped by selection, even if the resulting outcomes are not initially adaptive. So while adaptive plasticity followed by assimilation remains the most plausible and likely common path for genes-as-followers evolution, even highly divergent morphologies that aren't part of the normally produced developmental repertoire can manifestly be assimilated. Waddington himself marveled at the striking resemblance between the four-winged flies he produced and the typical four-winged condition of insect orders such as Odonata (dragonflies) or Ephemeroptera (mayflies), which have two functional pairs of wings. He said "If such a change occurred in Nature, it would certainly rank as a 'macro-evolutionary' step. It seems doubtful whether in any other selection experiment such a large change has been produced so rapidly from a normal wild type stock" (Waddington, 1957, p. 174–175). Presumably for an abnormal part of the developmental repertoire to become fixed in nature would require an unlikely combination of both unusual developmental conditions and unusual selective conditions, but it illustrates a potential mechanism for rapid and large-scale evolutionary change.

4.12 What does plasticity-first evolution mean for well-supported explanations of organismal form?

The next chapter illustrates that the strength of adaptationist evo-devo is its rich contribution to maximally supported explanations of why organisms have the form that they do. Considerations regarding plasticity contribute to these explanations in some potentially important ways: here are a few of these.

One is that a trait under selection very often will exhibit a plastic range, and the plastic range itself can be regarded as under selection. As we saw in Section 4.2, this fits perfectly well into a standard adaptationist scenario. There is intrapopulational variation between individuals in aspects of the plastic response—its amplitude, the morphologies expressed along the response, the morphology–environment associations, the thresholds required for a given response, and so on. This variation is heritable, such that individuals with a given response tend to have offspring with similar responses. Finally, individuals with certain plastic responses tend to have higher survivorship, mating success, or fecundity as compared to others.

Another is that plasticity and accommodation can interact in ways that often make it difficult to categorize as one phenomenon or the other. The plastic phenomenon of etiolation mentioned in Section 4.10—plant stems growing longer when denied access to light—provides an example. Etiolation can be regarded as plasticity because it represents a phenotypic difference, longer stems with little photosynthetic pigment, under certain environmental conditions, darkness. On the other hand, it could be interpreted as accommodation, developmental buffering of an environmental challenge in a way that maximizes the likelihood of producing a normal outcome. Plasticity and accommodation are helpful terms, but are expressions of the same buffering characteristic of ontogeny, hence two sides of the same coin or parts of a continuum.

Perhaps the most outstanding challenge is that in any given case of apparent adaptation it can be hard to say whether it has proceeded via plasticity-first or genes-first evolution. Both should lead to the same result of a fit between form and function, so no necessary difference in outcome exists that would help diagnose which process was responsible. Moreover, in selection experiments under differing environments, it is technically difficult to determine whether selection on standing variation in plastic ranges is occurring. Resolving which mode really is the preponderant one in evolution remains an empirical priority (more detail on this and similar debates in Section 5.7.8). For now, it is time to see how all of these considerations—genes-as-followers evolution, phenotypic plasticity, developmental systems, developmental potential, and standard adaptationist approaches—must be woven together in constructing explanations of organismal form.

5

Building adaptationist evo-devo explanations of organismal form

This chapter provides guidance for constructing evolutionary explanations. As scientists, we are faced with patterns in nature that demand explanation: Why do we observe this trait and not another? Why does one variant persist in these conditions while another fades away? The goal of explanation is to account for these observed regularities and deviations. There is a lot of debate, particularly in philosophy of science, about the terminology and structure of explanations: how to distinguish models from theories, theories from hypotheses, and how different kinds of theoretical constructs function in practice (Bailer-Jones, 2008; Love, 2013). These discussions can be useful for thinking carefully about the structure and function of scientific reasoning. That said, for working biologists engaged in explaining concrete phenomena, it's often enough to think of theories and hypotheses alike as potential explanations, frameworks that identify mechanisms, whether known or postulated, that account for observed patterns. While textbooks sometimes emphasize sharp differences between terms like "theory" and "hypothesis," in practice, there doesn't seem to be any meaningful distinction, certainly not for a practicing scientist. For our purposes here, we will treat these terms interchangeably, recognizing that both serve the same core function: offering candidate explanations for patterns in the world.

In evolutionary explanations, such as those produced by adaptationist evo-devo, complementary layers of empirical evidence, historical assumptions, and the observed data, along with mutual confirmation among them, make up an explanation (Olson and Arroyo-Santos, 2015). Biologists can come up with many potential explanations, and the one or ones with the best mutual fits among evidence, patterns to be explained, and assumptions are the preferred ones; we will explore this process below.

Even though the main goal of science is to come up with explanations, as scientists, we generally spend little time talking with one another about how to go about building these explanations and what we think constitutes a good one. And a lot of what we do say to one another is clearly incorrect. For example, many evolutionary biologists claim that they are Popperian deductivists, and that their efforts provide law-like proof for their hypotheses (see catalogue of statements to this effect in Olson and Arroyo-Santos, 2015; also De Santis, 2021a, 2024). However, no science proceeds entirely deductively, but instead by reasoning that goes by names such as Bayesian inference, abduction, or inference to the best explanation (Sterelny and Griffiths, 1999; Ladyman, 2002). What we will do in this section is to dissect the structure of an evolutionary explanation that scientists regard as well-supported. We will take this as the basis for how to construct our own explanations. After that, we will then turn to

Adaptationist Evo-Devo. Mark E. Olson, Oxford University Press. © Oxford University Press (2026).
DOI: 10.1093/9780197831533.003.0006

highlight some unnecessary burdens that we scientists can bring into scientific practice that get in our way in constructing explanations for organismal form. We will start by reflecting on the different sources of empirical evidence and the assumptions that make up an explanation, before seeing how they fit together.

5.1 The sources of empirical evidence: comparative, populational, optimality, and developmental potential

This section explores the main sources of empirical evidence used by evolutionary biologists in building explanations of organismal form. In this context, "evidence" is understood in contrast to assumptions: evidence consists of data that can support a particular explanation, whereas assumptions are the parts of the story that scientists need to interpolate to complete the explanation. The key theme is that each approach, comparative, populational, optimality-based, and developmental, offers distinct empirical strengths in ways that complement one another. The subsections that follow examine each in turn, describing how they operate, the types of data they yield, and how they contribute to the construction of explanations of organismal form.

5.1.1 Source of empirical evidence: comparative method

Perhaps most of what biologists think we know about specific adaptations comes from various applications of what can be called the comparative method for the study of adaptation, the "comparative method" for short. This approach adduces evidence for adaptation through apparent cases of convergent evolution. Convergent evolution refers to the independent evolution of similar organismal features under similar selective conditions, from distinct ancestral starting points. There are important elements of this definition to unpack and subtleties to add, and we'll begin with the most straightforward situation (the comparative method is also used to explore developmental potential, including the discovery of developmental impossibilities, and this application will be treated below in Section 5.1.4.3).

Obvious examples of convergence include fins and streamlined forms in swimming animals, the globular shapes of dryland succulents, and wings in flying animals; we'll use the latter as an example. All flying animals, insects, birds, bats, and pterosaurs have or had wings. Wings clearly participate in flight and have similar structure-function relations, so wings meet the "similar organismal features" criterion. An aerial lifestyle involving powered flight meets the "similar conditions of selection" criterion. The ancestors of pterosaurs, sister to the dinosaurs, obviously evolved wings from a different ancestor than insects or bats did. For that matter, the ancestral starting point was different from that of birds, too. From different organismal substrates, functionally similar if structurally different features evolved in similar selective contexts. In this case, the physics of powered flight are unforgiving, especially the larger an animal is: only a narrow range of wing configurations, wing-body relations, musculatures, and metabolic profiles will permit heavier-than-air flight. Forget for the moment that the physics of flight is well understood (the optimality perspective explored in Section 5.1.3), or that traits relevant to flight can be modified in model systems from pigeons to fruit flies (population biology, Section 5.1.2). Even setting that aside, the comparative observation that all powered fliers have wings supports the inference that selection favored wings repeatedly, regardless of the anatomical starting point.

This last point is key—even though the data are across-species, convergence is evidence of adaptation, a population-level process. With regard to the evolution of flight, say in insects, it implies that ancestral, flightless individuals varied in the propensity to fling themselves into the air, in their body mass, musculature, and the presence of proto-wing extremities. Because this is a scenario involving selection, this within-population variation had to have been heritable if the scenario is correct. At some point, some of the more leapy and wingy ancient insect variants began to have higher survivorship, say due to better predator evasion, than their more sedate conspecifics. Or perhaps they reproduced more, if proto-wings improved aerial courtship displays or boosted metabolic efficiency enough to allow production of more young. We'll never know the exact details. But if selection favored wing presence in insects during the transition to powered flight, then something along these lines necessarily must have occurred: a population of variably structured individuals in which this variation was heritable, exposed to consistent selective pressures, such that more winged forms outperformed others across generations. Analogous stories must apply to the ancestors of pterosaurs, birds, and bats as well. If we are taking the presence of wings in flying animals as consistent with selection acting in similar ways, then we are necessarily committed to a richly structured populational history. The comparative pattern is taken as the visible trace of that populational process. Just because the pattern is across-species does not make convergence a "macroevolutionary" phenomenon (see Section 1.12.1 on macroevolution that isn't). Convergence is evidence of selection, and there is nothing "macro" about selection, the paradigm standard population-level process.

Examples like the independent evolution of wings in insects and mammals are classic cases of convergence, in which both the organismal structures and the starting conditions differ dramatically. But in practice, evolutionary biologists often accept less extreme differences in states as sufficient for convergence when those ancestral states are functionally significantly distinct from the trait in question (Felsenstein, 1985; Martins and Hansen, 1997; Martins, 2000; Giannini, 2003; Hansen, 2014). Return to the example of femur length and body-mass scaling in terrestrial mammals from Section 2.3.8 and shown in Figure 1.2. At first glance, this seems a different scenario from wings in bats, birds, and insects, because all these mammals are relatively closely related. But if the primary predictor of femur length is body mass, not phylogenetic proximity, then the available evidence suggests that the appropriate bone length will evolve in its corresponding selective context from any ancestral starting point. In other words, if knowing a species' body mass tells us more about its femur length than knowing who its relatives are, scientists take the pattern as suggesting convergence driven by selection (Orzack and Sober, 1994, p. 200).

When trait values respond predictably to selective pressures, regardless of lineage, then this fits the definition of convergence, the evolution of similar organismal characteristics under similar conditions of selection from functionally significantly different ancestral states. Marchini *et al.* (2014) present a graph of tibia length against body mass, with various species helpfully labeled (Figure 5.1). There, we can see that species fall on the graph alongside others with their same body masses, not necessarily with close relatives. For example, the two rodents labeled are the guinea pig and the capybara, and they fall in the areas corresponding to their masses. Five primates, the tarsier, the sifaka, the saki, the indri, and the human all fall distributed along the

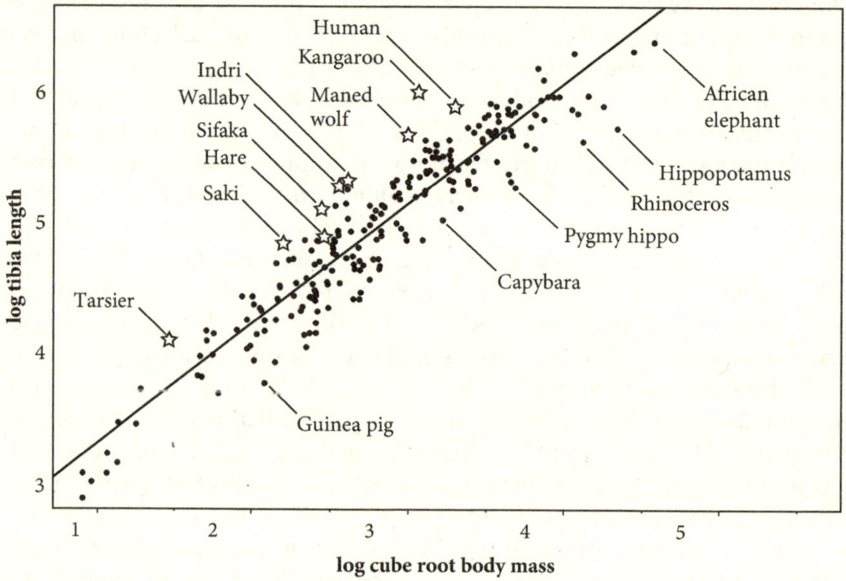

Figure 5.1 Convergence and statistical relationships between variables across species: tibia length scales with body mass across mammals. In this comparative dataset (Marchini *et al.*, 2014), tibia length is tightly predicted by body mass. Species fall along the same scaling line regardless of their phylogenetic affinities. This implies that ancestral state predicts little of the variation in present-day tibia length, and that different species routinely evolve similar trait values from very different starting points. From this perspective, the pattern is consistent with convergence, trait similarity shaped by shared functional demands independently of ancestral starting point. Although the phylogenetic comparative method aims to take into account non-independence among species (e.g. between wallaby and kangaroo, between the two hippo species, or between capybara and guinea pig), most variation in tibia length is explained by body mass alone. More important than any particular statistical correction is understanding why the analysis is interpreted as evidence for convergence in the first place: because tibia length is predicted mostly by body mass and practically not at all by ancestry, the regression line could in principle have a different slope or intercept under different conditions of selection, and the repeated evolution of similar tibia length-body-mass proportions is consistent with the population-level process of adaptation. Figure adapted from Marchini *et al.* (2014), licensed under CC BY 4.0, creativecommons.org/licenses/by/4.0/

scaling line, and not all bunched up in a cluster. The same for Christiansen's femur length data depicted in Figure 1.2: If femur length is shaped primarily by body mass, rather than being correlated with the ancestral state, then this comparative pattern is taken as consistent with convergence. In such a situation, apparently any state could be ancestral to any other, with the result that similar functional demands lead to similar outcomes. Elephants, for instance, occupy the far right of Figure 1.2 with body masses exceeding 4,000 kg. Their closest relatives, hyraxes, fall near the center with masses around 4 kg—three orders of magnitude smaller—yet each species has femurs proportionate to its size. Other clades show similar evolutionary lability: bovids range from 19 kg dorcas gazelle to the 440 kg African buffalo. Hippos from the 250 kg pygmy hippo to the 2400 kg common hippo; canids from the 1.2 kg fennec to the 29 kg gray wolf; felids from the 7 kg lynx to the 230 kg Siberian tiger. The population-level scenarios for bone–body-mass comparative patterns are laid out in Sections 1.11.1 and 2.3.8. That femur length tracks mass consistently and apparently from any ancestral starting point is empirical evidence consistent with repeated evolutionary responses to similar selective pressures, a classic pattern of convergence, even within clades.

The so-called "phylogenetic comparative method" incorporates this reasoning. For example, in a regression such as femur length against body mass, in one way or another, it attempts to incorporate phylogenetic information into the regression (Felsenstein, 1985; Martins, 2000; Blomberg *et al.*, 2003; Giannini, 2003; Hansen, 2014). The "phylogenetic" component of these methods refers to adjustments for the statistical complication that closely related species often resemble each other more than expected by chance. The worry motivating these methods is that failing to account for this similarity can violate assumptions of independent residuals in regression analyses. Phylogenetic comparative methods model shared ancestry as predicting correlation among species. In the case of Figure 5.1, some of the residual variation, that above and below the regression line, does seem likely to be predicted by ancestry. For example, both species of hippo have conspicuously short tibias, and the kangaroo and wallaby have long ones. However, body mass predicts practically all of the variation in tibia length, so any of this residual variation that would be accounted for by a phylogenetic comparative method is not a big deal to say the least. Regardless of statistical treatment, and the empirical accessibility or inaccessibility of the assumptions that these methods invoke notwithstanding, the logic behind them is the same as that sketched in the previous paragraph: if knowing X is all that is necessary to determine Y, and not ancestry, then this situation is compatible with convergence, and apparent cases of convergence provide evidence compatible with adaptation.

Much more important than the particular statistical method employed is understanding the logic of convergence that these methods aim to capture. When it comes to identifying convergence, what matters is whether the traits in question could plausibly have arisen from ancestrally distinct states, distinct enough that, under a different selective regime, a different pattern (say, a different slope or intercept) could have emerged (how much difference counts as "significantly different" leaves some wiggle room; see Section 5.9.3). This reasoning is the same kind of search for similar organismal responses to similar selective pressures across species as in classic qualitative examples like the independent evolution of wings in flying animals, the essence of the comparative method.

5.1.2 Source of empirical evidence: population biology

Selection emerges as the result of heritable variation with fitness consequences among the members of populations; population biology is the subdiscipline that examines these foundational variables directly. Whereas the informative variation in the comparative method is between species, in population biology, it is between individuals within species. Population biology takes many forms. It includes, among other studies, empirical analyses of genotype and phenotype frequencies within populations of a single species, measures of heritability, response to selection, quantitative genetics, and mathematical modeling of intergenerational genetic change within populations (Pigliucci, 2008; Templeton, 2021). Population genetics provides the theoretical core around which the Modern Synthesis, and thus most of today's evolutionary biology, is built (Dickins and Dickins, 2018). The mathematical rigor of population genetics, especially in its expression as quantitative genetics, is held often up as a gold standard for integration into comparative biology (Steppan *et al.*, 2002; Bégin and Roff, 2004; Butler and King, 2004; Auerbach *et al.*, 2023; Schraiber *et al.*, 2024). Physiology experiments can be categorized as essentially applications of population biology. Physiology experiments examine the ways that variation in organismal structure or metabolism affects functional performance. Physiology experiments that examine variation in performance or fitness across the members of a species go a long way toward explaining why heritable variation is associated with different tendencies to survive or reproduce. Population biology, then, broadly construed, provides key evidence regarding the very raw material of selection, in the form of direct study of intrapopulational variation, the heritability of this variation, and the performance and fitness consequences of this heritable intrapopulational variation.

As an example of the contribution of population biology to an explanation, body size affects everything from metabolic rate to reproductive output (West *et al.*, 2003). So, it is reasonable to think that natural selection could act on body size, and the only way to examine this notion directly is with the tools of population biology. In *Drosophila*, researchers have found that body size does indeed vary across the members of a population (Flatt, 2020). What is more, they have found that this variation is heritable, with progeny tending to resemble their parents in body size (Lafuente *et al.*, 2018). As might be expected, this heritable variation involves many genetic loci (Oldham *et al.*, 2000; Turner *et al.*, 2011) and has different impacts on fitness given different selective contexts. All of this evidence together shows that in a given population of *Drosophila*, body size does indeed meet all of the criteria for evolution by natural selection and can evolve adaptively across generations. Even if this information is from just a single species, it is valuable because it is consistent with across-species patterns like Bergmann's rule (selection favoring heat retention explains larger body sizes at colder latitudes). Of all evolutionary approaches, only population biology provides such direct empirical access to the very raw material that evolution by natural selection is made of—intrapopulational variation, heritability, and fitness.

5.1.3 Source of empirical evidence: optimality models

Which variant or variants prevail in a given case of selection makes biophysical sense. "Biophysical sense" means that there are always mechanistic reasons, often involving biomechanical or metabolic efficiency, but sometimes reflecting other physical or physiological interactions, that explain why one variant within a population has higher survivorship or reproduction than another. In other words, the association between the biophysics of a variant and its fitness is not random. Optimality modeling is the approach that most closely explores this expected fit between biophysics and fitness.

Optimality models build on basic biophysical principles, taking into account tradeoffs and selection acting in two or more opposing directions. These opposed factors mean that two or more functions cannot be simultaneously globally maximized. In plants, cell walls provide stiffness, so more cell wall material means greater mechanical stiffness. But cell wall material invades the cell lumen space required for storage or water conduction (Figure 2.2), so there is a necessary resource allocation tradeoff, where finite cell wall materials and space can't be simultaneously assigned to the same spot (Pratt and Jacobsen, 2017). The exact allocations to mechanical stiffness, storage of water and sugars, and conduction of water in any given species represent the results of a tug-of-war "negotiation" between these opposing functions. Selection acts in one direction until it is opposed by selection on one of the other functions, within the bounds of what is possible given the resource allocation tradeoff.

Optimality models include such negative relationships to find the configuration or configurations that maximize some performance currency (Parker and Maynard Smith, 1990; Vincent and Brown, 2005). Performance currencies can include factors such as fitness or one of its components, or minimization of carbon cost or metabolic output. Optimal foraging models, for example, balance the opposing investments of time spent intensively searching for food in a small area versus ranging widely and just eating the most readily accessed resources. The performance currency typically reflects energetic food intake, given energy or time investment, with selection tuning the balance between intensive and wide-ranging foraging.

"Optimality" is a term that frequently confuses, because it can imply all traits are somehow at their maximal imaginable values. But because there are resource allocation tradeoffs and adaptive tugs-of-war between two or more opposing functions, none of the functions in an "optimal" solution is at its maximal value (Orzack and Sober, 1994, p. 199) (this points to the difference between trait optimality and fitness optimality discussed in Section 2.3.10). For example, trees never have the maximum wood stiffness possible. This is because some amount of the wood is always taken up by the cell lumen, which is dedicated to storage and other functions. Eliminating all of the intracellular space artificially, by compressing the wood, leads to a material much stiffer than that found in any tree (Song *et al.*, 2018). At the same time, no tree has its entire wood volume dedicated to storage. In even the most succulent trees, such as baobabs, at least some volume is taken up by cell walls. The solution favored by natural selection in any given tree species, the "optimal" one for that species, is some compromise between support and storage (see Garland *et al.*, 2022 on flavors of "tradeoff"). It is therefore important to recognize that "optimality" does not imply that functions are separately maximized. Instead, it can be helpful to think of optimality models as finding the "least bad" combination given competing functions (cf. Potochnik, 2009, 2017).

In understanding how to use optimality models, it is helpful to think of them as just another piece of equipment in the laboratory. Statements and expectations about optimality models can be evaluated in this light. Biologists routinely complain that a given optimality model is incorrect or inaccurate (see, for example, Kozłowski and Konarzewski, 2005). Replacing "optimality model" with a piece of lab equipment, such as "microscope," reveals the misplaced expectation inherent in declaring this or that model "incorrect." Saying "this microscope is incorrect" is obviously an incomplete statement. It is incomplete because it misses the key qualifier of "for the task that I would like to perform." By the same token, if a researcher is interested in studying the effects of hurricanes in shaping plant trait distributions, then a model that does not include hurricanes is obviously not the correct tool, just as the too-coarse resolution of the light microscope would make it the wrong tool for studying the structure of the Golgi apparatus. But if we wish to examine major poles of natural selection applying to all of life, then including hurricanes would be absurd. Instead, it is more helpful to think about appropriate tools for different applications. In using any tool of discovery, it is important to understand where its inferential power lies.

As part of the process of discovering the factors shaping a trait of interest, optimality models are valuable not only for what they include but also for what they leave out. Adding or subtracting variables and comparing model outputs to real-world patterns is a way of testing ideas about what factors matter. A model's scientific value lies not in achieving universal accuracy, but in how it clarifies which components are necessary to explain a given pattern and which are not. This makes optimality modeling a fundamentally iterative process, akin to adjusting experimental apparatus to suit a particular task. Because all models are simplifications, no model can be globally "correct" or "realistic" in an absolute sense. Just as microscopes are instruments of discovery, optimality models are one important source of empirical evidence. In explanations of organismal form, they are the main resource for exploring the biophysical reasons for the prevalence of one variant over another within populations.

5.1.4 Source of empirical evidence: developmental potential

What puts the "devo" in "adaptationist evo-devo" is its focus on developmental potential. Just as there is a wide array of approaches to study adaptation—the comparative method, population biology, and optimality modeling—there is also a wide battery of approaches to explore developmental potential (Olson, 2012a). Exploration of developmental potential is essential for constructing explanations of organismal form. This is because some developmental outcomes are likelier than others (Fusco, 2001; Minelli, 2009; Badyaev, 2011; Lala *et al.*, 2024). In some cases, even plausibly imagined outcomes seem to be impossible. If a given configuration is truly cannot be produced in development, then natural selection cannot be invoked to explain its rarity. Therefore, determining whether a given "empty space" configuration can be produced but it has low fitness, or whether it apparently can't be produced—and might have higher fitness than the common configurations if it could be—are key empirical first steps. Chapter 2 showcased numerous examples of exploration of developmental potential. The methods they illustrate can be systematized into three broad strategies, as follows.

5.1.4.1 Manipulation

A very widely used strategy is manipulation. Instead of waiting for natural variation to present itself, biologists create the variants of interest, and then examine their performance or fitness relative to the common variants within populations. Expressions of this strategy include the following.

5.1.4.1.1 Laboratory selection

Laboratory selection experiments test the expectation that, if "empty space" variants are eliminated by selection, then it should be possible to move populations into those spaces under the right conditions of selection. The study in Section 2.3.2 of Conner *et al.* (2011) created, in just a few generations, flower part proportionalities never before seen in the focal species, wild radish. An excellent example in animals is the study of Frankino *et al.* (2007) in Section 2.3.11. In that study, the authors dramatically altered wing proportionalities in a species of butterfly, creating variants with lowered mating success compared to the wild type. Artificial selection provides an ideal test of the expectation that, if "empty space" is due to selection, then variants corresponding to those spaces should be ontogenetically possible, that these variants should be heritable, and that they should have lower fitness compared to the wild type under wild-type conditions.

5.1.4.1.2 Domesticated organisms

Many species have been subjected to selective breeding for centuries, leading to drastic expansions of their ranges of variation. These species provide superb study systems for seeing whether the "empty space" that surrounds the wild type is developmentally accessible or not, and if accessible, whether the variants are of lower fitness than the wild type. The study of domesticated dogs by Drake and Klingenberg (2010) in Section 2.3.3 provides a vivid example. They showed not only that the space of domestic dog skull shape is far greater than that of the wild ancestor of dogs, but that it is wider, even, than the space of all of Carnivora. The study of Sol (2008) on domestic versus feral pigeons in Section 2.3.5 is another example that leverages the strong divergence from the wild type in domesticated organisms. Trejo *et al.* provide an example from plants; despite nearly two centuries of intense selection, the domesticated Christmas poinsettia (*Euphorbia pulcherrima*), which is shorter, bushier, and leafier than its wild ancestor, still obeys the same total leaf area-stem volume relationship (Trejo *et al.*, 2018). This suggests a potent case of phenotypic accommodation maintaining a metabolic balance between leaves and stems, even as domestication has radically reshaped other traits. The case of domesticated goldfish treated in Section 2.3.4 is perhaps the most remarkable of all, given that some domesticated goldfish have two tails, an apparently unique condition in vertebrates. Domesticated organisms thus often provide veritable counterfactual clouds of developmentally possible variation that is manifestly heritable and that can be compared to the wild type.

5.1.4.1.3 Surgical intervention

When standing variation is too low, and biologists are too impatient for selection experiments, they can create the relevant variation artificially. If selection has extinguished variation in a given direction, then surgical intervention is one way of creating these variants. Surgical intervention not only can produce large displacements into "empty" space, but it's also quick—no need to exert generations of artificial selection to get the populations in those quadrants of trait space. Once created, it is then possible to compare the performance of the created variants against the wild type, to test the expectation that the "empty space" variants have lower performance or fitness than the wild type. The example of Polak and Rashed (2010) in Section 2.3.7 is one such example. The variation in male *Drosophila* genital spikes that the authors created was consistent with the notion that selection favors long spikes in male mating success. The "allometric engineering" of *Anolis* lizards by Sinervo and Licht (1991) provides another such example. By manipulating egg size before laying, they created mothers bearing fewer eggs for the same total egg mass. The large eggs either rupture in the oviduct or lead to eggbound mothers. Eggbound mothers have low fitness in the form of low survivorship. Mothers producing large eggs that rupture have low fitness in the form of low fecundity. Sinervo and Licht provide evidence consistent with the hypothesis that, given intrapopulational variation in egg size, selection favors a given egg mass and number as a balance between hatchling survivorship, maternal survivorship, and maternal fecundity. No mysterious "constraint" prohibits the construction of viable eggs that are larger than the ones seen in the usual clutches of four to six eggs; they are simply associated with lower fitness and not favored by selection. Along similar lines, Alexandra Basolo (1990b) used manipulation to examine female preference for swords in swordtail fish (*Xiphophorus*). Some species of *Xiphophorus* bear swords and others don't; swordlessness is ancestral with respect to sword-bearing. In sword-bearing *Xiphophorus*, females prefer males with larger swords, especially males with artificially enlarged ones (Figure 5.2). When she sewed artificial swords on the tails of swordless *Xiphophorus*, the females of *those* species preferred males with swords over the (normal) swordless males of their own species (1990a). Her use of manipulation suggests that the cognitive bias that led to a preference for males with swords predated the evolution of swords themselves, and that selection favoring swords exploited that preexisting bias.

5.1.4.1.4 Hormonal manipulation, directed mutagenesis, and manipulation of developmental conditions

These techniques are an additional powerful means of creating variation quickly. In Section 2.3.1, we saw the example of Camara *et al.* (2000) in *Arabidopsis*. To ask whether selection favors certain combinations of trait values over others, they induced mutations to produce an array of variants around the commonly observed one, showing that these alternative configurations are developmentally possible but likely of lower fitness than the common variant under wild conditions. In Section 2.3.6, Verna *et al.* (2019) and Linh and Scarpella (2022) manipulated hormone transport systems to generate leaf vasculatures never observed in wild plants (and with very low fitness). Also in that section, in a study of poplar trees, Rodriguez-Zaccaro *et al.* (2021)

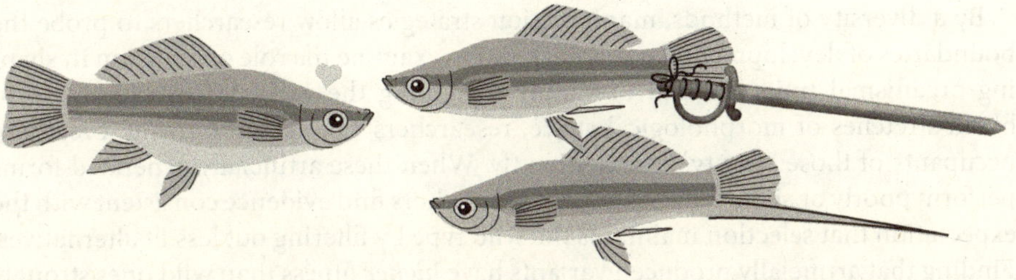

Figure 5.2 Swordtail manipulation. In the small, Mesoamerican freshwater fish *Xiphophorus*, the males of some species bear elongated fin rays on the ventral side of the caudal fin, forming a "sword." When Alexandra Basolo (1990b) sewed on artificially elongated swords, females preferred those males (the paper does not explicitly state that she did not use swords from martinis). When she sewed swords onto males of species of *Xiphophorus* that lack them, females preferred those males, which suggests that the preference predated the evolution of the sword.

used mutagenesis to generate a range of relationships between tree water conducting conduit diameter and tree height much wider than that observed in wild poplars. The temperature shock experiments of Edwards *et al.* on guinea pigs in Section 3.9 are another example of the manipulation of developmental conditions to see how the resulting morphologies fill developmental space. Further examples, of hormonal manipulation, can be found in Ketterson and Nolan (1999) in birds and Johnson *et al.* (2018) in plants.

5.1.4.1.5 Constructed organisms and models, digital and otherwise

Another expression of manipulation is constructed organismal variation, whether in the form of physical models or computer simulations. These models allow researchers to isolate and test the effects of specific variables that are otherwise entangled in natural systems. For instance, Jordan and Harder's (2006) use of painted Eppendorf tubes to mimic inflorescence architectures, summarized in Section 2.3.12, permitted a test of how flower arrangement alone shapes pollinator behavior, something not possible using natural variation. Taylor *et al.* (2025) pushed this idea further with 3D-printed insect models interpolated between real hoverflies and wasps, enabling precise control over shape, color, and pattern to test predator response to Batesian mimicry. These experimental systems serve as powerful stand-ins for natural forms and permit finely resolved tests of the adaptive significance, or lack thereof, of specific traits. Theoretical morphospaces offer a related strategy; by creating spaces defined by precise variables, such as coiling dimensions in mollusks (Raup, 1967) or plant architectural traits like stem taper and branching angle (Niklas, 2009), researchers generate complete formal spaces that allow visualizations of occupied versus unoccupied forms. This, in turn, makes it possible to investigate the causes of variation in the density of occupation or unoccupied zones in nature. Constructed variation, whether physically modeled or simulated, thus allows for a very fine population of variation.

By a diversity of methods, manipulation strategies allow researchers to probe the boundaries of developmental possibility and to examine the role of selection in shaping organismal traits. Instead of simply imagining the variants corresponding to blank stretches of morphological space, researchers can examine living, breathing occupants of those "empty" spaces directly. When these artificially generated forms perform poorly or show reduced fitness, researchers find evidence consistent with the expectation that selection maintains the wild type by filtering out less fit alternatives. Finding that artificially produced variants have *higher* fitness than wild ones strongly points to the possibility that some aspect of development impedes access to those parts of developmental space, and to the need to determine what those aspects are. Manipulation strategies, therefore, provide a crucial toolkit for exploring the causes of "empty" space.

5.1.4.2 Embryology

The embryological strategy takes up where the usual focus of population biology leaves off. Population biologists, as in artificial selection experiments, are generally content to work with the range of variation observed close to the population mean. It is probably safe to say that it is rare for population biologists or artificial selection experiments to make efforts to include sick, abnormal, or teratological individuals. The embryological strategy expands the focus to encompass the full range of variation within a species, including the rare, the abnormal (= very rare), and the teratological (=very very rare) (Alberch, 1989). One example is the intensive study of centipede segment number in Section 2.1.1. Combing hundreds of individuals has only turned up one with an even number of segments, an abnormal configuration for that group if ever there was one. The stubbornness of centipedes in producing only odd segments led to the discovery of the developmental patterning underlying segmentation in that group. In a plant example, Flores-Renteria *et al.* (2011) asked why pine trees have separate male and female cones. They patiently combed hundreds of pine cones in the conveniently low-statured dryland *Pinus johannis* to find a handful of teratological functionally bisexual ones. Once found, they were able to demonstrate that the bisexual cones are perfectly functional, producing viable seed and pollen, consistent with the notion that bisexual cones could be produced if selection were to favor them. The fact that cones in pines are normally male and female separately is therefore consistent with the likelihood that selection favors unisexual cones. Finally, the study of cervical vertebra number variation of Galis *et al.* (2006) in Section 2.3.13 is likewise based on variation from the normal to the abnormal. So, the embryological approach examines not just the normal individuals but also those that are usually disregarded, as being "sick" or "deformed," as evidence of what is possible in development.

5.1.4.3 The comparative method

The comparative method does not just provide evidence of the population-level process of natural selection, but at the same time provides information about individual-level developmental potential, again despite comparative data being across-species, including from extinct ones (e.g. Poe *et al.*, 2021; Jablonski, 2022). The allometric scaling line described by terrestrial mammal long bone-body mass relations is remarkably tight, with large empty spaces above and below the scaling line. The studies cited in

Section 2.3.8 show that, at least under some situations and in some groups, mammals can indeed occupy those empty spaces—mammalian tissue can indeed produce those variants in development. This occupation is consistent with the expectation that in terrestrial mammals, selection eliminates variants that extend too far into the empty spaces above and below the terrestrial mammal scaling line (Christiansen, 1999; Swartz and Middleton, 2008; Olson, 2012a). Similarly, most plant species have leaf veins that go from point A to point B as directly as possible, consistent with the notion that selection should minimize transport distances (West *et al.*, 1997). Finding that in some situations, as in some species of *Cistanthe*, veins can be longer than necessary (Figure 2.10) is consistent with the expectation that alternatives to the common configuration should be developmentally possible but not favored under most situations. The observation of pentadactyly across extant tetrapods, as explored in Section 2.1.2, is another example of the comparative strategy. It seems unlikely that five digits (max) is exactly the right number, favored independently by selection in species as diverse as lizards, whales, and humans. Instead, this across-species pattern suggests limits to developmental potential shape variation in a way distinct from standard adaptation accounts that include abundant variation, such as those outlined in Section 1.11. The comparative method thus provides information not just on convergence but also on what is possible in the intimacy of development in single individuals.

5.2 Generative assumptions and complementarity between sources of empirical evidence

With the wide range of methods available for gathering empirical evidence on organismal form, we can now turn to a less frequently examined ingredient of evolutionary explanation: the assumptions that make these methods relevant in the first place.[1] In what follows, we'll distinguish two types, "generative" and "historical" assumptions. In the previous sections, we reviewed the major sources of evidence for explanations of organismal form: population biology, the comparative method, and optimality models, and the diverse ways developmental potential can be investigated. What ties these approaches together is that each relies on a set of generative assumptions, ideas that simultaneously justify the method's relevance to evolutionary explanation and offer testable points of contact with empirical work.

Generative assumptions do the bridging work when individual approaches reach their explanatory limits. The central message of this book is that no single subdiscipline is sufficient on its own to build a robust evolutionary explanation. What allows these approaches to speak to one another, to support or challenge each other's conclusions, are their generative assumptions. Generative assumptions are the conditions that must surround the results of a particular approach, whether population biology, the comparative method, optimality modeling, or studies of developmental potential, in order for those results to contribute to an evolutionary explanation. These assumptions are not testable using further evidence from the same approach.

[1] some text in this section adapted with changes from Olson (2024). Is population genetics really relevant to evolutionary biology? *Evolutionary Biology* 5: 235–243, http://creativecommons.org/licenses/by/4.0/.

Instead, they are tested through findings from other methods; for example, the generative assumptions of population biology are evaluated not by population biology itself, but by comparative and other approaches.

These assumptions are "generative" because, by giving each approach its explanatory relevance, they breathe life into method; they establish the conditions under which explanatory connections become possible, as outlined in Table 5.1. The assumption that intraspecific patterns of variation and selection can, over time, give rise to interspecific patterns is "generative" because it allows population biology findings to be interpreted as contributions to evolutionary biology broadly. The assumption that comparative patterns reflect population- or individual-development-level processes is generative for the comparative method: it gives those observed regularities explanatory bite rather than reducing them to mere description. Generative assumptions are not afterthoughts; they generate explanatory relevance, hence the term. They allow, indeed demand, models, experiments, and empirical patterns from different approaches to be stitched together to achieve a narrative of biological form. In doing so, each method underwrites the claims of the others. The following sections examine this interdependence in more detail.

5.2.1 Populational weakness, comparative strength

Take a moment to contemplate the tiny scale of a population biology experiment. The population biologist takes a handful of individuals and describes their genotypes and phenotypes in meticulous detail. They subject the individuals of just a single species to differing conditions of selection and document the microscopic changes that their interventions provoke. On its own, a population biology experiment is a painstaking description of a handful of individuals from a micro-moment in the life of a single species. Nothing more. If the aim of science is to produce generalizations, then population biology, on its own, is barely science.

The actual content of population biology experiments provides no means of testing the generative assumptions that give the field broad relevance to evolutionary biology. A *Drosophila* experiment provides data only on the flies examined. It provides no direct empirical evidence whatsoever regarding unexamined populations of flies contemporaneous with the ones studied, and certainly not the uncounted and unobservable ancestors that precede the studied flies. Left alone with only their study population, a population biologist has nowhere to turn to justify the generative assumptions that give population biology results their relevance beyond the study system. Enter the comparative method.

The chief means of testing population biology's generative assumptions comes from comparative biology. The main generative assumption of population biology is that processes more or less similar to those observed in population biology, such as selection, can be extrapolated to explain patterns of variation across species. The assumption that processes observed in the narrow windows of population biology can be extrapolated to life at large is tested by looking for patterns across species that match what those processes would be expected to produce. Take population biologists studying variation in natural populations in the proportions of fork-tailed flycatcher wings (Carvalho Provinciato *et al.*, 2018), or the results of centuries of artificial selection on pigeon morphology and color (Domyan and Shapiro, 2017). These biologists will tell confident stories about similar heritable variation, with its indisputably dramatic performance consequences, as shaping the diversity of bird lifestyles across species. But within the confines of their population-level studies, they have no means of testing the assumption that similar processes shape bird wings at large.

The comparative method does just this. Witness albatrosses, which soar aloft for months, with their enormously long and slender wings. Witness forest-dwelling jays, with their short, broad wings ideal for flapping through dense woods. Witness the broad wings and fingertip-like extremes of vulture wings, ideal for lolling all day on rising air. These patterns of association are exactly what would be expected if processes very similar to those observed in population-level studies of selection on bird wings shape bird wings broadly. The gaping hole in population studies—its complete lack of empirical evidence of relevance beyond the studied organisms—is filled exactly by the comparative method. By a happy symmetry, the main gap of the comparative method is filled by population biology.

5.2.2 Comparative weakness, populational strength

Comparative studies are often accused of not providing "definitive evidence." A much-cited example is Leroi *et al.* (1994), who note that "the adaptive (or nonadaptive) nature of traits cannot be determined from most comparative data" and deride "false hopes about the ability of comparative studies to resolve the processes of evolution" (p. 397). Arnold (1983, p. 350) states that "comparative studies. . .do not provide direct evidence of selection or adaptation." Grandcolas and D'haese (2003) trot out the admonition that "Patterns cannot be taken as a direct indication of the occurrence of processes" (p. 484). Fuller *et al.* (2005) note that the "comparative approach can identify the possible selective agents but will not often be conclusive" (p. 392), implying the existence of a non-comparative source of information on evolutionary processes that is "conclusive." Kluge (2005) says that "phylogenetic patterns are limited in their ability to confirm evolutionary processes" (p. 656) and presents "laboratory and population studies" as somehow being the ultimate arbiters of adaptation hypotheses. Maddison and Fitzjohn (2015) say that comparative biology "cannot give definitive tests of evolutionary mechanisms" (p. 127), again implying that some true source really can provide "definitive" tests. Uyeda *et al.* (2021) complain that "observational datasets cannot definitively prove causal links between traits—correlation does not equal causation" (p. 1097). These critics are right—while comparative biology provides unparalleled generality, it provides little mechanistic detail—but that's as far as their being right goes.

These critiques are misplaced because they imply that some single source of data can provide, all by itself, a complete explanation. Instead, just as the weaknesses of population biology are addressed by the comparative method, the blind spots of the comparative method are addressed by population biology. Sections 1.11.1 and 2.3.8 provided examples of the comparative method in studies of mammalian long bone–body mass scaling. Those studies suggest that variation in long bone–body mass proportions should be developmentally possible, heritable, and have fitness consequences. It does not, however, directly examine any of these predictions. Population biology, however, can.

Within-species data from multiple sources show that various aspects of the limb-length–body-mass relationship readily respond to selection, from laboratory experiments to domesticated animals. Working in Campbell Rolian's lab in Calgary, Marta Marchini and colleagues (2014) showed that in mice, just 14 generations of artificial selection produced significantly longer tibias relative to body mass compared to control mice (Figure 5.3) (see also Sparrow *et al.*, 2017). In a move making any evo-devo adaptationist jump for joy, they then drilled down on the developmental mechanisms underlying greater limb growth in their selected lines. They found that increased length was not due to simple prolongation of the usual wild-type growth rate, which would result in longer limbs. Instead, they found that bone growth rates were actually *accelerated* relative to the wild type, leading to longer limbs in a similar developmental time frame (Marchini and Rolian, 2018). In a distinct but equally detailed series of laboratory evolution experiments in mice, Garland and Freeman (2005) showed

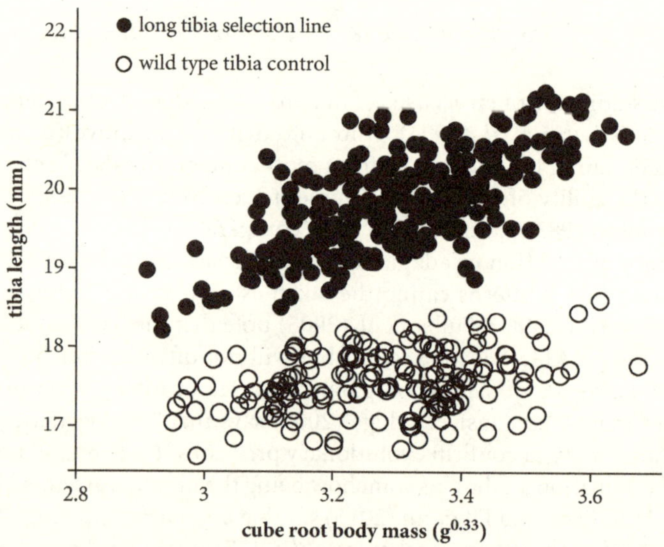

Figure 5.3 In mice, the long bone–body mass relationship responds readily to selection. Marchini *et al.* (2014) showed that in just 14 generations of selection, mice produced much longer tibias (black points) as compared to the wild type (open points), incurring far into space that appears empty based only on variation observed in normal populations. Results such as these are consistent with the interpretations that the comparative pattern in which long bone dimensions are predicted by body mass across terrestrial mammals is due to selection not favoring the "empty space" variants, not because they are not possible developmentally. Adapted from Marchini *et al.* 2014. Impacts of genetic correlation on the independent evolution of body mass and skeletal size in mammals. *BMC Evolutionary Biology* 14, licensed under CC BY 4.0, creativecommons.org/licenses/by/4.0/

that selection favoring high endurance running reduced the amount of asymmetry between hindlimbs that is the norm in mammals. Hindlimbs of similar length require less metabolically costly gait compensation, so exceptionally long periods of running would lead to less metabolic output per duration of running. Domesticated mammals also give evidence of large responses to selection within species. Domestic dogs show that long bone dimensions are highly malleable under artificial selection. Racehorses like thoroughbreds have been bred for long, slender limbs in maximizing stride length and speed, while draft horses like Clydesdales have shorter, sturdier limbs reflecting selection for pulling power. Though each of these examples gives data from within a single species, these results are exactly congruent with the expectation emanating from comparative data that heritable variation in the long bone–body mass relation not only must have existed at one time but remains present in contemporary populations and can respond readily to selection. The strength of populational data is that it directly examines intrapopulational variation, heritability, and fitness, directly complementing the strengths of the comparative method. This back-and-forth between comparative and population evidence highlights a key principle of building evolutionary explanation, which is the complementarity of different sources of evidence.

5.2.3 Essential complementarity of methods

Not just between comparative and population biology, but the blind spots in all four sources of empirical evidence are admirably complemented by the strengths of the others. Table 5.1 illustrates this complementarity between the subdisciplinary approaches for studying adaptation, the comparative method, population biology, optimality models, and exploration of developmental potential. Population biology provides great mechanistic detail regarding variation, heritability, and fitness differences. However, population biology provides practically no generality. If the goal of evolutionary biology is to explain the diversity of life, then an individual population biology study contributes little to this goal. The comparative method, by spanning tens or even hundreds of species across broad scales of size, habitat, and evolutionary divergence, offers precisely the kind of generality needed. It does, however, provide only the fuzziest view of intraspecific variation, the heritability of this variation, or the fitness consequences of this heritable interpopulational variation, exactly the detail provided by population-level studies. By the same token, neither populational methods nor the comparative method directly examines the biophysical expectations regarding which variant or variants should prevail over others within a population in a particular selective context. This gap is filled exactly by optimality models.

Relevant to our interest as evo-devo adaptationists is the complementarity between the three traditional adaptationist approaches and the study of developmental potential. The study of intrapopulational variation falls under the purview of population biology. Traditional population biology often focuses on standing variation within populations and generally does not emphasize the uncommon, the abnormal, or the teratological. Dovetailing seamlessly with this emphasis, adaptationist evo-devo study of developmental potential picks up where traditional population biology tends to leave off. This vital complementarity means that no subdisciplinary approach is superior to the others.

5.2.4 A (spurious) case can be made for the supremacy of any method

A cultural bias potentially fettering across-subdiscipline integration, such as that described above, is a tendency to put some kinds of data above others. A longstanding discussion in biology is whether "experimental" (population biology, including quantitative genetics; physiology, etc.) data are more rigorous, somehow more "causal," or more informative than comparative findings (Mayr, 1982; Cleland, 2001; Parke, 2014; Olson and Arroyo-Santos, 2015; Currie and Levy, 2019).

Experiments do indeed offer unparalleled precision in manipulating variables, no mistake there. The mistake arises when biologists or philosophers interpret this methodological strong point as a license to claim superior contributions to evolutionary explanation. Experimental and population-based approaches are valuable, of course, but their relevance rests on comparative and optimality-based methods, each of which brings explanatory resources that experimental work cannot hope to provide (see Table 5.1). One could just as easily argue for the special status of comparative methods or optimality models.

To turn the tradition of holding experimental and population biology as superior for a moment, a biologist could take the strengths of the comparative method from Table 5.1 and set them alongside the limitations of the other approaches to argue that the comparative method is, in fact, the privileged method of evolutionary inference. Its advantage begins with its unparalleled generality, the repeated observation of trait-environment associations across species. When lineages with different ancestral starting points independently evolve similar traits under similar selective conditions, the pattern of convergence provides compelling evidence that shared selective pressures shape phenotypes in similar ways. No other method provides this sweeping evidence, certainly not any single population biology study, with its navel-gazing emphasis on a handful of individuals of a single species.

But generality alone doesn't explain the comparative method's power. Unlike experiments, which rely on artificial settings, comparative studies examine naturally occurring specimens, real organisms shaped by the real processes of evolution. These are not the laboratory artifacts or theoretical constructs of population biology: they are the very outcomes evolutionary biology seeks to explain. The comparative method is the only approach that stretches the minuscule scope of a population biology study and the elegant simplifications of theory into the domain of real evolutionary history. It is the bridge between the short-term tinkering of selection in lab populations and the grand-scale patterns of life on Earth that evolutionary biology is charged with accounting for. Without the comparative method, we remain trapped in local anecdotes and elegant descriptions of evolutionarily insignificant temporal and spatial scales. With the comparative method, we gain access to the sweeping regularities of real-world evolution. Its power and scope clearly make the comparative method the privileged method for testing hypotheses regarding evolutionary processes.

Table 5.1 Vital complementarity between approaches in the study of form, plus the importance of assumptions (dashes indicate no empirical evidence).

Component of explanation	Population biology	Comparative method	Optimality models	Studies of developmental potential[a]
Variation is developmentally possible in current populations	Standing variation within populations, quantitative genetics	—	—	Studies of natural and induced variants, from typical to teratological
Variation is heritable in current populations	Quantitative genetic measurements of heritability, pedigrees, twin studies, response to selection	—	—	—
Variation in fitness in current populations	Direct studies of fitness, response to selection	—	—	Experimental manipulation to compare performance of "empty space" variants to wild type
Population-level or individual-level developmental processes plausibly produce interspecific patterns	—	Putative convergent evolution	—	Individual-level factors that bias or impede access to some parts of trait space; do species in uncommon selective contexts fall in otherwise "empty" space?
Differences in performance, fitness, are consistent with predictions from engineering logic or functional theory	—	—	Link biophysical generalities to fitness	Predict the range of morphologies that should be observed. May predict "holes" in morphospace

continued

Table 5.1 *continued*

Component of explanation	Population biology	Comparative method	Optimality models	Studies of developmental potential[a]
Is the trait of interest a "module," i.e. an organismal subset that can vary to some extent independently of the rest of the organism?	Trait (co)variation across individuals, G matrix	Trait covariation/ independence across species	Explicit focus on functionally coupled traits, tradeoffs	Direct studies of development show quasi-independence between parts
Contribution of trait to current variation in fitness ("current utility")	Fitness variation across the members of a population	Comparisons of species with different character states in comparable situations	Generate expectations regarding performance differences between variants	—
A trait evolved with its current function in its current selective context ("historical genesis")	—	Comparisons of species with the derived trait in current selective context relative to ancestral trait	—	—
Interindividual variation existed in past populations	*Accepted as part of assumptions*	*Accepted as part of assumptions*	*Accepted as part of assumptions*	*Accepted as part of assumptions*
Variation heritable in past populations	*Accepted as part of assumptions*	*Accepted as part of assumptions*	*Accepted as part of assumptions*	*Accepted as part of assumptions*
Variation in fitness in past populations	*Accepted as part of assumptions*	*Accepted as part of assumptions*	*Accepted as part of assumptions*	*Accepted as part of assumptions*
Unobserved contemporary conspecifics have similar biology	*Accepted as part of assumptions*	*Accepted as part of assumptions*	*Accepted as part of assumptions*	*Accepted as part of assumptions*

[a]The myriad approaches for exploring developmental potential exemplified in Chapter 2 and summarized in Section 5.1.4 address at heart a population biology issue, the range of intrapopulational variation that selection can act upon. They are separated here to emphasize the importance of their contribution; the study of developmental potential often examines ranges wider than those in traditional population biology.

Of course, this argument is simply to illustrate that, because they all have strengths that the others lack, a case can be made for the supremacy of *any* method.

Such arguments about the supposed primacy of one approach over others achieve nothing, change nothing, and contribute nothing. Yes, experiments and population biology offer detail; comparative methods offer scope; optimality models connect to first principles; developmental studies reveal how developmental potential is accessed or isn't accessed. If our goal as evolutionary biologists is to understand why organisms have the traits they do and why patterns in nature take the forms they do, then no single method can claim preeminence. Table 5.1 illustrates that each offers something essential, and none can substitute for the others. A real benefit of recognizing the structure of evolutionary explanation is that it allows us to build stronger evolutionary explanations by working together.

Pooling together multiple approaches helps offset the blind spots of any one method. Studies with limited empirical evidence are widely viewed as weak; if anything qualifies as a robust explanation, it is one built not only on abundant evidence but also on the mutual reinforcement of distinct sources of empirical evidence. However, this does not mean that all blind spots are compensated. This leads us toward the other main protagonist in scientific explanations, historical assumptions.

5.3 Assumptions and Just-so Stories

Adaptationist explanations are frequently dismissed as "just-so stories," a phrase borrowed from Rudyard Kipling's (1902) *Just so Stories*, a collection of playful, imaginative tales for children that offer fanciful explanations for animal traits (Smith, 2016). Each story whimsically explains how some animal got its distinctive feature, like "How the Leopard Got His Spots" or "How the Elephant Got His Trunk." In evolutionary biology, the term is used pejoratively to suggest that an explanation is speculative, untestable, and accepted more for its surface plausibility than for its empirical grounding (Lennox, 1991; Durrant and Haig, 2001). The label gained traction through influential critiques by Gould (1977, 1978, 1997, 2002), who used it to criticize supposed cases of storytelling unmoored from rigorous testing. Though the "just-so stories" label is recent, concern over lightly supported evolutionary explanations has been present for essentially the entire history of evolutionary biology (e.g. Bateson, 1894). Just-so stories are useful here because they illuminate the structural features of evolutionary explanations that scientists have long considered poorly supported—plausible-sounding narratives based on little evidence.

Let's take the classic example of the giraffe's long neck. There's a longstanding debate about its adaptive function. The most familiar version, often found in textbooks, is that it was favored by selection because it allowed giraffes to browse leaves from tall trees. Other proposals suggest it was favored by virtue of its predator-spotting advantages. Still others argue that it was favored in male-male combat, with longer necks producing greater torque in head-swinging fights over mates (Simmons and Scheepers, 1996). Whatever the scenario, the structure of the account is the same: a functional advantage is proposed, and that advantage is used to explain the trait's presence. For the explanation to work, certain elements must be in place. It assumes that, in some ancestral population, individuals varied in neck length; that this variation was heritable, so that offspring tended to resemble their parents; and that longer necks conferred a fitness advantage, whether through survival, mating success, or fecundity. Together, these elements form the backbone of the adaptationist story: variation, heritability, and fitness leading to change and maintenance across generations. The details may shift, but the narrative logic remains the same.

What makes the giraffe account a just-so story, at least as presented, is its scant empirical evidence. All we have to go on is that giraffes today do indeed browse from

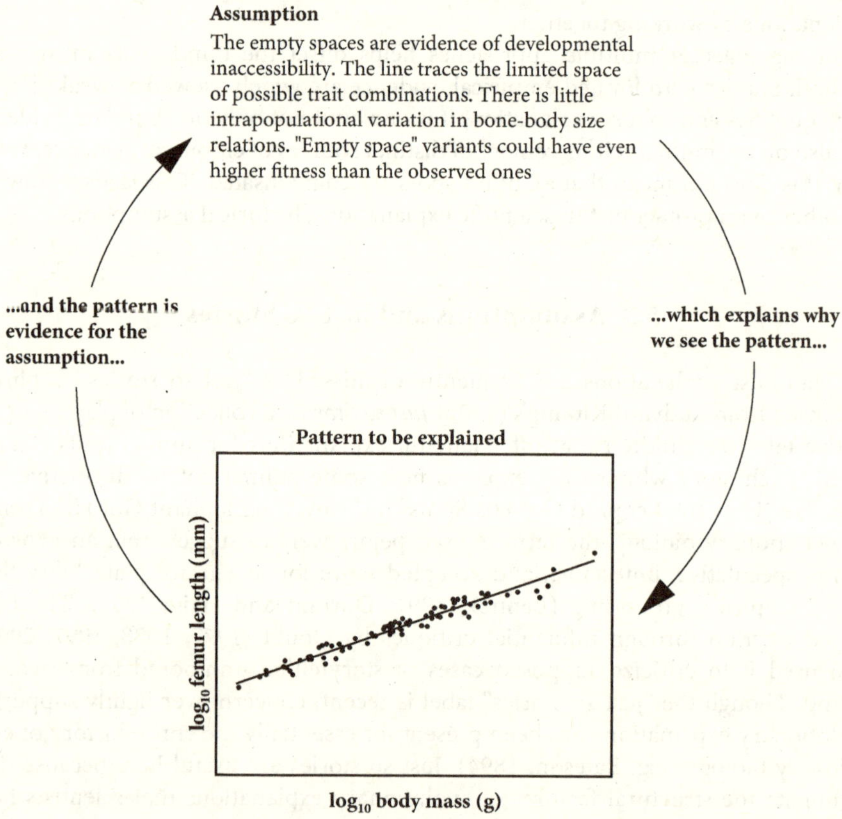

Figure 5.4 "Just-so stories" as preliminary explanations. In evolutionary biology, explanations with little empirical support are often dismissed as "just-so stories." In this example, the only evidence offered is the very pattern that needs explaining: the absence of points above and below the bone-body-mass scaling line is taken as evidence that such variants cannot occur. This kind of non-adaptationist just-so story appeals to unexplained limits in developmental possibility. The skepticism that just-so stories receive serves as a useful guidepost: biologists prefer explanations grounded in abundant empirical evidence over those with very little. Still, many elements of the scenario shown here are testable, making it a plausible starting point rather than a finished explanation.

tall trees, survey the landscape for predators from atop their lofty perches, and males do engage in head-swinging battles. As presented here, that's the only information for accepting any one of the adaptive scenarios given in the previous paragraph. In other words, the data themselves are the reason for accepting a given adaptive story, and the adaptive story explains the observed data. This mutual justification of the pattern to be explained and the explanation certainly sounds like circular reasoning.

Circular-sounding just-so stories are not just adaptationist. It's just as easy to come up with structuralist, internalist, development-centered, nonadaptive "constraint" just-so stories (Figure 5.4). To return to the remarkable pattern of the scaling of femur dimension with body size from Section 1.11.1, a circular-sounding explanation is that the empty spaces above and below the scaling line are caused by some sort of "constraint" that makes those trait combinations developmentally inaccessible. This is just an assumption, and at this point, this assumption constitutes the entire explanation for the pattern. And now comes the closing of the circle. We accept our assumption for the time being because it would explain the observed data so well if it were true: the spaces are empty because they are inaccessible and we know that they are inaccessible because they are empty.

Such circular-sounding explanations often cause anxiety among biologists. Circular reasoning occurs when the conclusion is inherent in the premises. Accepting assumptions that explain the data on the basis of the data themselves can sound awfully circular, and earn "just-so story" status (Holcomb, 1996; Smith, 2016). What we will see in a moment is that just-so stories aren't really so bad because piles of empirical evidence are available to test them.

5.4 Moving from assumptions to evidence

What distinguishes a just-so story from a well-supported explanation is not so much its content as the proportion of the story that is represented by empirical evidence and how much of the story remains as assumptions. Adaptationist and "constraint"-based stories alike can begin as just-so stories; science doesn't leave them there. The daily work of evolutionary biology involves using the diverse methods we examined at the beginning of this chapter to test those assumptions, to narrow the pool of competing explanations, and to move as large a portion as possible of the initial conjectures into the category of empirical support. This process is illustrated in Figure 5.5.

Returning again to the example of mammalian long bones, the pattern of scaling between femur dimensions and body size, introduced in Section 1.11.1, might initially seem explainable by a vague appeal to "constraint." But what once lived as a background assumption can be tested from multiple angles. Population biology studies reveal variation in bone-body size relationships within species. Comparative analyses extend this to a wide range of mammals, including the lifestyles of bats and aquatic species, which demonstrate that mammals can indeed occupy the apparently empty space above and below the terrestrial mammal scaling line. Optimality models help understand which limb proportions should perform best under different biomechanical demands. Evo-devo research, including studies of teratologies, shows that a much wider range of bone shapes and sizes is developmentally

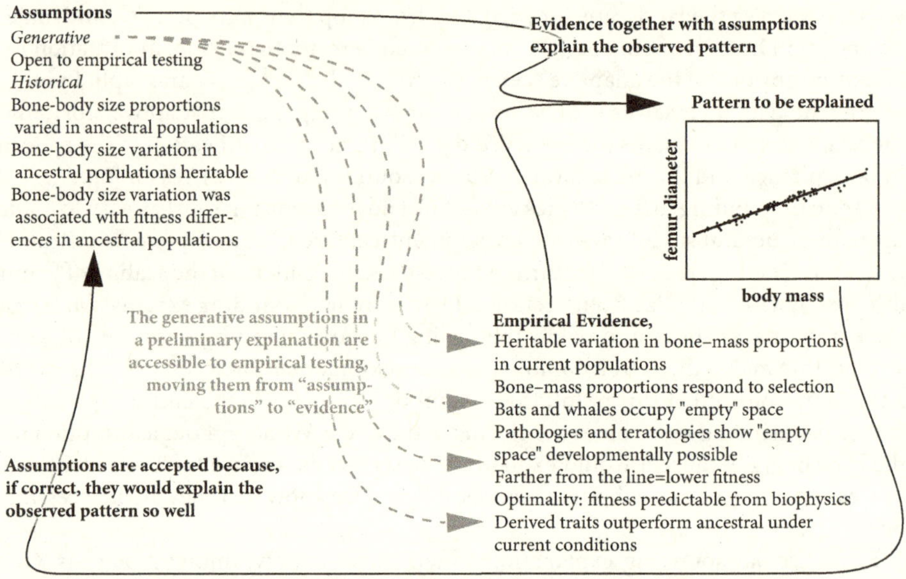

Figure 5.5 From just-so stories to well-supported explanations. Many of the aspects of a "just-so story," understood as a preliminary explanation based on little empirical evidence, are accessible to testing. Moving aspects from the "assumptions" category to the "evidence" category is the day-to-day work of scientists. However much empirical evidence is adduced, though, explanations always retain assumptions about unobserved individuals and populations. These assumptions are accepted not because they can be empirically confirmed but because, among the plausible alternatives, they provide the most coherent account of the data.

possible than typically observed. Together, these approaches transform the original just-so "constraint" claim into one rich with empirical evidence. The pool of viable explanations narrows as the evidence accumulates.

Even for cases in which much less empirical evidence is available, such as the giraffe's neck, the story is not doomed to remain a just-so tale. It may begin with assumptions about variation, heritability, and fitness consequences, but many of those assumptions are, in principle, testable. We can examine standing variation in neck length and its fitness correlations in extant populations. We can investigate the biomechanical plausibility of combat hypotheses. We can look for comparative patterns across groups that vary in neck length. While the available evidence may not identify just one out of the pool of competing explanations, the possibility for testing remains. And as evidence accrues, the pool of compelling explanations can narrow further.

It is entirely natural that the pool of potential explanations should remain large when less empirical evidence is available. Bumping up against limited sets of evidence are efforts to explain unique features, such as the long necks of giraffes or many attributes of humans. In humans, especially, obvious ethical limitations restrict experimental approaches such as breeding experiments or surgical manipulation. Traits of interest, such as language, may lack close analogs in other species, making comparative data sparse (Smith, 2016). When less evidence is available, there is less resolving power, and the pool of coexisting potential explanations will naturally remain larger. In these situations, these pools of competing explanations are a completely expected consequence of the inverse proportionality between the size of the pool and the amount of empirical evidence that can be brought to bear. This challenge simply underscores the point that the day-to-day work of scientists is moving considerations from the "assumptions" category to the "empirical evidence" one (Figure 5.5).

5.5 Untestable limits and historical assumptions

Some components of evolutionary explanations resist being moved from the "assumptions" category to the "empirical evidence" one. Philosopher of science Paul Griffiths refers to these as "historical" assumptions. They are claims about past populations that are, by their nature, inaccessible to direct observation (Griffiths, 1996b; Sterelny and Griffiths, 1999; Amundson, 2001). These assumptions concern features of ancestral populations that are simply lost to time. Unlike generative assumptions, which can be tested through present-day evidence, historical assumptions are tethered to past events for which no direct evidence remains.

Take again the case of giraffe necks. An adaptationist hypothesis might propose that longer necks were favored because they were associated with greater access to food. Embedded in that hypothesis are two distinct kinds of commitments: first, that there might exist in present-day giraffes heritable variation in neck length with associated fitness differences; and second, that this same pattern of variation, heritability, and selection existed in ancestral populations stretching back to the origin of long-neckedness. It's only by historical accident that we can test the first but not the second. The ancestral populations are gone, their members unobservable. That we cannot test these assumptions isn't due to their logical structure, but to a contingent absence of access.

Therefore, the difference between "generative" and "historical" assumptions isn't a deep difference in the content of the assumptions. It simply reflects what kinds of evidence we happen to have access to. If we could somehow look directly into the past—if we could outrun light and watch the history of life unfold, or if every individual from every past population had fossilized—then many of the assumptions we now call "historical" could be tested. As things stand, they're just parts of our explanations that remain out of reach. The contrast between historical and generative assumptions, then, is one of contingent testability. "Generative" assumptions can, under the right conditions, be subjected to empirical scrutiny, even if they originate as background commitments. "Historical," in Griffiths' sense, refers to those that cannot, even in principle, be examined using available data. Both types are assumptions, but their vulnerability to testing, and therefore the reasons for accepting them, are distinct.

5.6 Assumptions are part of even the best-supported evolutionary explanation

Historical assumptions about unobserved and unobservable past populations are not optional add-ons to evolutionary explanation but remain essential. Without them, an explanation collapses. If our explanation for the long necks of giraffes is an adaptationist one, then we must invoke a story involving heritable variation in neck length and differences in fitness across generations. Whether the selective advantage was browsing, predator detection, or in male-male combat, the logic is the same: if long necks are the product of natural selection, then past populations must have included variation on which selection could act, and some selection must have driven necks to become longer over time. The explanation fails if we strip out these assumptions about ancestral populations, because the entire claim that long necks are a product of selection depends on them. Without heritable variation in neck length, selection can't occur. It also can't occur without fitness differences associated with variation in neck length. And without a historical sequence of populations in which these populational phenomena occurred, there is no way to account for the presence of the trait from the point of view of adaptation. In short, remove historical assumptions, and what remains is not an evolutionary explanation but a description of a trait with no postulated story of how it came to be. But if those populations are lost to time and the assumptions are untestable, why do we accept them?

This is the surprising move that Griffiths highlights as key in scientific practice. Biologists justify historical assumptions by appealing to the very data those assumptions help explain. The presence of long necks in giraffes today is taken as evidence for the ancestral conditions that would, if they had existed, account for that outcome. The assumptions are accepted because they would explain the pattern so well if they were true. This kind of reasoning is not rare or exceptional. It is routine. Biologists regularly accept assumptions about ancestral variation, heritability, and fitness consequences—assumptions that no one will ever directly test—because no other explanation that anyone has been able to come up with accounts for the observed pattern as well. We will never observe the ancestral population in which this variation occurred. We will never recover the continuum of populations linking ancestral and modern giraffes. Even the richest fossil record leaves most individuals and generations, and the fitness impacts of variation, undocumented. Historical assumptions are what fill those gaps. In the just-so story of the giraffe, assumptions are doing a great deal of work.

But even the best-supported, most robust, and widely accepted explanations of organismal form still include historical assumptions that are accepted because of how well they explain the observed data if the assumptions were true (Figure 5.6). If wings are favored in flying animals as adaptations in an aerial lifestyle, then this implies many states of affairs. It implies that, at least at some point in the past, the degree of winginess varied among the members of populations. This variation was heritable and associated with differences in gliding or flying performance. The means of testing these deductions are limited only by scientific creativity and by which data are actually empirically accessible.

To test these scenarios regarding the evolution of wings, abundant evidence is available from the categories of empirical evidence mentioned above. Comparative

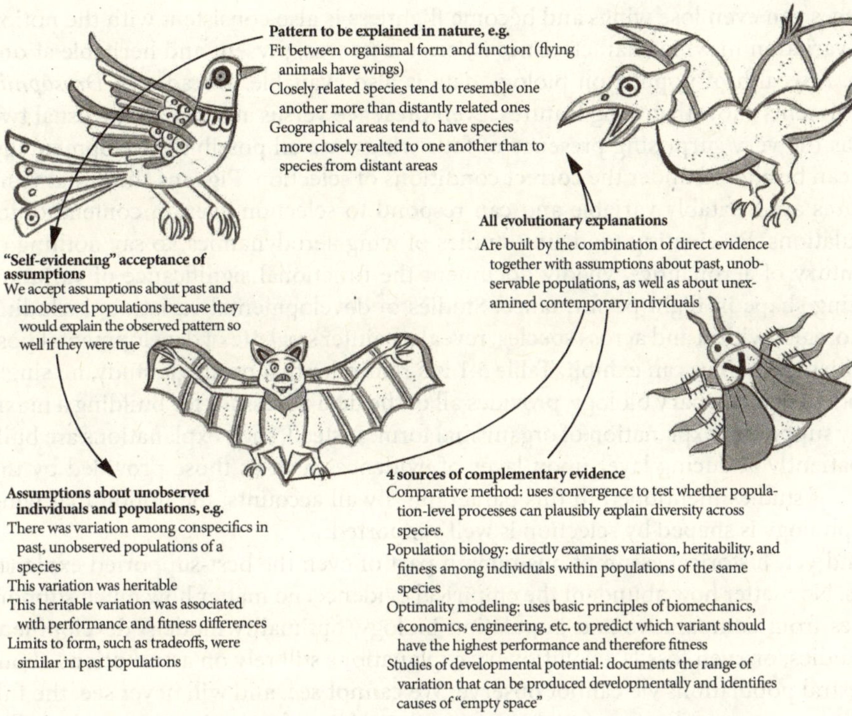

Figure 5.6 "Self-evidencing" explanation. Scientific explanations, including evolutionary ones, are "self-evidencing": they involve a back-and-forth between evidence and assumptions that are accepted to the extent that they would, if true, make sense of the evidence. In explaining the presence of wings in aerial animals as the result of selection favoring wings out of a much wider array of developmental possibilities, scientists can bring a great deal of evidence to bear from the four sources of empirical evidence: populational, comparative, optimality, and studies of developmental potential. This evidence, together with assumptions about unseen and unseeable populations, together makes up the explanation. So, the evidence and the assumptions together explain the pattern to be explained, and the assumptions are accepted given their good fit in the explanation. Philosopher Paul Griffiths calls this structure the "adaptationist abduction," abduction being a kind of inference in which what we accept as explanatory is shaped by how well it would make sense of the empirical evidence. Illustrations by Eduardo Martínez Díaz.

approaches include not just the observation that all flying animals have wings, but also that flying animals with similar lifestyles tend to have similar wing features, such as aspect ratios, even though they have evolved from ancestors with different aspect ratios. Comparative data show that wing features vary across species in ways that fit form to function, consistent with the action of selection on heritable variation. At the same time, the very wide range of wing characteristics across species is consistent with a very wide array of developmental potential in wing characteristics. The fact that lineages can even lose wings and become flightless is also consistent with the notion that variation in wing characteristics is, or at least was, present and heritable at one time. A wealth of population biology data is also available. Decades of *Drosophila* experiments show that wing features, even presence versus absence or the usual two versus the very surprising presence of four wings, are all possible developmentally, and can be altered under the correct conditions of selection. Pigeons show that wing features are heritably variable and can respond to selection even in contemporary populations. Fascinating modeling studies of wing aerodynamics, to say nothing of a century of aeronautics, vividly document the functional significance of variation in wing shape in flight performance. Studies of developmental variation, including teratologies within and across species, reveal a painter's palette of developmental possibilities that wings can exhibit. Table 5.1 is a reminder that no single study, no single school of evolutionary biology, provides all of the data necessary for building a maximally supported explanation of organismal form. Instead, such explanations are built by patiently adducing layer upon layer of evidence, such as those provided by the types of studies mentioned in this paragraph. By all accounts, the notion that wing morphology is shaped by selection is well supported.

And yet, historical assumptions remain part of even the best-supported explanations. No matter how abundant the empirical evidence, no matter how much support comes from comparative and population biology, optimality models, developmental studies, or even fossils, evolutionary explanations still rely on assumptions about ancestral populations we cannot observe. We cannot see, and will never see, the full continuum of populations stretching from the origin of proto-wings to today's flying species. Even in explaining wing evolution via natural selection, we must assume that wing traits varied in ancestral populations, that this variation was heritable, and that some forms were favored over others due to differences in survival or reproduction. We must assume that the kinds of associations between heritable variation and fitness that we study in present-day populations also obtained in the past. These assumptions are not optional—they are required. No evolutionary explanation, however well-supported, escapes the need to invoke claims about unobservable past populations.

The goal, then, is not to eliminate assumptions, but to reduce their weight relative to empirical evidence. A "just-so story" is simply a hypothesis resting heavily on assumptions that haven't yet been tested. What transforms a just-so story into a well-supported explanation is evidence, layer upon layer, drawn from multiple approaches, converging to support the story's claims. At that point, the historical assumptions that remain no longer seem controversial, but are accepted because no better explanation accounts for the data more consistently.

5.7 Key clauses and loopholes in the fine print of evolutionary explanation

Even the most thoughtful evolutionary explanations can come with subtle design quirks. These are familiar features of the explanatory landscape, often reasonable in context, but occasionally prone to getting in the way of thinking about evidence, postulated mechanisms, and assumptions. This section walks through a set of patterns that crop up from time to time and shape how evolutionary ideas are framed and, in turn, how they guide the construction of explanations for organismal form.

5.7.1 No smoking guns

Scientists often seem to think that evolutionary explanations should be built on smoking guns (Thomson, 1997). A "smoking gun" is a single piece of evidence so compelling that practically no other information is needed to reach a conclusion. The expression refers to a scenario in which the police arrive at a person lying dead from a fresh bullet wound, and another person is holding a gun that has smoke coming out of it, indicating that it has been fired recently, and they regard it as practically definitive evidence that the live person shot the dead one. In constructing evolutionary explanations, biologists often act as though a single source of information, such as comparative or population biology, or even a single study, should provide definitive evidence. With well-supported explanations consisting of complementary layers of direct evidence from comparative, populational, and optimality sources, plus adaptationist evo-devo exploration of developmental potential, and let's not forget bundles of historical assumptions, there are no smoking guns in real evolutionary explanations (Cleland, 2013; De Santis, 2021b). This realization helps align expectations about the role that any single piece of evidence can play in an evolutionary explanation.

5.7.2 The "adaptationist abduction" provides a healthier view of "circularity"

It is fairly common in the evolutionary literature to find a biologist detecting a whiff of "circularity" and deciding that the argument in question is invalid (Olson and Arroyo-Santos, 2015). For example, Waters and Craw (2006) worry about circularity in inferences of the origin of New Zealand's biota. The practice that concerns Waters and Craw is the following. Biologists notice shared organismal lineages between New Zealand and continents that are known to have formed part of Gondwana. On this basis, biologists infer that New Zealand also made up part of Gondwana. The circular formulation is thus:

> The presence of shared lineages between New Zealand and Gondwanan continents shows that New Zealand was once part of Gondwana;
> The fact that New Zealand was part of Gondwana explains why it has shared lineages with Gondwanan continents.

The "New Zealand was, once a part of Gondwana" scenario is a just-so story, in that, as presented, the only evidence supporting the potential explanation is the pattern to be explained itself. The story can be considered a preliminary explanation, one in which its assumptions are ripe for testing via the gathering of additional empirical evidence. Waters and Craw are quick to provide suggestions about empirical evidence that can be feasibly gathered to test the Gondwanan New Zealand idea. In particular, they note that geological evidence is not consistent with the expectation of continuous terrestrial habitats between New Zealand and Gondwana during the relevant geological period. They argue that it's likely that New Zealand was below water well after it separated from the other landmasses (this is controversial; Berridge, 2023), and if it was underwater, any shared biota can't be explained by it having been on New Zealand ever since it rafted away. The fact that empirical evidence can be adduced to test, and in this case likely reject, the implications of a given just-so story suggests that there is nothing wrong with this form of "circularity." As scientists adduce more and more empirical evidence, the relative importance of assumptions becomes less and less, and the reasons for accepting them greater.

This non-noxious form of "circularity" is integral to all scientific explanations. It is found in strategies of reasoning that go under names such as abduction, inference to the best explanation, and Bayesian inference (Holcomb, 1996; Ladyman, 2002; Lipton, 2008; Olson and Arroyo-Santos, 2015). This "circular" quality has been called "self-evidencing" (Lipton, 2001) because it involves accepting assumptions based on how well they would explain the data if they were true, which is precisely our case (Figure 5.6). All of these forms of reasoning involve some degree of back and forth between assumptions and the data to be explained. Philosopher Paul Griffiths has termed this explanatory structure, in which empirical evidence, the data to be explained, and historical assumptions interact, the "adaptationist abduction" (Griffiths, 1996b; Sterelny and Griffiths, 1999), referring to a type of inference known as abductive inference.

In the case of the giraffe's neck or of Gondwanan New Zealand, there is very little empirical evidence beyond the pattern to be explained. In consequence, the parts of the explanation that are simply assumed bulk large relative to empirical evidence. Authors such as Waters and Craw are entirely correct that these are poorly supported explanations. But as explained in Section 5.5, even very well-supported explanations still involve the acceptance of historical assumptions as a function of how well they would explain the data if they were true. So, simply finding these loops of reasoning that are part of any scientific explanation does not automatically disqualify an explanation. Any time additional empirical evidence can be adduced to test the implications of a given hypothesis, then the likelihood is that any loops of reasoning detected are simply part of normal self-evidencing scientific explanations that accept assumptions as a function of how well they would explain the data if they were true.

Such acceptance of assumptions occurs across science, not just in evolutionary biology (Olson and Arroyo-Santos, 2015). For example, in astronomy, inference of atmospheric composition always involves self-evidencing reasoning. When an exoplanet makes its transit across its sun, instruments are able to measure the way the atmosphere of a planet modifies the passage of radiation through it. On this basis, astronomers infer the contents of the exoplanet's atmosphere. This reasoning has a structure like, "To be consistent with the spectrographic data, the atmosphere of exoplanet Zarc-77 must contain methane and hydrogen sulfide." We know that the atmosphere contains methane and hydrogen sulfide because of the spectrographic data, and the spectrographic data are explained by the presence of methane and hydrogen sulfide in the exoplanet's atmosphere (Kreidberg *et al.*, 2014). This self-evidencing acceptance of assumptions is part and parcel of all scientific explanation.

Truly pernicious circularity is relatively rare in science. Pernicious circularity occurs when, in contrast to conventional self-evidencing explanations, a conclusion is shielded from the accumulation of further empirical evidence, very different from situations such as wings in volant animals, fins in aquatic animals, Gondwanan New Zealand, or the atmosphere of Zarc-77, all of which are very much accessible to testing via many complementary layers of empirical evidence. A potential example of pernicious circularity is when a bioinformatician proposes a method to infer evolutionary processes in the deep past. The method includes specific assumptions about the evolutionary process. The bioinformatician then simulates a dataset, say of DNA sequences across species, using the same assumptions about evolution that are built into the method. The method then correctly reconstructs the evolutionary process that produced the simulated set of DNA sequences. Such a procedure implies the following circular reasoning:

Assume that X process generates Y pattern
Simulate Y pattern using X assumptions
Infer X process from the simulated Y pattern

Our faith in the validity of the method could be excessive, with its performance appearing artificially successful, given that the same assumptions were included in the method as in the simulated dataset. However, even this potential example of pernicious circularity is accessible to testing. It is quite common for simulated datasets used in the validation of novel methods to be generated using a wide array of plausible scenarios. This array of plausible scenarios that a method can be pitted against shows that even in this situation, it is often possible to find ways to break pernicious circularity.

Another possible example of true circularity is afforded by the notion of "juvenile wood" in the study of trees. Tree biologists use features such as wood cell characteristics to delimit "juvenile" wood from "mature" wood in trees. Juvenility is diagnosed by wood cell characteristics, and the presence of those wood cell characteristics is explained by the juvenility of the wood. The circular formulation is as follows:

I know that this wood is juvenile because it has X cell characteristics.
This wood has X cell characteristics because it is juvenile.

While the notion of juvenility in wood does provide a good example of facile circular reasoning in biology, its effect in actual biological use seems quite harmless. Tree biologists don't spend their time testing the hypothesis that a given wood is juvenile or not, for example. Instead, the notion of juvenile versus adult wood is mostly used by foresters who want consistent wood characteristics for wood products. The juvenile-adult distinction helps them reach this pragmatic goal by only including "mature" wood in their products (Zobel and Sprague, 1998). The notion of "juvenile wood" does not, however, seem to play any mischievous role in the formulation of biological theory or even affect empirical studies to any great degree.

Proponents of intelligent design and other critics of evolutionary biology often like to present loops of reasoning from self-evidencing explanations as fatal flaws of evolutionary science. Perhaps the most classic example is the criticism that natural selection explains the survival of some individuals within a population by appeal to their higher fitness, and we know those individuals have the highest fitness because they are the ones that survive. The straw man "circular" formulation in these criticisms takes the form:

> These individuals survive, so they must be the fittest ones.
> I identify them as the fittest ones because they survived.

As an evo-devo adaptationist familiar with the adaptationist abduction and the real structure of evolutionary explanation, in which scientists accept historical assumptions on the basis of how well they would explain the observed data if they were true, you're in a position to refute these criticisms.

Evolutionary explanation is built not just by the pattern to be observed and by assumptions, but also by a dense network of layers of empirical evidence. Scientific explanations are not built on a single datum, the way the "survival of the fittest" criticism implies. Instead, many layers of empirical evidence are brought together, and it is their general pointing in the same direction that gives confidence in a given explanation. A similar criticism of evolutionary biology on the part of creationists has to do with the dating of fossil strata. The criticism goes that the age of fossils is inferred from the age of the strata, and the age of the strata is inferred from the presence of those same fossils. Obviously, in ascertaining the age of a fossil, here too, piles of empirical evidence, such as isotopic dating, are available to bring to bear.

The concern that circularity poses a major threat to evolutionary biology thus seems largely exaggerated. Most supposed instances of circularity merely highlight self-evidencing components within broader, well-supported explanations. Other cases are preliminary hypotheses for which ample empirical evidence has yet to be gathered, but this evidence is accessible, allowing for testing of the explanations. Even in cases of allegedly pernicious circularity, there are often additional sources of evidence or methods that can strengthen the inference. Some possible cases, in the form of "phylogenetic inertia" and "niche conservatism" are treated in Section 5.7.11; even their effect is likely relatively minor. This reframing of "circularity" shifts it from a disqualification of an evolutionary explanation to a call for seeking empirical evidence that reduces the role of assumptions, guiding biologists away from mutual disqualification and toward mutual regard and collaboration.

5.7.3 The structure of a robust explanation—descriptive, not prescriptive

There is a long tradition within evolutionary biology of referring to the mid-20th-century philosopher of science Karl Popper's views as somehow normative regarding what good science is.[2] In this context, there is a useful distinction that can be made between what we might call prescriptive guidelines for good science versus descriptive

[2] Some text adapted from Olson and Arroyo-Santos (2015). How to study adaptation (and why to do it that way). *The Quarterly Review of Biology* 90: 167–191. Reused with permission under the University of Chicago Press republication policy.

standards. Prescriptive guidelines take forms such as "good science includes X, Y, and Z characteristics because Karl Popper said so." Descriptive standards are very different, taking the form "scientists tend to regard explanations that include X, Y, and Z as representing solid science."

Popper was formulating his views about how science works in Europe between World Wars I and II. Theoretical physics, including Einstein's work, was rightly dazzling the world with its content that went so far beyond the intuitive and that made very daring predictions. For example, Einstein had predicted that gravity should bend light twice as much as Newtonian gravity allowed. This is a very specific prediction, and one that would be very easy to falsify. These sorts of specific predictions impressed Popper, who saw scientists boldly following the deductive consequences of their theories into venturesome predictions. The more specific a prediction, the easier it is to falsify, to find data that can potentially reject the prediction. No one likes being wrong, so Popper saw profound courage in scientists willing to take risky stances on the basis of their theories. The role of powerful formal (mathematized) theories in generating bold predictions was not lost on Popper, who noticed the power of the mathematical deductions involved in making the riskiest predictions. On this basis, Popper depicted good science as involving the generation of daring predictions using formal theories, and emphasized the importance of the deductive steps in the generation of these predictions.

The deductive vision of science has been controversial in the philosophy of science since the birth of the field in the early 20th century with the Vienna Circle. The Vienna Circle was formed by a group of philosophers who are regarded largely as defenders of a deductive vision of science (Uebel, 2024). A deductive inference is one in which the conclusion follows inevitably from the starting conditions and the rules of interaction between the participants. So, if we know what "one" means and we know what "plus" means, we know without any doubt that "one plus one" equals "two." The conclusion "two" provides no feedback or confirmation of any sort regarding our notions of "one" and "plus." The idea of science as deductive had its crowning moment with the deductive-nomological (DN) model developed by Carl G. Hempel and Paul Oppenheim in the late 1940s. The "deductive" part of the DN model specified that scientific explanations are deductive arguments in which the phenomenon to be explained, the *explanandum*, is the logical consequence of a set of premises, the *explanans*. The "nomological" aspect of the DN model said that the *explanans* must contain at least one law of nature, nomos being Greek for "law." Accordingly, for any valid deduction, accepting the premises of an argument implies accepting the conclusion. The DN model was criticized from the outset. Not all philosophers believed that all of science could really be fit into a deductive mold (Uebel, 2024). In addition, it is not entirely clear what a "law of nature" actually is, or how to tell one from a generalization or a model (Salmon, 1989). Another criticism was that perfectly valid deductions can result in perfectly invalid explanations, for example, the conclusion that the height of a flagpole is caused by its shadow (Bromberger, 1966). Consequently, philosophers quickly got over the notion that science has to be entirely deductive (see, for example, Scriven, 1959; Salmon, 1989; Ladyman, 2002).

Even though scientific explanations are not purely deductive, many biologists in the late 1960s firmly embraced describing their methodologies as deductive, largely attributing this stance to Popper, a tradition that continues to influence scientific practice today (though see Scriven, 1959). Some prominent examples include Ghiselin (1966) and Medawar (1967), both of whom stated again and again that biology is built on deduction. Their approach was mirrored exactly by Mayr in his influential work *The Growth of Biological Thought* (1982). Gould (1980b) even entitled a paper stating the aspiration of paleobiology to become a "nomothetic" discipline, in reference to the DN model. Ghiselin said, echoing Hempel and Oppenheim, that "Biology ... is a hypothetico-deductive, predictive, deterministic, and nomothetic science" (Ghiselin, 1984, p. xiii). These exact arguments are repeated by biologists to the present, perhaps most prominently by Ayala (2009).

Following a similar trajectory as statements about deduction, there is a long tradition of biologists saying that what they do is "Popperian" (Panchen, 1992; Holcomb, 1996; Sterelny and Griffiths, 1999; Haig and Durrant, 2002; Ladyman, 2002; Rieppel, 2003; Morange, 2009; Lancaster, 2011). Along these lines, biologists go to great lengths to show that Popper supports one or another position (e.g. Wiley, 1975; Platnick and Gaffney, 1978; Jaksić, 1981; De Queiroz and Poe, 2003; Ayala, 2009). The essence of Popper's vision was a deductive one. But because scientific explanation involves a back-and-forth between assumptions and empirical evidence (Figure 5.6), science in general and evolutionary biology in particular cannot be Popperian. It is not clear why Popper has been made a "patron saint of science" (Ruse, 1979, p. 287), but Ruse (1979, 2005) suggests that it is the simplicity of Popper's scheme, that the scheme makes scientists look daring and clever, and that it fulfills desires to see progress in science.

When scientists attempt to justify their work by showing that it conforms to Popper's vision, they are using Popper's views prescriptively, but this really isn't justified. Using Popper's views prescriptively implies that they are not merely observations about what scientists consider good science but authoritative standards that define what should be accepted as good science and rejected as bad science. But there isn't any position from which to make normative statements in science regarding what good science is. Such a position would require knowing what objective reality is and then determining which scientific procedure best approximates this reality. Obviously, no one has such a perspective. The philosophy of science, of course, informs science (two good examples are Pigliucci and Kaplan, 2006; Griffiths and Stotz, 2013), for example, by offering conceptual tools missing from science, by identifying conceptual issues that need addressing, or in the present case, identifying the structure of evolutionary explanation as self-evidencing and definitely not Popperian. Attempts to justify some scientific ideas or procedures because they are Popperian and to reject others because they are not Popperian is an appeal to the prestige of a given philosopher, not scientific reasoning (Arroyo-Santos *et al.*, 2015).

Instead, the best that scientists can hope for is to select explanations of organismal form that we consider satisfactory and then dissect the structure of these explanations. If a given structure yields an explanation that is well-regarded in the field, then that becomes a starting point for building studies of organismal form, regardless of whether they conform to a given philosopher's stipulation of what science ought to be (cf. Rieppel, 2003; see also Bird, 2010). These explanatory structures turn out to have the back-and-forth evidence-and-assumptions structure discussed above.

This structure is not offered prescriptively; no one is insisting that evolutionary explanations *must* follow it. Rather, it is presented descriptively, because these are the kinds of explanations that tend to be respected in the field, as opposed to those dismissed as mere just-so stories. That said, there appears to be no alternative structure that yields well-supported evolutionary explanations.

5.7.4 Putting the "deductive" in the hypothetico-deductive method

Having seen that the structure of scientific explanation is not deductive, it's important to flag the very real role that deduction does nevertheless play in everyday evolutionary biological inference. One of the most important scientific practices, though it sounds easier than it is, is to set out an idea and then follow its consequences without losing track of the original idea. The basic structure of this practice is "if X, then Y." For example: "My clade includes 95% of its species on one side of the basal bifurcation. Is it an adaptive radiation?" According to Guyer and Slowinski (1993), an adaptive radiation is a clade in which ≥ 90% of the species are on one side of the basal bifurcation. In "if X, then Y" terms: if an adaptive radiation meets that ≥ 90% condition, and my clade has 95% on one side, then my clade does represent an adaptive radiation.

This kind of reasoning depends on holding the premise steady long enough to explore its consequences. One helpful way to think about this is to imagine putting on a lab coat, an intellectual uniform that signals you're stepping into a particular stance entailed by the definition or hypothesis under consideration. You don't have to believe, deep down, that a given position is correct. What matters is that, for the sake of the exercise, you take its implications seriously and follow them through to their logical conclusions. In Chapter 1, we donned our adaptationist lab coats: what must be the case if this trait was shaped by natural selection? Tomorrow, you might switch to a coat that takes a different perspective. Every hypothesis invites a different set of questions, and by mentally "putting on your lab coat," you discipline your reasoning to stay within the boundaries of the chain of reasoning you are trying out. The coat can be taken off again. This flexibility is what allows scientists to explore competing explanations rigorously. Strictly speaking, what we are calling generative and historical "assumptions" are predictions, deductive consequences that must follow if the explanation being tested is correct.

This approach was particularly helpful in our analysis of the human female orgasm in Section 1.11.2. By putting on our adaptationist lab coats, we could ask what would necessarily have to be true if the trait were shaped by selection. That means checking whether the three conditions for selection are met: variation, heritability, and fitness differences (more in Section 5.1.2). In this case, the data didn't fit very well. For example, there isn't strong evidence that variation in the presence or intensity of orgasm is heritable, nor is there evidence that any such variation is associated with increased mating success or fecundity. Only by carefully setting out what must be the case, deductively, if the human female orgasm were shaped by selection, do the mismatches with empirical data become obvious.

Most scientists describe their practice as hypothetico-deductive. But if evolutionary explanations are not purely deductive, then why is it described this way? The deductive part comes precisely in a "lab coat" step. In attempting to explain a given pattern in nature, scientists generate scenarios based on available data, background knowledge of a universe of natural processes, and often some intuition. They are often motivated to generate new explanations because existing explanations failed to explain all of the patterns that they would like to explain. Therefore, they attempt to generate new explanations that account for everything that the old explanations accounted for, plus the additional observations that concern them (Barseghyan, 2015). Natural selection is the differential survivorship or reproduction of certain heritable variations within populations between generations, as the result of the interaction between the differing biophysical characteristics of the variants and their selective context. This definition of natural selection is analogous to "1 + 1" in our example of deduction in the previous section. Knowing what natural selection is means that it must be the case that is, it deductively entails that there have to be differences in survivorship or reproduction between heritable variants within populations.

The deductive step of saying "if my scenario is correct, then X, Y, and Z observations must obtain in the natural world" is at the very heart of science. The step tells a scientist what to look for in testing potential explanations. It is what sends scientists to the laboratory and to the field on the empirical quests that take up most of their time and money, with scientists often subjecting themselves to physical danger and certainly to fatigue. Given these stakes, no scientist, in their calm moments, would ever hope to be on a wild goose chase, expending their finite resources, time, and energy on avenues of research that for sure will not contribute to testing their hypothesis. Avoiding such detours requires applying the "if X is true, then Y" deduction carefully and explicitly, without losing track of which lab coat one is wearing, and without forgetting that the overall structure of the explanation is not itself deductive.

5.7.5 Ceteris paribus

Another key and sometimes underappreciated step in generating scientific predictions is the assumption of "all else being equal." This assumption is often referred to by the phrase *ceteris paribus*, which means "other things being equal." Invoking *ceteris paribus* is a key step in generating scientific predictions because it allows us to isolate factors one by one. The presence of *ceteris paribus* in science can seem so pervasive and obvious that it would not even seem necessary to highlight it as an overlooked and important resource. For example, what could be a more obvious example of *ceteris paribus* than laboratory experiments in which all conditions are kept uniform except for variation in a single experimental factor?

However, there are often many overlooked opportunities for using *ceteris paribus* to sharpen the predictions that emanate from our hypotheses. Generating expectations regarding the components of fitness is a good example. For instance, it is common for students to postulate that variants with greater survivorship should have higher reproductive output because they live longer and therefore have the opportunity for longer lifetime reproduction. Or, they will postulate that individuals with higher mating success will have higher fecundity because of the greater number of matings. All of these expectations might be true, but by mixing expectations across different components of fitness, researchers risk missing key predictions. We can ask, *given the exact same survivorship*, what variation in reproductive output might we expect given the variation at hand? We can ask, *given the same mating success*, how might the variation at hand potentially affect fecundity? *Ceteris paribus* is a powerful but sometimes overlooked tool for generating the precise expectations that flow from a given hypothesis.

5.7.6 Teleology, licensed and not

Teleological language, invoking purpose or goal-directedness, is a common source of ambiguity in how biologists explain patterns in nature. Teleology is manifest in implications suggesting that traits exist "in order" to serve some function, often phrased in terms that imply intention or agency on the part of the organism. For example, one might say that "pine cones open their scales in order to expose the ovules to pollen so that they can be fertilized" or that "coral polyps evolved to capture zooxanthellae so that the algae can photosynthesize and nourish the polyps." While such phrasing is often meant to gesture toward natural selection, it tends to obscure the actual causal claims being made. When the evolutionary mechanisms behind a trait are not clearly specified, the resulting hypotheses are less clear and the predictions weaker.

It can be tricky to manage teleology in day-to-day evolutionary biology because there are legitimate and not-so-legitimate forms that it can take. Organismal features do seem to have characteristics that are their "proper functions," as opposed to just incidental ones (Griffiths, 1993; Sterelny and Griffiths, 1999; Garson, 2016). Pumping blood is the proper function of the heart; other effects, such as making noise or having heart attacks, are just incidental effects. It also seems to make sense to talk about malfunctioning structures, such as hearts with ventricular congenital defects, as ones that can't perform their proper function. Talking about proper functions or malfunctions has clear teleological content. This teleological content can be naturalized; it can be expressed as processes in nature that do not depend on human imagination.

The proper function of a biological feature is the one favored by selection, such that it explains the presence of the feature. So, scenarios that depict variation in past populations and blood-pumping explain the presence and characteristics of hearts in present populations, whereas noise-making does not. The sense of "function" is thus naturalized in the action of natural selection. This is a legitimate and interesting expression of teleology in biology (see, for example, biologists and philosophers who explore whether or not the metaphor of organisms having "agency" is useful or not (Walsh, 2015)).

Not so helpful is anthropomorphic teleology, expressions that seem to imbue organisms, or parts of organisms, with conscious volition (Gregory, 2009; Varella, 2018). Portraying pine cones or coral polyps as acting deliberately to bring about foreseen outcomes may serve as an evocative metaphor, but it risks hiding the underlying biology. In most cases, such phrasing is just a vague shorthand for natural selection, and translating it into standard adaptationist terms makes the predictions much clearer. The prevalence of teleological language in biology, leading to unclear postulates about cause, is one of the reasons that we spend so much time on standard adaptation scenarios in Chapter 1. In the case of female pine cones, if alternative scale configurations are developmentally possible and that variation is heritable, then natural selection can favor cones that open just enough to trap pollen. Cones that open too widely may lose pollen; those that never open block fertilization altogether. These predictions are plausible and testable. Yet they're effectively buried away by the phrase that "pine cones open their scales in order to capture pollen." One straightforward way to make the content of evolutionary scenarios clearer is to be alert to this kind of anthropomorphic teleology and replace it with more explicit causal reasoning about variation, heritability, and fitness consequences.

5.7.7 "Progress" in phylogenies

A close cousin to teleology is evolutionary "progress." One expression of "progress" thinking is to read phylogenetic trees as though the across-tips axis conveyed information about relative ancestry or even about the traits that should be observed. In a phylogenetic tree, one axis does contain information. The axis from the tips to the deepest divergence (Figure 5.7A) contains unambiguous information. The divergences lower in the tree, closer to the deepest divergence, occurred earlier in time than ones closer to the tips. The perpendicular axis runs across the tips. This axis contains no information at all. An easy way to convince yourself that this axis contains no information is to see that it is possible to rotate the branches about the nodes so as to produce many different arrangements along the across-tips axis (Figure 5.7B). All of these different arrangements contain the same information.

However, certain phylogenetic patterns lend themselves seductively to thinking in terms of progress (see the excellent guide to the correct use of phylogenetic trees in Crisp and Cook, 2005). The most seductive such pattern is a small clade sister to a large clade. In the flowering plant phylogeny shown in Figure 5.7A, a single species, *Amborella trichopoda*, is sister to all the rest of the flowering plants. Biologists know that the *Amborella*-rest of the flowering plants divergence is a perfectly symmetrical bifurcation (Figure 5.7C). At this bifurcation, we have on one side the rest of the flowering plants, and on the other side of the bifurcation, *Amborella*. It's impossible to say that one side of a bifurcation is older or younger than the other. However, regarding *Amborella* as somehow ancestral to the rest of the flowering plants often proves irresistible, even in the scientific literature, as when Xi *et al.* (2014, p. 919) refer to the possibility that *Amborella* is "the first lineage of extant angiosperms" (see Crisp and Cook, 2005). In a similar vein, one author hailed the sequencing of the genome of the platypus as giving the animal "top billing at the end of evolution's tree" (Brown, 2008). The notion that a tree has an "end" reflects reading progress into the axis along the tips of a tree, when in fact, the divergence between the monotremes and the rest of the mammals, like the divergence between *Amborella* and the rest of the flowering plants, is a bifurcation. So, phylogenetic topology alone does not give any reason to think that the monotremes are ancestral to the rest of the mammals, or that *Amborella* is ancestral to the rest of the flowering plants.

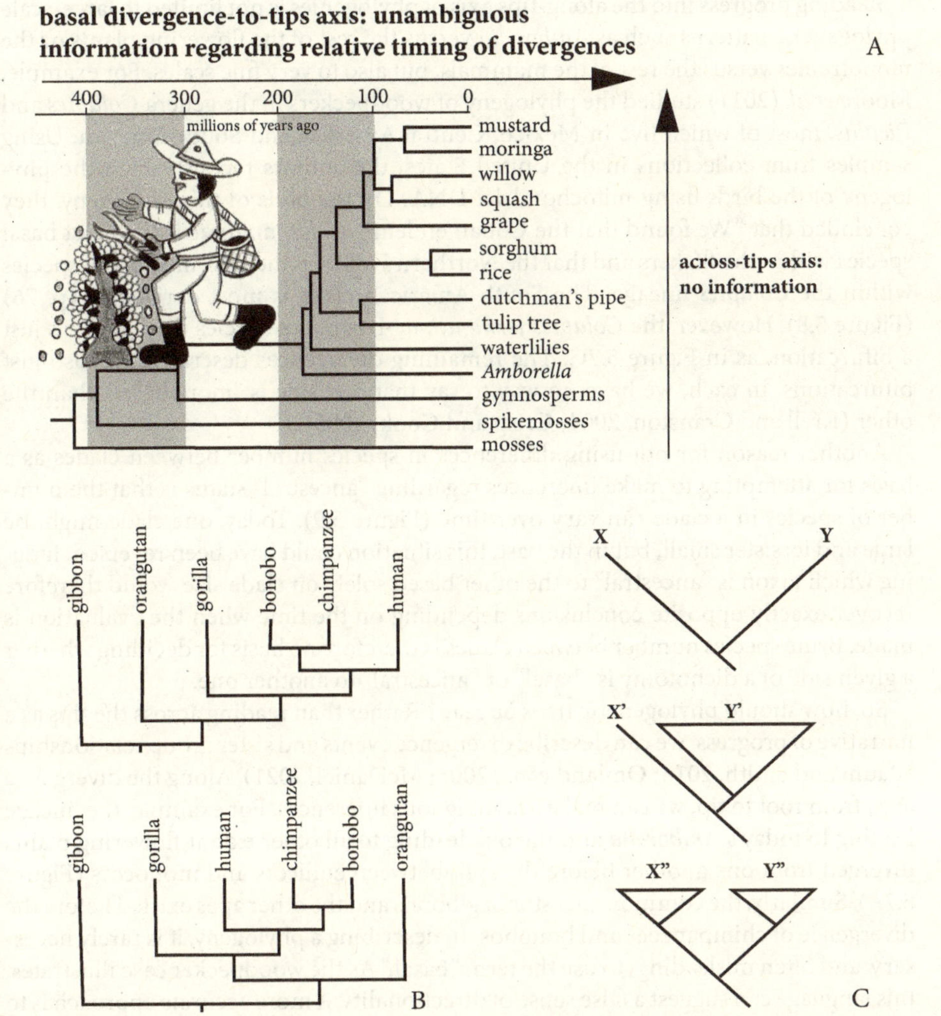

Figure 5.7 Reading phylogenies. (A) The axis along the nodes of a phylogenetic tree (in this image, the left-to-right axis) contains unambiguous information regarding the relative timing of evolutionary divergences. Divergences closer to the base of the tree happened before those that are found more toward the tips. The relative positions of divergences thus trace an unmistakable temporal axis in a tree. The other axis, across the tips of a tree (in this image the vertical axis) contains no information. (B) It contains no information because the same tree can be drawn in different ways. Both trees here are identical, but the left-to-right order of the tip taxa differs between them. (C) In each phylogenetic tree, see if you can tell which taxon is the basal one, which one should have the largest number of primitive characteristics, and which one branched off first. In the figure, a triangle means "a large clade with many species." The answer to these questions is that all of these trees provide exactly the same information because they all represent simple dichotomies. Because the divergence between each pair of taxa is a simple

Reading progress into the along-tips axis of phylogenies is not limited to large-scale phylogenetic patterns such as *Amborella* versus the rest of the flowering plants or the monotremes versus the rest of the mammals, but also to very fine scales. For example, Moore *et al.* (2011) studied the phylogeny of woodpeckers in the genera *Colaptes* and *Piculus*, most of which live in Mexico, Central America, and South America. Using samples from collections in the United States, the authors reconstructed the phylogeny of the birds using mitochondrial DNA. On the basis of this phylogeny, they concluded that "We found that the Cuban endemic *C. fernandinae* is the most basal species within the flickers and that the Northern Flicker is the next most basal species within the *Colaptes* lineage. The South American clade is most derived..." (p. 76) (Figure 5.8). However, the *Colaptes fernandinae*-rest of the species divergence is just a bifurcation, as in Figure 5.7C. The remaining divergences described are also just bifurcations. In each, we have no way to say that one side is "more basal" than the other (Krell and Cranston, 2004; Crisp and Cook, 2005).

Another reason for not using differences in species number between clades as a basis for attempting to make inferences regarding "ancestral" status is that the number of species in a clade can vary over time (Figure 5.9). Today, one clade might be large and its sister small, but in the past, this situation could have been reversed. Judging which taxon is "ancestral" to the other based solely on clade size would therefore recover exactly opposite conclusions depending on the time when the evaluation is made. Brute species number between clades is therefore no basis for deciding whether a given side of a dichotomy is "basal" or "ancestral" to another one.

So, how should phylogenetic trees be read? Rather than reading across the tips as a narrative of progress, we can describe divergence events and sister group relationships (Baum and Smith, 2014; Omland *et al.*, 2008; McDaniel, 2021). Along the divergence axis, from root to tip, we can make unambiguous inferences. For example, the lineage leading to today's *Amborella* and the one leading to all other extant flowering plants diverged from one another before the split between eudicots and monocots (Figure 5.7A). Similarly, the common ancestor of gibbons and the other apes existed before the divergence of chimpanzees and bonobos. In describing a phylogeny, it is rarely necessary, and often misleading, to use the term "basal." As the woodpecker case illustrates, this language can suggest a false sense of directionality. A more accurate approach is to describe the pattern of nested sister group relationships. In the tree shown in Figure 5.7B, the deepest divergence separates gibbons from a clade including orangutans, gorillas, humans, chimpanzees, and bonobos. Within that larger clade, orangutans form a sister group to the rest. Gorillas are then sister to the human-chimpanzee-bonobo clade, and within that, humans are sister to the chimpanzee-bonobo clade.

Figure 5.7 *(continued)* dichotomy, both diverged from one another at exactly the same time. Consequently, it is impossible to say that one side of the dichotomy is older than the other. The X'-Y' pair is drawn to resemble the situation in angiosperms, with *Amborella* sister to the rest of the flowering plants. *Amborella* is simply the sister taxon to the rest of the angiosperms, not basal to them and certainly not ancestral to the angiosperms. The (hypothetical) ancestor of both is represented by the lowermost node in the tree. The X-Y and X"-Y" pairs provide exactly the same information as the first tree regarding the divergence between the two taxa.

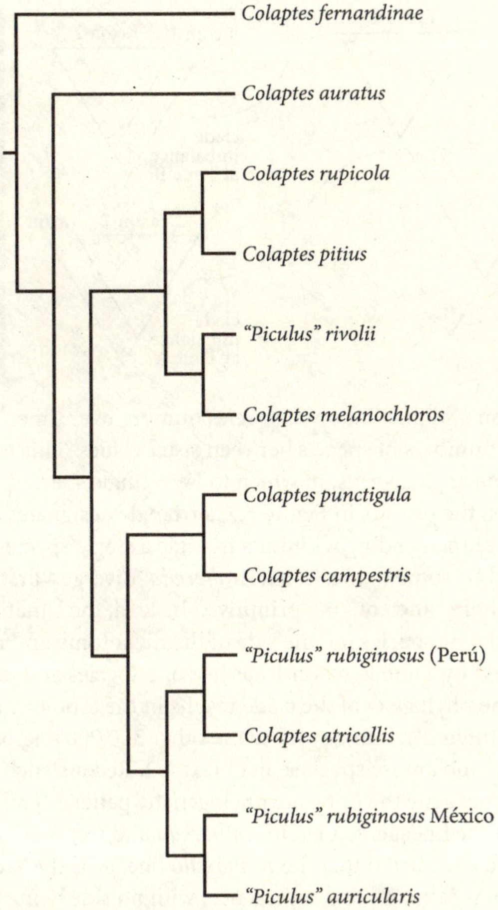

Figure 5.8 Phylogenies and the challenge of descriptive economy without overinterpretation. Moore *et al.* reconstructed the phylogeny of the *Colaptes* group of woodpeckers using mitochondrial DNA sequences. In this phylogeny, *C. fernandinae* is sister to the rest of the *Colaptes*+ "*Piculus*" clade. The *C. fernandinae*-rest of the genus divergence is a bifurcation, as in Figure 5.7C. Therefore, it is impossible to say that one side is ancestral to, basal to, or more ancient than the other.

Each of these splits represents a bifurcation, not a step in a linear progression. That phylogenetic trees can support multiple, sometimes conflicting interpretations is just one example of a situation called underdetermination.

5.7.8 Underdetermination

Another bugbear that can bedevil evolutionary debates is underdetermination (Turnbull, 2018). In the sense most useful to working scientists, underdetermination arises when multiple explanations are compatible with the same pattern or dataset. Giraffes having long necks is a pattern in need of explanation. One long-standing

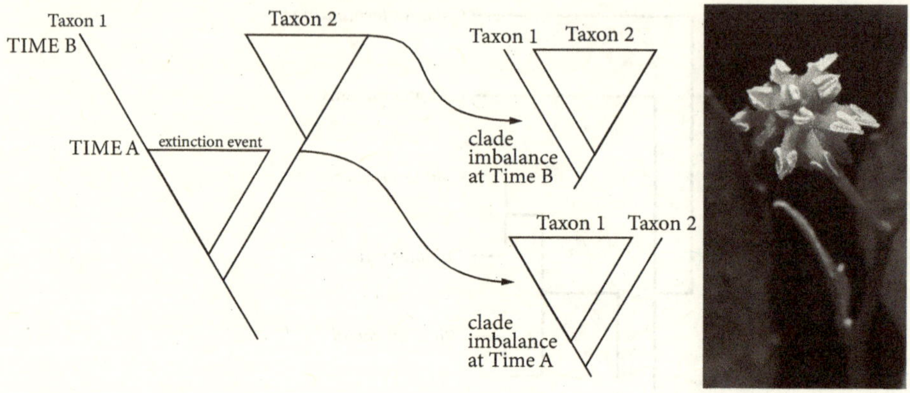

Figure 5.9 Extinction and fluctuation in species number over time. "Clade imbalance" refers to the relative numbers of species between sister clades. This example draws on the pattern seen in the angiosperms, in which today a single species, *Amborella trichopoda*, is sister to the rest. As in Figure 5.7, a triangle designates a large clade with many species, and a branch ending without a triangle a single species. This phylogenetic pattern is interpreted by some to mean that *Amborella* "diverged first" or is somehow "basal," "ancestral," more "ancient," or "primitive." Instead, the situation is simply one of clade imbalance, with few species on one side of the dichotomy and many on the other. Currently, represented by Time B, taxon 1 has just one species and taxon 2 has many. So, reconstructing the phylogeny of the clade results in the topology at upper right, as with one species of *Amborella* being sister to the other 300,000 angiosperms. But at Time A, taxon 1 was much more speciose than taxon 2. Reconstructing a phylogeny at time A would recover exactly the opposite phylogenetic pattern. This would correspond to 300,000 species of the lineage that led to *Amborella* and just one in the rest of the angiosperms, and the conclusion that the *Amborella* lineage is the "advanced" one. Instead, a bifurcation is just a mutual divergence, with no side being primitive and the other advanced. It is important to keep an eye out for this mistaken tendency to interpret small sister taxa as "early diverged," "basal," or "ancestral." At right, a male flower of *Amborella* in a New Caledonian cloud forest. Cladograms from Olson (2023). A skeptic's guide to Sherwin Carlquist's inferences of xylem function. *IAWA Journal* 1–17. Licensed under CC BY 4.0, creativecommons.org/licenses/by/4.0/

account holds that natural selection favored longer necks by virtue of their high-branch access. A second explanation proposes that selection favored longer necks in males because those with longer necks led to greater mating success after prevailing in head-swinging contests, and that shared developmental pathways limit the degree of neck-length dimorphism between the sexes. The challenge is that the available data, namely, that giraffes today have long necks, browse from tall trees, and engage in male-male combat, are equally compatible with both hypotheses. These explanations are therefore "underdetermined" by the data.

One way to think of science as a whole is as an ongoing effort to reduce underdetermination: to design studies and generate evidence that can distinguish among competing explanations and narrow the field of viable contenders. Underdetermination is a scientist's constant companion, an ordinary feature of scientific practice that fuels scientists' ongoing push to distinguish between plausible alternatives with greater precision.

The problem with underdetermination is that it can smuggle subjectivity into this process of inference to the best explanation (Andersen and Rocca, 2020). The most conspicuous and consequential cases of manipulation of underdetermination are outside of science. In any scientific explanation, there will always be some incompatible data. One reason is that there can be errors in the data. Another more important reason is that scientific explanations always involve some level of generalization and abstraction, such that they cannot cover every single case in nature. Variables that are outside a given explanation will influence the variation in the data to be explained. So, members of society who are invested in current levels of energy use can always argue that "not all the data are in yet" regarding the anthropogenic nature of climate change (Glazebrook and Goldsby, 2018). CO_2 is a known greenhouse gas, and an increase in temperature is consistent with an increase in global CO_2 concentration being anthropogenic, given liberation of carbon previously stored in fossil fuels and in forests. But scientific explanation only points to the best explanation or explanations from among the pool of potential explanations. It does not provide "conclusive proof." So, even though climate change being anthropogenic is the explanation uncontroversially with the best fit to the data, the debate is easily manipulated by insinuating that the criterion for choosing the best explanation is that it be associated with "definitive proof." If scientific explanations need to produce "definitive proof," then the difference between anthropogenic and natural climate change is indeed underdetermined. In this way, underdetermination provides a side door through which bad-faith actors can introduce deliberate distortions.

The possibility for manipulating underdetermination also appears within the scientific community itself (Dietrich and Skipper, 2007). Much of the resistance to ideas that processes like "genes-as-followers" evolution, epigenetic inheritance, or the contributions of nongenetic resources to development lead to what can be called "relative importance debates" (Lloyd and Gould, 1993; Kovaka, 2021; Kovaka and Novick, 2024). Relative importance debates center not on the correctness or not of two or more possibilities, but instead about whether one is a predominant or frequent phenomenon. These debates can be useful as part of candid explorations, but just as often, they can be vulnerable to subtly taking advantage of underdetermination.

The useful aspect of relative importance debates is that, as biologists, we really do want to know what the most important processes are, which the less important ones are, and why we observe these relative proportions. For example, genes-as-followers evolution relative to genes-as-leaders evolution: it is extremely important to know which is the primary mechanism of adaptive change. Likewise, it would be very informative to know whether heritable variation in nongenetic developmental resources really does have long-term evolutionary impact. Or, alternatively, can everything nongenetic for all practical purposes be safely backgrounded for describing the reconstruction of parental phenotypes? In these debates, biologists generally don't negate either side as being impossible or even implausible. They do, however, often suggest that there is so little evidence for one side as to make it generally unimportant.

Their arguments motivate determining these relative importances, so in this sense, these debates stimulate useful empirical research.

Another kind of maneuver in relative importance debates proves less helpful: using the scarcity of evidence for a hard-to-study process as if it were evidence that the process is less important than an easier-to-study one. Arguments about the primacy of genes-as-followers evolution often hinge on this disparity of empirical evidence. It's easy to make a strong case that genes-as-followers evolution could be widespread, based on the simple logic that plastic responses to a given environmental change should often occur faster and in more individuals than de novo mutations associated with the same magnitude of change (Section 4.11). But at the same time, a strong case can be made for the prevalence of genes-as-leaders evolution, given how often researchers observe genetic changes that track phenotypic evolution. Referring to genes-as-followers evolution, De Jong and Crozier (2003, p. 17) ask, "How plausible is all this? Not very," dismissing genes-as-followers evolution not as impossible but just not the "dominant evolutionary factor." However, most lab-based or artificial selection experiments involve applying selection in a fixed environment where all individuals face the same conditions, and researchers decide who reproduces. These setups are structurally incapable of detecting genes-as-followers evolution.

Genes-as-followers evolution requires changes in environmental conditions that provoke plastic responses, some of which end up influencing survival or reproduction. Because both genes-as-followers and genes-as-leaders evolution manifest as intergenerational phenotypic change, standard selection experiments collapse these mechanisms into the same observable outcome. As a result, based on current data, the prevalence of one over the other must be considered underdetermined; that is, the data don't actually distinguish which is the prevalent mode of adaptive evolution. Teasing apart these scenarios requires detailed developmental and genetic investigations, well beyond standard measurements of heritability or fitness. It's therefore no surprise that genes-as-followers evolution has left a lighter empirical footprint than genes-as-leaders.

But this disparity in evidence plausibly reflects research tractability, not real-world frequency. As Houle (2024) notes in a frank reflection on the realities of research design, the reason some phenomena remain understudied may have less to do with intellectual resistance than with the sheer logistical difficulty of studying them. If some phenomena being harder to study than others influences the amount of empirical data that support them, then ranking the importance of evolutionary processes by how often they're observed risks confusing ease of study for prevalence in nature. Relative importance debates are most productive not when they confuse logistical accessibility with evolutionary relevance but when they motivate empirical inquiry.

These relative importance debates illustrate that while addressing underdetermination is a routine part of scientific practice, it is problematic when ambiguity is treated as confirmation of one side over another (Kovaka, 2019), or when the absence of definitive proof is taken as grounds for dismissing an explanation entirely. In cases like the genes-as-followers hypothesis, where available data cannot yet resolve the issue, good scientific practice calls for allowing alternative narratives to coexist, awaiting data that can genuinely arbitrate among them. Situations like this, where apparently opposing explanatory models compete for attention, are common in biology. Given how frequently these binary framings arise, it's useful to pause to consider dichotomies themselves and how they function in scientific practice.

5.7.9 False dichotomies

Biology is populated with dichotomies, and it is worth keeping an eye on them to make sure that they are taken with a grain of salt. Section 4.3 examined the "genetic-plasticity" dichotomy. There, we saw that there's a common notion in the literature that variation in organismal features can either be fixed and adaptive or vary markedly as mere plasticity (Pigliucci, 2005). That section showed that plasticity itself is, in fact, an adaptation. To the extent that adaptation by natural selection involves genetic inheritance, then plasticity is of course "genetic." Moreover, biologists usually aren't really positing that absolutely all individuals in a species are for some reason completely genetically uniform in the loci related to a trait of interest, as the "plastic" end of the "genetic-plastic" dichotomy implies. Nor are they positing situations in which there is absolutely no plasticity in the trait of interest within individuals, as the "genetic" end implies. Instead, the "genetic-plastic" distinction usually reflects a concern about the relative breadths of plastic ranges within individuals as compared to the range across individuals: given a wide range across individuals, are the plastic ranges within individuals wide, such that most individuals cover most of the across-individual range, or are within-individual plastic ranges narrow, such that each covers just a small stretch of the total range? Stating this concern is much clearer than an untenable "genetic-plastic" dichotomy.

Likewise, Chapter 3 examined the genotype-phenotype dichotomy, highlighting that nucleic acids are essential participants in cellular processes but not the sole determinants of phenotypes. From that point of view, the "genotype" does not cause phenotypes on its own, but instead makes up part of developmental systems. There is therefore no alternation between generations of genotype, phenotype, genotype, phenotype, as depicted in traditional evolutionary biology. Instead, developmental systems are transmitted wholesale from generation to generation, and evolution is a change in developmental systems (Figure 3.9). It is surely possible to split living things into genomes versus everything else, but it is also possible to split them into lipids versus everything else, or water versus everything else (Griffiths and Stotz, 2013; Archer and Lavie, 2023). Simply being able to make such splits does not mean that any one categorization is more biologically meaningful than another. When biologists split living things into two categories like this, they are at least implicitly claiming that one resource has a more important role as compared to the others. As explored in Chapter 3, though, the genome is, of course, a crucial resource, but it can't be elevated above any of the other key participants in development. Development will fail just as dependably in the absence of water, oxygen, or the right temperature as it will in the absence of a genome. The genotype-phenotype dichotomy makes it seem, though, that the genome is more important than the rest of the organism, rather than a part of it. The "genes-aren't-everything" folks would argue that to the extent that the genotype-phenotype dichotomy mischaracterizes developmental processes, then scientific understanding of the causes of phenotypes will remain shaped by that distortion.

Chapter 3 also touched on the "organism-environment" dichotomy (especially in Section 3.14), showing that any "abiotic" variable can be shown to be influenced by organisms, and that "the environment" is itself defined in relation to the features of organisms. For this reason, the "biotic-abiotic variable" dichotomy has so far resisted any non-arbitrary definition. Dichotomies like biotic-abiotic variables can be useful handles in that they evoke a wide array of resources, but the distinction never adds any clarity to the description of a biological process beyond what can be said by simply enumerating the specific variables.

5.7.9.1 Dichotomies as tools of convenience

Dichotomies can be useful in casually indicating the breadth of scientific concerns—surely all biologists are interested in nucleic acids and in the outcomes of ontogeny, and in organisms and environments. But like many terms in biology, the usefulness of these dichotomies only goes so far. Many authors have explored the limitations of building evolutionary biology on the notion of a separation between genotype and phenotype (Depew and Weber, 1997; Oyama *et al.*, 2003; Jablonka and Lamb, 2005; Griffiths and Stotz, 2013), for example. Likewise, baking conclusions about the relative importance of biotic versus abiotic factors into a scientific scenario is bound to cause confusion because no one has found a way to distinguish between biotic and abiotic variables unambiguously. Including false dichotomies in places where their reality is required for a hypothesis to hold true builds on shaky ground. But dichotomies can be useful shorthands, and the conclusion at the end of the book will return to affirm the pragmatic usefulness of a central false dichotomy. But first, we turn to another major set of scientific tools that, like so many dichotomies, are useful but not literally true.

5.7.10 Managing metaphors

Metaphors are another case of excellent servants but poor masters. Like everyday language, biology is replete with metaphors and arguably even requires them (Black, 1981; Boyd, 1993; West-Eberhard, 2003; Ball, 2011; de Lorenzo, 2011; Boudry and Pigliucci, 2013; Archibald, 2014; Archetti, 2015; Falkner, 2016). Metaphors such as ecological "niches," molecular "clocks," adaptive "radiations," or "landscapes," phylogenetic "trees,"—to say nothing of natural "selection" (and debates about whether it is merely "filtering" or whether it is "creative")— all have clear benefits.

The most obvious benefit is evocatively inspiring research. A great benefit of metaphors is that they allow scientists to project features of a familiar source system onto a biological system of interest, inspiring research questions. Inspired by electromagnetic "radiation" as the source, biologists can ask what features might be observed in nature. Electromagnetic radiation proceeds at the speed of light, inspiring questions about how fast biological radiations are. Electromagnetic radiation streams from a point source, inspiring questions about whether biological radiations include all descendants from a common ancestor or not. Electromagnetic radiation emanates in all directions, inspiring research on how wide or narrow the diversity in lifestyles and morphology is in biological radiations. Metaphors unquestionably inspire research by projecting aspects of a familiar source domain onto biological systems, often bringing otherwised unasked questions into view.

A less conspicuous but just as important benefit of metaphors is that they bring scientists of diverse backgrounds together. Studies of "adaptive radiation" bring together scientists who study functional morphology, rates of molecular evolution, phylogenies, and behavior, just to name a few, all under the expansive umbrella term, in conferences, in collaborations, and in edited volumes. Clear benefits of the "adaptive radiation" metaphor are thus that it not only inspires research in new directions but also fosters dialogue and cross-pollination that would seem unlikely to occur in the absence of the metaphor.

Like dichotomies, metaphors are not so useful when scientists take them too seriously (Olson *et al.*, 2019a). One way of taking them too seriously is when scientists attempt to operationalize them via quantitative definitions. One definition of "adaptive radiation" is a situation in which 90% or more of the species in a phylogenetic bifurcation are found in one clade and 10% or less in the sister taxon (Guyer and Slowinski, 1993). Under this definition, a situation in which 89% of the species are in one clade and 11% in the other would not qualify as an adaptive radiation, whereas one in which 91% of the species are in one clade and 9% in the other would qualify. This is obviously arbitrary. The scientist could just as well have chosen 5% or 15%. The choice of 5%, 10%, or 15% threshold is not based on some salient biological discontinuity in nature.

Giving a quantitative expression to a vague metaphor in this way runs into the fallacy of reification. Reification is from the Latin "res," which means "thing," with reification referring to "making something a thing." "Testing" the "hypothesis" that a clade represents an adaptive radiation is simply holding up a situation in nature to an arbitrary threshold that some scientist invented as a reification of a vague metaphor. So, by all means, use metaphors to inspire research and to bring scientists together. But make sure to avoid treating the metaphors as though they were real phenomena, as exemplified by testing "hypotheses" that a given situation in nature coincides with an arbitrary implementation of a vague metaphor.

5.7.10.1 Metaphor diagnostics

At this point, a scientist might be doubting that some key scientific terms really could be metaphors, so it's helpful to have some metaphor diagnostics at hand. Could terms as widely used as "adaptive radiation" really be metaphors? To help address these questions, here are just a few of the diagnostics of metaphors that were devised by MIT philosopher Stephen Yablo (Yablo, 2000), precisely to help in identifying whether a term is a metaphor or not.

The first is *expressiveness*. A metaphor evokes associations in a way that a non-metaphorical term does not. As discussed above, the "radiation" in adaptive "radiation" evokes notions of rapidity, origin from a central point, and divergence in all directions. Compare this evocativeness with "nucleic acid," which invites no associations beyond its technical meaning, reflecting its non-metaphorical status.

A second diagnostic is *paraphraseability*. Paraphraseability means that when the metaphor is replaced with alternative terminology, there is an increase in precision. By "adaptive radiation," some biologists mean "suddenly elevated speciation rate," whereas others mean "statistically significantly higher disparity among the members of a clade with respect to its sister clade." Note the way that replacing the metaphor instantly intensifies the precision of expression. "Nucleic acid" can be replaced too, say with its definition ("biopolymer composed of nucleotide monomers, which are linked together by phosphodiester bonds. . ."). But replacement of non-metaphorical terms does not lead to an increase in precision because we simply replaced the term with its known meaning. Instead, the replacement of a metaphor suddenly unveils the real interpretation being given in any specific application, but which was hidden behind the vague term.

A third diagnostic, *silliness*, sounds frivolous but is very important. It highlights that for all their expressiveness, metaphors also have aspects that don't successfully project to the target system. So, while rapidity, point origin, and divergence of the "radiation" metaphor do apply evocatively to biological systems, other aspects do not. The silliness criterion highlights these non-applicable aspects with statements such as "we detected adaptive radiations with a Geiger counter" or "the Galápagos finches are a radiation because they simultaneously behave like waves and particles." These aspects that don't project to the target system are sure-fire diagnostics of a metaphor.

If a term is expressive, in evoking ideas beyond its literal definition; if replacing it results in a marked increase in precision; and if there are clearly aspects that don't apply to the target system, the term in question is very likely a metaphor. Testing these diagnostics on a couple of scientific metaphors will demonstrate the usefulness of Yablo's diagnostics.

5.7.10.2 The "adaptive landscape" metaphor

Unlike most metaphors in science, the adaptive "landscape" one has been subjected to excellent analysis (Kaplan, 2008; Dietrich and Skipper, 2012; Pigliucci, 2013; Fusco *et al.*, 2014; Serrelli, 2015), so we are on firm ground here. Let's take each of Yablo's three criteria in turn. First, the image of "landscape" is unquestionably evocative, meeting the *expressiveness* diagnostic. It evokes notions of certain allele or other trait combinations as being associated with low fitness, and others associated with high fitness. It evokes the challenge facing populations moving from one peak to another through a valley of low fitness (as alluded to by Wright himself, 1932). It evokes the notion that arriving at point B from point A requires traversing the series of intervening points, or even teleporting between points or landscapes with inaccessible "holes" in them (Gavrilets, 1997). It invites thinking about the ruggedness of landscapes, how many peaks there are, how close they are to one another, and to what degree the slope of the landscape at any given point can be followed up to the highest peak or not (Kauffman, 1996). Given these and many other examples of biological analogies that the metaphor evokes, the notion of "adaptive landscapes" unquestionably meets Yablo's expressiveness diagnostic.

"Adaptive landscape" also meets the *paraphraseability* diagnostic. The paraphraseability diagnostic is the increase in precision that obtains when the vague metaphor is replaced by what is actually meant. Saying "certain allele frequencies correspond to higher or lower fitness" is much more precise than just saying "adaptive landscape." Upon hearing "adaptive landscape," one biologist might think about how fitness varies with different combinations of allele frequencies, while another might think about how fitness varies with different phenotypic traits. Other scientists might think of the variation among different protein structures, with some structures being more functional or stable than others (Poelwijk *et al.*, 2007). Many more interpretations of "adaptive landscape" could be cited (e.g. Fusco *et al.*, 2014). Notice the way in all cases, replacing the vague "adaptive landscape" metaphor results in a dramatic increase in clarity. This increase in clarity upon paraphrasing away a term is a good sign that it is a metaphor.

Finally, "adaptive landscape" meets the *silliness* diagnostic. This diagnostic is the obverse of the expressiveness diagnostic. Expressiveness highlights that there are aspects of a metaphor that do make sense when projected from the source system onto the target system. Silliness highlights that if a term is a metaphor, then there will be aspects that don't make sense. For example, setting up camp is a normal part of traversing real landscapes, but it makes no sense to say "the population set up camp halfway up the adaptive peak, complete with tents and a campfire." Likewise, statements such as "the species surveyed the terrain before deciding on the best path to the peak," "the population decided to enjoy the scenic route around the adaptive landscape," "the species used a GPS to find the best route to the fitness peak," or "the population called for helicopter rescue when stuck in the fitness valley," all of which make sense in the source system of real landscapes, but are obviously not projectable to the target system of biological evolution. Its expressiveness, paraphraseability, and aspects that don't project clearly reveal evolutionary landscapes as metaphorical.

5.7.10.3 The "genetic blueprint" metaphor

Another metaphor that has been explored well is that of the genetic "blueprint" (Mahner and Bunge, 1997; Morse, 1998; Oyama *et al.*, 2003; Noble, 2015; Held, Jr., 2021; McKenna *et al.*, 2022; Ball, 2023). According to Yablo's *expressiveness* criterion, this metaphor evokes a precise, pre-specified design, DNA as the comprehensive schematic of the organism, with its adult form laid out in all detail in the genome. With respect to the *paraphraseability* criterion, the metaphor is easily dispensed with: we can simply say things like "DNA is a vital component of developmental systems, contributing to ontogeny and metabolism, participating as a template in the synthesis of proteins and regulatory RNAs." The blueprint metaphor shines when it comes to the *silliness* criterion, inviting vivid absurdities that expose the many areas where "blueprint" does not apply to living systems. One could imagine saying, "the chameleon looked sleek in the representation in its genetic blueprint but turned out clunky in real life," as if the genome contained a depiction of the adult form. Or, "the genome called for the heart to be 8 centimeters left of the sternum, but the embryo didn't have a ruler handy," when instead there is no fixed coordinate system in development, but structures arise from relative positions and local interactions. Or "due to a mutation in the scale bar on its DNA blueprint, the mouse was accidentally built to the size of a cat," when the genome contains no explicit global scale indicator for the overall size of the organism, organs, or limbs. Though maybe frivolous sounding, these mismatches do the very heavy lifting of highlighting the metaphorical nature of a genetic "blueprint." Given the pervasiveness and usefulness of metaphors in science, being able to recognize them and deploy them consciously of their strengths and limitations, as Erika Edwards did in Section 2.3.9, is a key skill in building evo-devo explanations of organismal form.

5.7.11 Pattern versus process and the descriptive substitution fallacy: "phylogenetic inertia" and "niche conservatism"

The distinction between pattern versus process and the admonition that pattern does not unambiguously diagnose process are so often repeated that they would seem unnecessary to emphasize. One such criticism is particularly tiresome, and this is that "comparative patterns do not provide definitive proof of adaptation" (discussed in Section 5.1.1). Even though biologists are all aware of the "pattern is not unambiguous evidence of process" principle, and even though they are entirely justified in using patterns as evidence of process as part of well-supported explanations, there are situations in which confusion regarding pattern and process nevertheless sneaks into practice within evolutionary biology. To define terms, "patterns" are the data that, as scientists, we wish to explain and the evidence we adduce, which include associations between organismal traits broadly construed, or trait-environment relationships, trait-geography relationships, etc. "Processes" are the mechanisms that participate in the narratives we construct to account for the observed patterns. Divergence from common ancestors together with limited dispersal is part of the process accounting for the pattern that species tend to be more closely related to members of the same group on the same continent, and more distantly to ones on other continents. Evolutionary biologists adduce pattern data in an effort to adjudicate between different process explanations, so it's easy to think that distinguishing pattern from process is always straightforward. And yet there are situations in which pattern and process can be confused fairly easily in current biology, with a whiff of circular reasoning thrown in too. Here are two examples, both having to do with the explanatory power of phylogeny.

It is very common for biologists to appeal to phylogeny to explain situations in which closely related species resemble one another in a given trait of interest more closely than distantly related ones. The tendency for closely related species to resemble one another is sometimes referred to with terms such as "phylogenetic effect," "phylogenetic conservatism," or the technicolor "phylogenetic inertia" (Orzack and Sober, 2001; Blomberg *et al.*, 2003). When confronted with this carnival of terms, a biologist could be forgiven for being impressed and thinking that the terms refer to some real process in nature, likely a very important one. Appeal to the explanatory power of phylogeny takes the form that "bonobo molar enamel thickness resembles that of chimpanzees more than that of humans because bonobos and chimpanzees diverged more recently from each other than either did from humans." Whether the trait in question is enamel thickness, the number of lumbar vertebrae, or the presence of stipules, the idea is that shared ancestry not merely predicts but explains similarity.

What is crucial to notice here is that this kind of explanation moves in a circle: it explains a pattern by appealing to another pattern. Start with the first pattern: bonobos resemble chimpanzees more than they resemble humans, at least in the particular traits a scientist is studying. Now add the second pattern: bonobos and chimpanzees are more closely related to each other than either is to humans. The resemblance and the closeness on the tree are both patterns. So when someone says that bonobos resemble chimpanzees because they are more closely related, they are not pointing to a process that produced the resemblance; they are just noting another regularity that happens to align with it. Coincident with these patterns of similarity are also traits that *don't* resemble one another more closely between related species.

Evolutionary biology seeks to explain patterns by identifying processes, not by pointing to yet more patterns. The fundamental question is: why do some traits show high resemblance among close relatives while others diverge? That's a pattern in need of explanation, one that must ultimately come from causal evolutionary processes, not just from the structure of a phylogenetic tree. Mahner and Bunge (1997, p. 365) say "all phylogenetic constraints are really developmental constraints." What they mean by this is that, in reality, it's not the phylogeny per se that's limiting change. Instead, when some feature remains conspicuously constant despite changes in selective context, there is likely some aspect of development that is resistant to change. So, "phylogenetic inertia" is not so much an explanation, but an invitation for adaptationist evo-devo exploration of the causes of conspicuous lack of variation.

A related example of attempts to explain patterns by appeal to pattern is "niche conservatism" (Crisp & Cook, 2012). According to Wiens and Graham (2005, p. 519), conservatism is "the tendency of species to retain ancestral ecological characteristics." There is some debate over how to define niche conservatism. The definition of Wiens and Graham can apply to a single species, with ecological preferences remaining constant between individuals across generations. It can also apply to species within a clade. Under one notion of the term, niche conservatism would be said to be present if closely related species resemble one another in ecological preferences more than distantly related ones do. This definition is essentially the "phylogenetic effect" of the previous paragraphs applied to ecological preferences. Whatever the definition, biologists appeal to niche conservatism to explain everything from why some species are microendemics to the latitudinal diversity gradient. If a species formerly occupied a wider range, and shifting climate contracted its range, then a lack of change in climate preference would mean that the species would become restricted to the tiny remaining patch of that climate. If the Earth had a wider latitudinal distribution of tropical habitats in the past, and these tropical habitat preferences do not change over time (i.e. the "tropical niche is conserved"), then this explains why diversity today is restricted to the relatively narrow tropical band in today's climate (Brodie & Mannion, 2023).

But niche conservatism, when defined simply as the tendency for ecological preferences not to change over time, is itself just a pattern (Losos, 2008; Crisp and Cook, 2012). The problem arises when this pattern is treated as if it were a cause. Biologists sometimes give the name "niche conservatism" to the observation that related species tend to share similar ecological preferences, and then use that name to explain why the preferences are similar. But that doesn't actually identify any process or mechanism. It simply restates the pattern in different words. Saying that species share similar niches because of niche conservatism is no more informative than saying that preferences stay the same because they haven't changed.

Explaining a pattern by reference to another or the same pattern, rather than by identifying a process, can be thought of as the "descriptive substitution fallacy." The term emphasizes that what is offered sounds like an explanation, but is in fact merely a rephrasing or substitution of the original description, not a true causal or mechanistic account. For example, saying "species A and B share similar traits due to niche conservatism" might sound insightful, but if "niche conservatism" merely restates the observed similarity without unpacking the underlying processes, such as similar results under similar conditions of selection, then it has simply swapped one description for another without advancing understanding. The descriptive substitution fallacy lives up to its name: it substitutes a new description for an explanation, leaving the original question unanswered.

Terms such as "niche conservatism," which are suggestive of mysterious processes, are so confusing and seductive that they seem either best avoided or at least used mindfully. The history of life is so complicated that using terminology that makes us evolutionary biologists vulnerable to confusion seems self-defeating. Reframing niche conservatism as "the tendency, of unknown cause, for species not to change in ecological preference" would help strip the term of its misleading explanatory allure. In their review of niche conservatism, Crisp and Cook (2012) reveal themselves once again as the forward-looking and generous scientists that they are (I have already cited their guide to reading phylogenies in Crisp and Cook, 2005). With admirable optimism, Crisp and Cook (2012) conclude that "niche conservatism" is a useful term because it draws attention to situations of similarity that require explanation (Wiens *et al.*, 2010 can be construed as making a similar argument). Another tricky term is "macroevolution"; because Section 1.12.1 focuses on where *not* to use it, it is important to mention more legitimate uses.

5.8 Real macroevolution

In Section 1.12.1, we saw that most uses of "macroevolution" are just confusing references to the comparative method for the study of natural selection (Olson, 2021). There are legitimate uses for the term macroevolution, and it is important to flag them here for two reasons. One is that they are almost certainly important for explaining certain patterns of organismal form. The other is that this flagging underscores just how unhelpful the macroevolution-to-comparative-method equivalence is. The criticism in Section 1.12.1 was that most uses of the term "macroevolution" in the evolutionary literature simply refer to evolution. No entity or process distinct from the usual population-level ones, or distinct from individual-level ontogenetic processes, is at work. Adding the "macro" prefix makes it sound, though, that some additional process is at work when there is not. There are, however, situations in which additional entities and processes could in fact be at work.

The most important of these for our purposes go by the names of macroevolution *sensu stricto* (or "species selection") and effect macroevolution (Jablonski, 2007). They involve both entities as well as processes that are distinct from the usual population-level ones (Fábregas-Tejeda and Vergara-Silva, 2018). Macroevolution in this sense is part of the "hierarchical expansion of Darwinian theory," which refers to asking whether natural selection can be observed in entities above and below the species level. It starts with the question "what if we regard species as individuals?" This question was posed in 1974 by biologist Michael Ghiselin (1974). There are some analogies to be found between species and traditional individuals. Like traditional individuals, species are made up of smaller component parts. Where traditional individuals are made up of cells and organs, species are made up of individual bodies and populations. Just as an individual is born, lives its life, and dies, a species has its origination in time, persists, and goes extinct. Just as a single individual is of finite size, species also have finite ranges. This analogy makes it easy to engage with the inquiries of biologists such as Steven Stanley, Stephen Jay Gould, and Niles Eldredge, who asked what processes above the traditional population level might occur if species are a kind of individual (Stanley, 1975; Gould, 2002).

If species can be thought of as individuals, then perhaps something like natural selection can occur at the interspecific level. To explore this possibility, we need to ask how each component of traditional, population-level selection might look when species constitute evolutionary individuals. In traditional natural selection, there is variation between individuals within a population. Analogously, at the macroevolutionary level, species within a clade vary in their traits. The second condition for selection, heritability, also has a clear counterpart. Macroevolutionary "heritability" refers to the tendency of species to give rise to descendant species that resemble them in the traits of interest. So, the first two conditions necessary for natural selection, variation and heritability, can be extrapolated to the species level without much strain. Fitness extrapolates only partially (Gould and Lloyd, 1999; Gould, 2002; Hautmann, 2020; Simpson *et al.*, 2024). The first component of fitness, survivorship, corresponds at the species level to how long a species persists in geological time. Some traits may increase or decrease the likelihood of persistence or extinction. The second component of fitness, mating success, has no macroevolutionary equivalent that anyone has been able to come up with. Finally, fecundity, the number of offspring produced, translates to a species' propensity to speciate. Some traits may increase or decrease the likelihood that a species will give rise to new species. This exercise shows that the three classical conditions for natural selection—variation, heritability, and differential fitness, can be meaningfully extended to the level of species. Taken together, these analogous properties mean that it is plausible to think that something like selection could occur at the between-species level.

The effects of traditional natural selection at the population level and macroevolution in this sense are very different. In traditional population-level natural selection, certain variants, in the form of traditional individuals within populations, have higher survivorship or reproduction, and therefore that variant comes to predominate within a population. This process leads to a fit between organismal characteristics and the selective context. Instead, at the macroevolutionary level, some species tend to go extinct sooner than others, persist longer than others, or speciate more or less than others, due to their characteristics. Through this process, species with certain characteristics become overrepresented relative to species with other characteristics. While traditional natural selection explains the fit between form and function within individual bodies and their selective contexts, macroevolution is one process that can lead to there being more or fewer species with certain characteristics.

An important distinction in macroevolutionary selection has to do with the level at which differences in macroevolutionary fitness "emerge," that is, are observed. In macroevolution *sensu stricto*, "fitness" differences between species are due to attributes that are observed or "emerge" at the species level. In effect macroevolution, the attributes that lead to differential tendencies between species to survive, go extinct, or speciate are observed at the traditional individual level (Jablonski, 2007; though see Grantham, 1995). Characteristics that are observed or "emerge" at one level of organization as a result of interactions between components at lower levels of organization are very common in biology. For example, the maintenance of body temperature in endotherms involves coordinated actions and feedback mechanisms that span multiple organ systems, such as the nervous system, endocrine system, the skin (including fur or feathers), and the cardiovascular system. Examining individual cells or organs in isolation cannot capture the interactions involved in maintaining stable body temperatures. So, thermoregulation in endotherms is an example of a feature observed at the whole individual level, not at lower levels. Properties that are observed or "emerge" at one level as a result of interactions between lower-level components are often called "emergent" properties. Emergent properties are in contrast to "aggregative properties," which are simply the sum of component parts. For example, it does not matter how you stack the rocks in a box; the total weight of the box of rocks will remain the same, so the weight of the rocks is an aggregative property (Wimsatt, 1997).

In macroevolution *sensu stricto*, the feature that causes differences in macroevolutionary fitness between species emerges at the species level. The example usually cited is species range size (Grantham, 2007). The geographical range of a species depends on the ways that individual organisms within the species are distributed across space relative to one another. This distribution has to do with individual body size, metabolic rates, and resource needs. It depends on the interactions among individuals that determine individual home ranges and, collectively, the total species range. Unlike the weight of a box of rocks, species range size depends on how individuals are deployed across the landscape relative to one another. Therefore, species range size is not simply the aggregate consequence of multiple individuals. It certainly can't be measured at the individual organism level. Instead, it can only be measured at the whole-species level. Species range size is an emergent property at the species level that tends to have profound consequences for species persistence or extinction. Species with large range sizes tend to have higher macroevolutionary "survivorship" than species with small range sizes, which tend to go extinct more quickly than species with wide ranges (Newsome *et al.*, 2020).

In the case of effect macroevolution, the feature that causes differences in macroevolutionary fitness between species emerges at the traditional individual level. A good example involves the nectar spurs of columbine flowers (*Aquilegia*)

(Hodges, 1995, 1997). These small wildflowers of the north temperate zone have, extending from the backs of their flowers, conical projections known as nectar spurs. Apparently, the insects that pollinate columbines are so picky that small changes in the lengths and curvatures of nectar spurs are enough to deter a species' usual pollinator while attracting others. So, a variant arising within a columbine population that has sufficiently different nectar spurs from the rest of its species stands a good chance of remaining reproductively isolated. Reproductive isolation, the scenario goes, leads to speciation. Therefore, the presence of nectar spurs in columbines confers a greater propensity to speciate than closely related lineages that lack spurs. Nectar spurs are observed at the level of individual flowers, not entire species.

Thus, macroevolutionary "fitness" can vary between species as a result of attributes that emerge either at the species level, as in macroevolution *sensu stricto*, or at the individual organism level, as in effect macroevolution (see Vinogradov (2004) for another empirical example). There is some debate regarding whether these two phenomena are distinct from one another. The debate hinges on the degree to which features apparently emergent at the species level really just reduce to individual-level properties.

What is clear, though, is that macroevolutionary processes in this sense cannot simply be reduced to standard microevolutionary, population-level processes (Grantham, 2007; Jablonski, 2007). "Macroevolution" is often used to grandly denote traditional population-level processes (microevolution) extrapolated to long time scales. If the processes being invoked are simply the usual population-level ones, then calling these patterns "macroevolutionary" does not add a level of helpful clarity because where a "short" time scale ends and a "long" one begins is up to the taste of each biologist. However, if macroevolution *sensu stricto* or effect macroevolution have been at work, these do involve entities and processes that clearly are distinct from traditional microevolutionary ones (Serrelli and Gontier, 2015; Fábregas-Tejeda and Vergara-Silva, 2018). These macroevolutionary processes certainly require consideration for research questions such as "why are there so many species with characteristic X compared to Y?" or "why are there so many more species in this clade than in its sister clade?"

Saying that macroevolution is qualitatively distinct from traditional population-level selection is not to say that it is incompatible with it. Something very much like the logic of macroevolution was present in Sewall Wright's shifting balance theory (Wright, 1931, 1932; Templeton, 2021). In the third phase of his model, "interdemic selection," some demes (subpopulations) happen to arrive at a favorable combination of alleles. These demes, composed of high-fitness individuals, contribute disproportionately to the migrant pool, thereby spreading their allele combinations to other demes. The result is a species-wide shift toward that peak. What's key here is that the trait combinations that spread are still individual-level traits, as in effect macroevolution, and these traits can be thought of as conferring a deme-level "fitness" advantage to the deme that bears them. In effect macroevolution, differential persistence or proliferation of species occurs because of differences in individual-level traits. In Wright's model, the entities are subpopulations rather than species, but it illustrates that thinking about selection acting between entities above the traditional individual level is not a total departure from traditional thinking.

While true macroevolution, understood as differential proliferation or geological persistence of species with certain characteristics over others within clades, may not help explain traits like "why do flying animals have wings?," it may contribute to explaining why features such as organismal modularity are so widespread (a process briefly alluded to in Section 3.19.2). Modularity, understood as the organization of organisms into quasi-independent parts, may influence the long-term persistence and diversification of lineages. Developmental architectures that allow different parts of organisms to vary semi-independently can result in a broader range of viable phenotypic outcomes. Over time, lineages with such architectures may be more likely to persist through environmental changes or to give rise to descendant species with divergent forms that are well suited to their selective contexts. In consequence, modular construction may become more common across clades not because it is directly selected within populations, but because lineages with this feature tend to be overrepresented among surviving or proliferating species. In this sense, the prevalence of modularity may reflect a macroevolutionary pattern shaped by differential lineage survival.

This macroevolutionary framing also gives us a way to revisit the possibility, introduced in Section 3.19.2, that selection can shape not only traits but the very structure of developmental systems. There, we saw that selection may favor systems at what Kauffman described as the "edge of chaos," where the number and arrangement of attractors lead to both stability and evolvability. From a macroevolutionary perspective, lineages with such systems would be more likely to persist over geological timescales because they reliably reproduce viable phenotypes across generations while still allowing for the developmental flexibility that permits adaptation to changing conditions. Lineages with overly rigid developmental systems may fail to adapt and go extinct, while those with excessively broad or shallow attractors may lack the required heritability. By contrast, systems that balance order and variability are more likely to survive and diversify. These network architectures may become overrepresented across clades through differential lineage persistence and proliferation. So, while true macroevolution doesn't offer an explanation for the detailed fit between organismal form and function, it does potentially contribute to explaining why some organismal features are widespread across clades. The topic of how biologists appeal to different processes in explaining different patterns invites a look at what biologists actually mean when they say that a trait is "explained by" selection or "explained by" constraint.

5.9 "Explained by selection" or "explained by developmental constraint"

Biologists often use language that suggests that organismal form might be "explained by selection" or "explained by constraint" (Schwenk and Wagner, 2004; Shanahan, 2008). Some of the case studies examined in Chapter 2 provide examples of what biologists mean by these expressions (see also Dwyer, 1984). For example, the scaling of long bone dimensions across terrestrial mammals is a pattern that biologists regard as essentially satisfactorily explained by selection. On the other hand, patterns such as geophilomorph centipede segment number, mammalian cervical vertebra number, or *Drosophila* sex comb tooth number are patterns that biologists, many of them at least, would explain by appeal to some sort of "constraint" that does not straightforwardly involve selection. Given that so many processes and phenomena go into explaining any given case of organismal form, it is important here to unpack what biologists mean when they say that a pattern is "explained by adaptation" or "explained by constraint." Before we do this, though, it's necessary to see what the problem with these formulations could be.

All explanations of biological form include natural selection as well as "constraint," however construed (Beldade and Brakefield, 2003; Schwenk and Wagner, 2004; Shanahan, 2008; Minelli, 2019). The studies of Malagón *et al.* on *Drosophila* sex combs, Galis *et al.* on mammalian cervical vertebra number, and Badyaev and colleagues on avian carotenoid coloration in Sections 2.3.13–15 illustrate this inextricability. Minimal cell size provides another example. The smallest naturally occurring cells are mycoplasma, just 3 μm in diameter (Moger-Reischer *et al.*, 2023). Far below this size, single-bilayer lipid vesicles of just 100 nanometers are known (Mueller *et al.*, 1983). Even though compartments of this size are possible, living cells 100 nanometers in diameter are not favored by selection. Given variation within a population, variants that contain all of the organelles, metabolites, cofactors, enzymes, nucleic acids, and other cellular contents crucial for metabolism would be favored over variants that lack them. The minimal set of cell contents of even the smallest mycoplasma would not fit within a "cell" 100 nanometers in diameter. Any organism with cells this size would have very low fitness, so selection almost certainly imposes a lower bound on cell size far above the biophysically possible minimum lipid bilayer compartment.

Analogous scenarios must apply to all situations of biological form: there is a range of the biophysically and developmentally possible, and within this range, however narrow or wide, lies the range of the favored in a given context. If the limits of the biophysically possible are designated as some sort of "constraint," then for this reason alone, all biological form is bound by both selection and "constraint." Resource allocation tradeoffs, selection acting in multiple directions in multifunctional structures, pleiotropy, and developmental bias favored by selection are involved in any given pattern. If these are considered "constraints," then this is all the more reason to regard every pattern as shaped by both selection and "constraint." As a result, at first glance, declaring a given pattern to be "explained by selection" or "explained by constraint" would seem nonsensical. But these expressions are shorthands that do reflect legitimate concerns of working biologists.

5.9.1 "Explained by selection"

What biologists mean when they say that a pattern is satisfactorily "explained by selection" is as follows. This biologist shorthand denotes situations in which *there is enough developmentally possible variation such that functionally significantly different configurations could be favored*, with this variation adjacent to the wild type and without conspicuous holes (see Orzack and Sober, 1994). Let's revisit some case studies to illustrate this position. In the case of mammalian long bone scaling, terrestrial mammals exhibit a tight fit between long bone length and body mass across species (Section 2.3.8). A wealth of evidence, including intraspecific (think the pigeons in Section 2.3.5 or among dog breeds or the mouse selection experiment in Figure 5.3) and pathological variation, as well as long bone proportions in bats and aquatic mammals, suggests ample developmental potential exists above and below the observed scaling line of terrestrial mammals. If this is true, selection could favor higher or lower intercepts, or steeper or shallower slopes than those empirically observed (Figure 5.10A, B). Different slopes or intercepts would lead to distinctly different functional outcomes. The long bone proportions in bats, terrestrial mammals, and aquatic mammals are markedly different in function. Developmental limits likely lie far beyond the range delimited by realized bone dimensions in living and extinct mammals. Therefore, in terrestrial mammal bone scaling, what determines the placement of slopes and intercepts does not appear to be any sort of developmental limitation. Instead, there is strong reason to suspect that slope and intercept positions are finely tuned by selection from among a much broader spectrum of developmental possibilities.

Another example "explained by selection" is Drake and Klingenberg's study of canine skull proportions (Section 2.3.3). They found that, starting from the wolf, taller and flatter skull heights have been selected in domestic dog breeds. This also applies to snout length, with domestic dog breeds exhibiting snouts much shorter and longer than the ancestral wolf configuration. Therefore, when biologists ask why wolves have the skull heights and snout lengths they do, the answer is not because alternative developmental options adjacent to the wild type are unavailable. Instead, there is evidently selectable variation adjacent to wolf skull height and length, leading to significant functional divergence from the ancestral wolf form. Hence, the reason why wolves have their current proportions, instead of the short snouts of a pug or the long snouts of a borzoi, is likely that wild conditions do not favor those manifestly possible configurations. Neither the example of long bone scaling nor that of dog skull proportions negates the existence of ultimate developmental limits beyond the observed range of developmental potential. Rather, in situations "explained by selection," these ultimate limits, while always present, lie far outside the realized range, and developmental potential is ample in the immediate neighborhood of the wild type.

In practice, situations "explained by selection" are, at least in shorthand terminology, well accounted for by selection within fields of functionally significant developmental possibility that are sufficiently wide that they don't obviously impinge on standing, favored variation. As Orzack and Sober (1994, p. 363) put it, selection "provides a sufficient explanation because taking other factors into account could not significantly enhance the predictive accuracy of the [explanation]." Studies such as those of Malagón *et al.* (2014), Galis *et al.* (2006), or Morrison and Badyaev (2016), illustrate situations in which developmental possibility is so structured that inaccessible configurations are directly adjacent to the commonly observed morphologies, leading to the other end of the continuum.

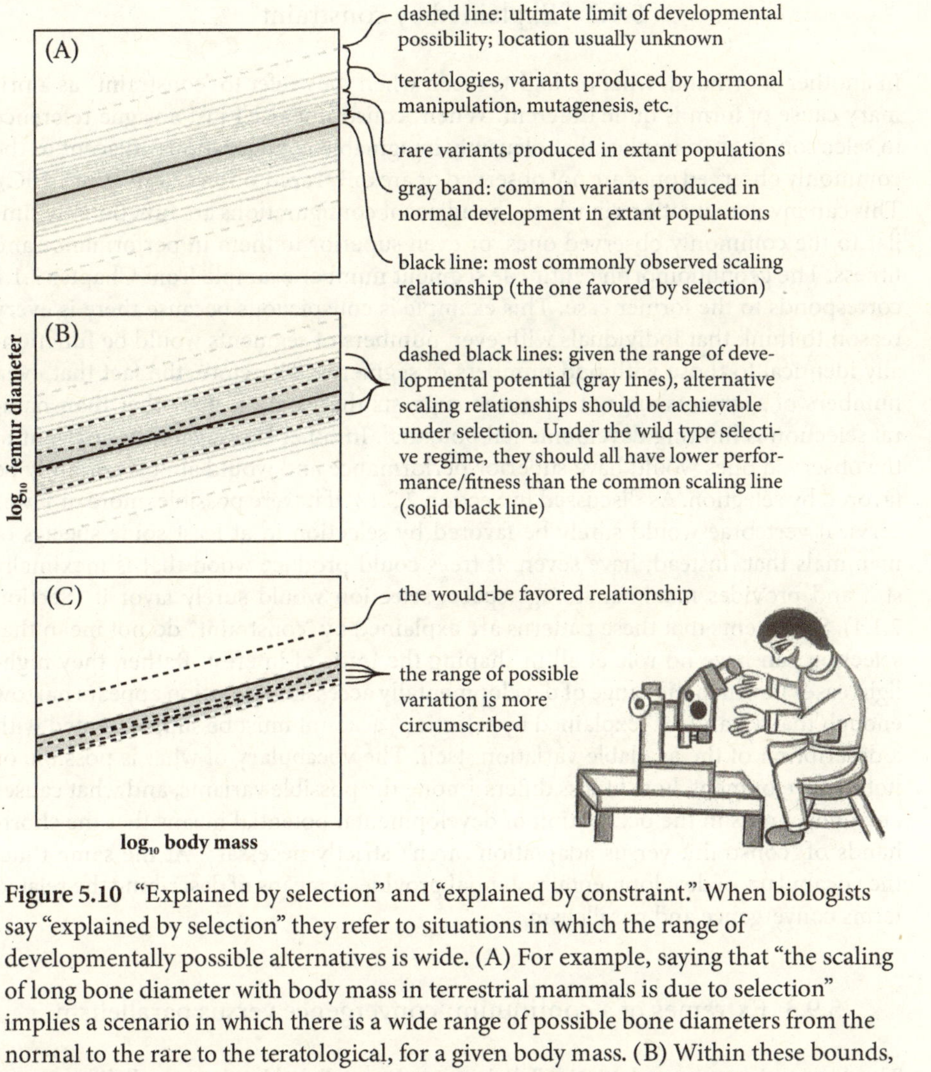

Figure 5.10 "Explained by selection" and "explained by constraint." When biologists say "explained by selection" they refer to situations in which the range of developmentally possible alternatives is wide. (A) For example, saying that "the scaling of long bone diameter with body mass in terrestrial mammals is due to selection" implies a scenario in which there is a wide range of possible bone diameters from the normal to the rare to the teratological, for a given body mass. (B) Within these bounds, there should be sufficient developmental variation such that different slopes and intercepts could be produced and potentially favored by selection. "Constraints," as in some sort of ultimate ontogenetic limits to bone dimensions, certainly exist, but the range of variation around the observed scaling line is so wide that the location of these ultimate limits is essentially irrelevant for why a given slope and intercept are observed and not others. So, while these and other sorts of "constraint" are involved in an explanation, their importance is low relative to the fitness differences between developmentally readily available alternatives. (C) By the same token, when biologists explain a pattern by "constraint," it doesn't mean that selection plays no role. It refers to situations in which the range of developmentally available alternatives is so circumscribed that the range of the available seems necessary to include in an explanation. Selection still can play a role in determining where a given population mean lies. In this case, the favored proportionality (black line) is as close to the

5.9.2 "Explained by constraint"

In another shorthand, what biologists mean when they refer to "constraint" as a primary cause of form is quite different. When "constraint" isn't just a vague reference to selection, it often implies that plausibly imaginable configurations adjacent to the commonly observed ones are not observed or are only rarely observed (Figure 5.10C). This can involve situations in which the adjacent configurations are functionally similar to the commonly observed ones, or even superior to them in performance and fitness. The geophilomorph centipede segment number example from Chapter 2.1.1 corresponds to the former case. This example is conspicuous because there is every reason to think that individuals with even numbers of segments would be functionally identical to those with odd numbers of segments. Therefore, the fact that even numbers of segments are not observed suggests that some factor other than natural selection is limiting developmental potential. In other cases, variants adjacent to the observed ones would have superior performance and would almost certainly be favored by selection. As discussed in Section 2.3.13, if it were possible, more or fewer cervical vertebrae would surely be favored by selection in at least some species of mammals that, instead, have seven. If trees could produce wood that is maximally stiff and provides maximal storage space, selection would surely favor it (Section 2.1.4). Statements that these patterns are explained by "constraint" do not mean that selection can have no role at all in shaping the traits of interest. Rather, they highlight cases in which the range of developmentally accessible variation appears narrow enough that a standard "explained by selection" account must be supplemented with a description of the available variation itself. The vocabulary of what is possible or not in development, how fitness differs among the possible variants, and what causes any unevenness in the occupation of developmental potential means that the shorthands of "constraint versus adaptation" aren't strictly necessary. At the same time, the vocabulary of developmental potential provides a means of describing the related terms convergence and parallelism.

5.9.3 Extremes of a continuum: convergence versus parallelism

The "constraint versus adaptation" dichotomy is paralleled by the "parallelism versus convergence" dichotomy. Tidy definitions make parallelism and convergence seem very different phenomena. Convergence, the definitions go, is a situation when, under similar conditions of natural selection, different species arrive at similar evolutionary outcomes from totally different starting points. Parallelism, instead, is a situation in which two species start from similar starting points and under similar conditions of

---

Figure 5.10 *(continued)* would-be favored one as is possible given the developmentally available variation. The shorthand is "due to constraint" not because selection isn't involved at all, but because of the high relative importance of factors that seem to impede access to plausibly imaginable alternative configurations. Scientist by Eduardo Martínez Díaz.

selection independently arrive at similar evolutionary outcomes. Similar outcomes can be produced from different starting points only if developmental potential is wide, so convergent evolution is often interpreted as evidence against some sort of limitation or bias to what is possible in development. Parallelism, on the other hand, is often taken as evidence of shared developmental tendencies. Two species, the account goes, inherit similar developmental possibilities from a common ancestor. When they evolve similar outcomes in similar environments, this is evidence of their limited developmental options. These definitions do highlight situations that differ in biologically important ways.

Demanding a clean difference between convergence and parallelism in practice, however, runs into the "how much is enough?" problem (Arendt and Reznick, 2008; Scotland, 2010, 2011; Powell, 2020). The "how much is enough?" problem refers to the difficulty of figuring out how to divide a continuum into two or more categories. No organism on Earth is completely unrelated to any other, so we all share some features, such as water being a fundamental solvent, the presence of lipids, proteins, nucleic acids, ATP, metabolism, homeostasis, etc. While organisms manifestly diverge markedly from one another, this divergence has clear limits, with none of us operating in a totally anhydrous environment or in a vacuum, for example.

The Australian thorny devil lizard (*Moloch horridus*) and the horned lizards of North America (*Phrynosoma*) are medium-sized dryland lizards that are covered with large spiny scales, move slowly, harvest moisture from rain and dew with their skin, and eat ants (Figure 5.11). *Moloch* is an agamid, a largely Old World group of lizards, and *Phrynosoma* are phrynosomatids, a largely New World group. Agamidae and Phrynosomatidae are part of the larger clade Iguania, which also includes the Old World chameleons, iguanas, anoles, fence lizards, and other familiar lizards.

An argument can certainly be constructed in favor of parallelism. The similar features of *Moloch* and *Phrynosoma* evolved within a deeply shared anatomical and developmental scaffold. All lizards have (maximum) four limbs, a similar body plan, epidermal scales, similar cranial and axial skeletal elements, and they share similar limb and digit structures. For all their variation, all Iguanians, the large clade *Moloch* and *Phrynosoma* are a part of, have four limbs, and mostly five well-developed digits. Other groups, such as amphisbaenians or pygopodids, have two or even zero limbs. So, even if the ancestors of *Moloch* and *Phrynosoma* diverged tens of millions of years ago, their starting material arguably wasn't wildly different. Both had the capacity to develop spiny ornamentation, cranial projections, and dermal traits that facilitate water harvesting. That shared foundation makes it plausible that the same types of traits could evolve repeatedly, not by the sheer force of similar selective pressure on endlessly malleable organisms but because they represent accessible solutions within similar spaces of developmental potential.

However, the ancestral lizards from which *Moloch* and *Phrynosoma* are descended were almost certainly more standard lizard-like, not squat, slow, spiny, and ant-eating. So, *Moloch* and *Phrynosoma* are often presented as a good example of convergence: from ancestral states clearly differing from both *Moloch* and *Phrynosoma*, they evolved very similar morphologies and lifestyles under the similar selective environments of drylands on distant continents. This is exactly the stipulation that

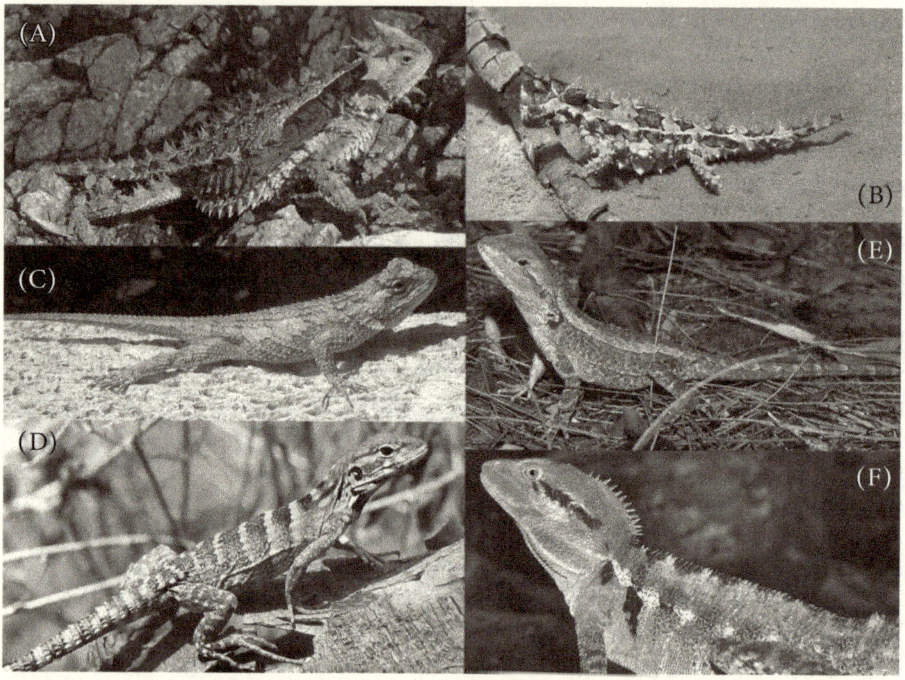

Figure 5.11 Convergence or parallelism? Lizards in the genus *Phrynosoma*, from warm drylands of North America, resemble *Moloch horridus* of central and western Australian deserts. (A) *Phrynosoma*, such as this *P. asio* from the Chamela research station on Mexico's Pacific coast, belong to the New World family Phrynosomatidae, whereas (B) *Moloch*, here in Australia's Red Center, belongs to the Old World family Agamidae. *Moloch* and *Phrynosoma* are similar-sized, slow-moving, covered with spiny scales, and eat ants. Because of the geographical and phylogenetic distances separating them, and because their ancestors were more gracile, less spiny, and had different diets, this similarity of morphology and lifestyle is often regarded as a classic case of convergence. (C) and (D) give examples of *Phrynosoma* relatives in the family Phrynosomatidae. (C) *Sceloporus melanorhinus*, a slow-moving "spiny lizard" that perches cunningly, hunting insects on tree trunks along the tropical Pacific of Mexico. (D) *Petrosaurus thalassinus* darts nervously on rocks and trunks in tropical dry forest of the tip of the Baja California peninsula, Mexico, where it hunts large insects and small lizards, and also eats leaves and flowers. Both of these lifestyles are very different from *Phrynosoma*. (E) and (F) are Australian representatives of Agamidae, the group *Moloch* belongs to. (E) *Diporiphora nobbi* in moist woodland on the temperate Australian east coast, where it forages for insects. (F) The habitat of *Intellagama lesueurii* could hardly differ more than that of *Moloch*, here shown on a rock in a lowland tropical rainforest stream, where it lumbers with little provocation into the water. No other taxon in either family closely resembles *Moloch* or *Phrynosoma*; it seems certain that the ancestors from which they evolved were certainly non-*Moloch*-like and non-*Phrynosoma*-like, underwriting convergence interpretations. However, both Phrynosomatidae and Agamidae are in the same larger clade Iguania, whose members share many characteristics. The substrate from which these similar morphologies evolved, therefore, was not *entirely* different, meaning that a case could be made for parallelism underlying their similarity.

convergence represents a situation in which functionally significantly different variants could be favored by selection (Figure 5.10B).

As with the adaptation-constraint distinction as depicted in Figure 5.10, an adaptationist evo-devo perspective provides the tools for making the convergence-parallelism distinction more informative than a simple dichotomy. Emphasis on what variation is possible developmentally, and of the possible and the imaginable, how variants differ in fitness, naturally locates any given case along a convergence-parallelism continuum, just as it highlights mutual contributions of selection and whatever flavor of "constraint" is being entertained. In a pattern such as long bone-body mass scaling, it is true that all of the species are from the same clade and can be argued to have the same developmental substrate. However, the finding that there is enough variation available, such that selection could perfectly well favor different intercepts and slopes, represents sufficient information on its own, whether we declare it convergence or parallelism or some combination. In the same way, we might declare passerine carotenoid space (from Alex Badyaev's work mentioned in Section 2.3.15) to be so structured that all species traverse it in more or less the same way, given what particular colors are favored in a given species and why. Here too, we could construct an argument for the relative contributions of convergence and parallelism, but as with adaptation versus "constraint," with a sufficient description not just of relative fitness between variants but also the space of developmental possibilities and the cause of its structure, this information to a large extent obviates the convergence-parallelism dichotomy in the same way it does the "adaptation-constraint" dichotomy.

5.9.4 "Contingency" versus selection

Another very close cousin of the adaptation versus constraint and convergence versus parallelism dichotomies is "historical contingency" versus "natural selection." As with the "adaptation versus constraint" dichotomy, scientists tend to lean toward "selection" as an explanation when there is clearly a large amount of variation adjacent to the wild type, and when variants farther from the wild type have lower fitness. "Contingency" refers to historical happenstance. One sort of contingency is that the space of alternatives was never broad to begin with, because of the way a developmental pathway happened to evolve. Another is that one possibility out of many got frozen, as by sheer sampling error. Whatever the reason, contingency in this sense implies that the explanation for a trait is not because selection winnowed from among a wide array of developmental contenders. Cases such as only-odd segment number in geophilomorph centipedes, seven cervical vertebrae in mammals, and perhaps the stubborn production of interpetiolar stipules regardless of habit or habitat in the coffee family are situations that send biologists toward wondering about such happenstance because it's hard to see how selection could explain them.

As with the "adaptation versus constraint" and "convergence versus parallelism" dichotomies, an argument can always be constructed to show that both selection and contingency, however that term is defined, are involved in any example of form. For one, a given outcome can always be construed as "contingent," given how far one stretches that term. Any outcome depends, at least in part, on the historical accident of the particular combination of gravity, pressure, temperature, and chemistry experienced on Earth; where the intercept and slope of the terrestrial mammal long bone-body mass scaling depends on all of these, and if conditions were different, those parameters would fall elsewhere. So, the dichotomy seems again to be best taken as useful guideposts, but as we saw in Chapter 2, especially Sections 2.3.13–15, with the examples of seven cervical vertebrae, *Drosophila* sex combs, and songbird carotenoid coloration, these examples illustrate that with enough information about the causes of phenotypes in hand, it's impossible to adjudicate entirely in favor of adaptation or entirely in favor of any sort of "constraint." The same seems likely true of "contingency." Much more thoughtful explorations of contingency can be found (Losos, 2017; Powell, 2020), and this setup is just to be able to ask the following question.

Whether a valid dichotomy or not, one thing that became clear in researching this book is that there are way more papers focusing on adaptation than on contingent aspects of organisms. One question is whether this proportion reflects the relative importance of different natural processes in shaping diversity. Certainly, as scientists, we would like at least to have an idea of whether our efforts are distributed in ways more or less proportional to the major causes in nature, or whether we are concentrated off in a corner because it most resonates with our interests. So, here's a thought experiment. First, suppose a clean dichotomy between contingency and selection as explanations for biological traits. On the contingency side, things just happened that way. On the other side are adaptations, features whose main explanation is that selection favors them, with a wide array of developmental possibilities surrounding the favored value on all sides.

Now, imagine that Mother Nature suddenly handed us a stone tablet revealing the exact percentage of biological features shaped by contingency versus those honed by adaptation. The tablet doesn't change the world, just our knowledge of it. In both scenarios that follow, the biological facts and the scientific literature to date remain exactly the same. If the tablet claimed that 99% of organismal traits are the result of natural selection and only 1% stem from contingency, then the adaptation-heavy focus of the literature and our daily research efforts would seem perfectly justified. But what if the tablet said the opposite, that 99% of features arose through contingency and only 1% through adaptation? Would our scientific focus change accordingly? Would we shift our attention toward historical happenstance? Or would we go on asking the same kinds of questions in the same proportions, because that's what we find most tractable, most satisfying, or most aligned with our intellectual habits? There are reasons to think that our efforts might not shift that much, for several reasons.

A low-hanging fruit reason is pragmatic manipulation, with adaptation being the actionable part of nature. In agriculture, understanding how plants adapt to poor soils, short growing seasons, or salinity allows us to breed crops that thrive where others fail. Human efforts to intervene in the natural world can only do so in these actionable parts, so even if the non-actionable parts are the bulk, they are irrelevant for our efforts to intervene in the natural world. So adaptation thinking is deeply tied to applied biology, and this is surely part of why it's so resilient in the scientific imagination. From this angle, there would be no reason to shift our focus.

Possibly more importantly, we are an engineering, problem-solving species with a cognitive tendency known as agency detection. This trait, widely thought to have been favored by selection (Kuhnen and Kitayama 2021), refers to our bias toward perceiving intention behind observed effects, even when none exist. One idea is that agency detection was favored by selection because, in uncertain environments, assuming agency ("something caused this") was better than ignoring it ("it's just random"). Whatever the reason we have it, one consequence is that humans are strongly drawn to causal stories with goals. Adaptation provides just such a story: a trait exists because it serves a function. It may be, then, that scientific attention tilts toward adaptation not only because it proves useful, but because it is cognitively motivating in ways that other types of explanations are not. Given this bias in our interests, perhaps we would change nothing because evolutionary biology provides not so much an isomorphic picture of the world but a reflection of human interests.

These considerations seem important given that the world really might be 99% contingency, and because biologists are well aware that we bring biases to our work. It's well known, for example, that we preponderantly focus on organisms large enough that we can see them, and we neglect microorganisms. In this way, the work of biologists and philosophers of science is far out of proportion to the disparity between macro- versus micro-organisms, given their species diversity, abundance, and contribution to the biosphere (O'Malley and Dupré, 2007). With regard to contingency, as Weiss and Buchanan (2011, p. 769) note, "most elements in the flux of life involve small or rare context-specific combinatorial effects that are historically ad hoc and heterogeneous or even unenumerable." The question is not whether biological features are in principle explainable, surely they are, but whether the subset we choose to explain as biologists mirrors the proportionality of processes in nature. Knowing whether our explanations approximate the actual distribution of causal processes in nature or instead reflect the structure of our investigative preferences seems important, if only to clarify the aims of our scientific practice.

5.9.5 The adaptation-constraint/contingency dichotomies as a question of scale

In the midst of these debates, Wilkins and Godfrey-Smith (2009) propose that the adaptation-constraint distinction is not just empirical, given that both selection and "constraint" are both always at play in the shaping of form, but also a matter of the level of focus of different scientists. They use a distinction between fine and coarse-grained perspectives. When examining evolution at a fine-grained level, such as the single-generation timescale or within a very small region of the adaptive landscape (the authors invoke this metaphor), phenomena such as interindividual variation, drift, and selection are highly salient. Evolutionary processes at this level often highlight local adaptation, with populations found at or near fitness peaks, reflecting the power of natural selection in shaping many aspects of species. As domestic dog breeds and the work of Drake and Klingenberg show (Section 2.3.3), bone proportions can vary dramatically across generations. There is enough heritable variation that people have successfully selected for extremely short-legged and long-legged dogs. At this close-up level, population-level processes such as adaptation are clearly important.

Zooming out to a coarse-grained perspective, taking in very broad patterns of morphospace occupation and long timespans, reveals markedly heterogeneous patterns of occupation. At this level, the metaphorical landscape occupied by species living and extinct appears largely empty, with organisms seeming to have explored only a tiny fraction of conceivable forms. The notion that these vast empty spaces are developmentally possible but simply not favored seems hard to believe. To return to vertebrates, for all the variability in limb length between dachshunds and greyhounds, the basic humerus-radius-ulna structure of the tetrapod forelimb is astonishingly ancient, dating back over 360 million years, to the earliest tetrapods in the Late Devonian, like *Acanthostega* and *Ichthyostega*. These early limbed vertebrates already had a recognizable upper arm bone (humerus) and paired forearm bones (radius and ulna). That three-part configuration has been conserved ever since across nearly all land vertebrates, including amphibians, reptiles, birds, and mammals. There seems to be, for example, no known extant or extinct tetrapod with a duplicated long bone proximal to the radius/ulna, no "double humerus" giving a true two-elbow configuration. Across tetrapods, from the strangest frogs to the most flamboyant pterosaurs, the ancestral 1-2-many limb pattern (stylopod-zeugopod-autopod) apparently remains constant. In this unimaginably long stretch of time, it's hard to believe that a significant deviation from the 1-2-many limb pattern wouldn't be favored in at least some selective context. So, from this zoomed-out perspective, saying that the power of natural selection appears limited is the height of understatement, with organisms restricted to small subsets of morphospace almost certainly due to the contingent ways that their developmental systems happened to evolve. Framing this apparent dichotomy as a matter of grain, with both adaptation and "constraint" or contingency playing crucial roles at different scales of observation, offers yet more common ground between biologists working from different perspectives.

Much of the interest in "constraint" over the past 40 years, plus some of the confusion around this term and what biologists should do about it, can be traced to the influence of one famous essay, so it's important to understand what that paper said, and what its actual intention was.

5.10 Understanding the *Spandrels* paper

Many of the biologists inspired to adaptationist evo-devo have drawn motivation from one particular paper, so it's important to understand its arguments (Rose and Lauder, 1996; Pigliucci and Kaplan, 2000; Wagner, 2014). The paper is Stephen Jay Gould and Richard Lewontin's 1979 essay "The Spandrels of San Marco and the Panglossian Paradigm: A Critique of the Adaptationist Programme" (Gould and Lewontin, 1979). The paper is one of the most widely cited in evolutionary biology and is the most-read evolutionary biology paper outside of our field. It has inspired at least one book dedicated to explicating its rhetoric, arguments, and style (Selzer, 1993). It is an engaging paper to read, and if you haven't read it, you should.

However, don't expect to find much clarity in it. If you have read it and weren't sure how to act on their suggestions, it's not your fault. The central analogy in the paper was intended to illustrate "constraint," but it instead illustrates standard natural selection. So, if you carefully read the paper looking for Gould and Lewontin's lessons about constraints, you almost certainly came away confused. Even more seriously, they did not actually provide guidance on what biologists should do when they return to the field or the laboratory after having read the paper (Dwyer, 1984; Goodwin, 1990). Here's some help in understanding the paper's message and navigating its central confusions.

First, the very valuable intent of the paper: Gould and Lewontin were worried that biologists relied too much on standard adaptationist accounts instead of considering the possibility that a given pattern was observed because developmental potential was limited in ways that had nothing to do with selection. In adaptationist evo-devo vocabulary, they were concerned that explanations of organismal form were too quick to assume very wide fields of developmental potential without empirically testing this assumption. Testing these assumptions is precisely a central goal of adaptationist evo-devo, and the *Spandrels* paper was an important motivation in turning the thinking of many biologists toward developmental potential. So, the central exhortation of the paper to include developmental potential in explanations of organismal form is one well taken. If you are happy to conclude that you have absorbed the adaptationist evo-devo mission to explore developmental potential as part of well-supported explanations of organismal form, and feel you don't need a deep dive into the guts of the *Spandrels* paper, then you can safely skip this section. If you do want some signposts for reading the *Spandrels* paper, read on, but take a deep breath or two first.

5.10.1 The central analogy of *Spandrels* was a perfect illustration of selection

The deep breath is helpful because the central analogy that Gould and Lewontin used was a bad one for illustrating "constraint." They focused on the huge curving triangular masses of masonry, known as pendentives, that gracefully bear the weight of a large dome, channeling its load into the supporting arches beneath. For whatever reason, Gould and Lewontin focused on the decorations in the Basilica of San Marco in Venice. They noted that the four masses of masonry offered great places for locating mosaics of the four apostles and the four rivers mentioned in the Bible. Gould and Lewontin's argument was that the resulting composition is so appealing that one would think the triangular masses of masonry were intentionally located there, in those shapes, because artists wanted more than anything some curved triangular surfaces on which depict four evangelists and four rivers. However, Gould and Lewontin warn, giant curving triangular masses of masonry are inevitable architectural consequences of locating a dome atop a square base. If "inevitable" means that there was no other configuration available, then the reason for their being there can't be as surfaces for locating evangelists.

Translated to biology, the message is well-taken: if developmental alternatives are impossible, then selection cannot act. Therefore, the limited developmental potential must be an important part of the explanation of the trait in question. Gould and Lewontin did not use the vocabulary of "developmental potential," or explicitly discuss the need to study and include it in explanations of organismal form. Perceptive scientists did manage to extract this core message of the need to study developmental potential from Gould and Lewontin's essay, as in the methods showcased in Chapter 2. But without a doubt, their message was missed by many readers, and understandably so.

It was missed because, in reality, pendentives are a paradigm illustration of selection, and not any type of "constraint" (Houston, 1997, 2009). It turns out that pendentives aren't an "inevitable consequence" at all, and that there are many ways of fitting a dome onto a square building in architectural traditions around the world. In some parts of the world, it is common for domes to be supported by brackets, as in horizontal shelves that bridge the empty space between the dome and the corners of the supporting walls (Figure 5.12A). In other places, domes are supported by squinches. Squinches are half-domes set into the corners beneath the main dome (Figure 5.12B). The realm of the architecturally ("developmentally") imaginable is much wider. A dome could be supported by a ring of columns, or even a circular wall, like the Pantheon in Rome, could be placed inside a church. A tracery of steel girders could traverse the interior of the church, or for that matter, the girders, walls, or columns could be made out of paper, or yummy candy. Far from being restricted, the realm of the possible is very wide indeed (Olson, 2012a). The variant actually observed in the Basilica of San Marco, the great masses of masonry mentioned in the previous paragraph, is the pendentive (Figure 5.12C).

Of all the architectural alternatives that medieval architects explored, pendentives were the option that most effectively transmitted the mechanical load of the dome to the square building and to the Earth below (well, in Venice, to the piles sunk into the floor of the lagoon) (Houston, 1997, 2009). "Most effectively" in this context means that it held up the dome for the longest period with the least amount of material and the least obstruction of the view in the cathedral's interior. Apparently, for large domes, it is the only alternative explored that holds up a huge dome for any amount of time. Favoring the alternative with the best performance from among a much wider field of alternatives is simply selection. So, when delving into the *Spandrels* essay, it is important to recognize that its central analogy is not any kind of meaningful "constraint" as they intended, but instead an analogy of completely standard natural selection.

5.10.2 True spandrels illustrate issues of trait delimitation, not "constraint"

It gets worse. Let's concede for a moment that instead of pendentives, Gould and Lewontin really were referring to "spandrels." They almost certainly used the term "spandrels" because "the pendentives of San Marco" doesn't sound at all cool. But pendentives and spandrels are very different things. Spandrels are leftover spaces in

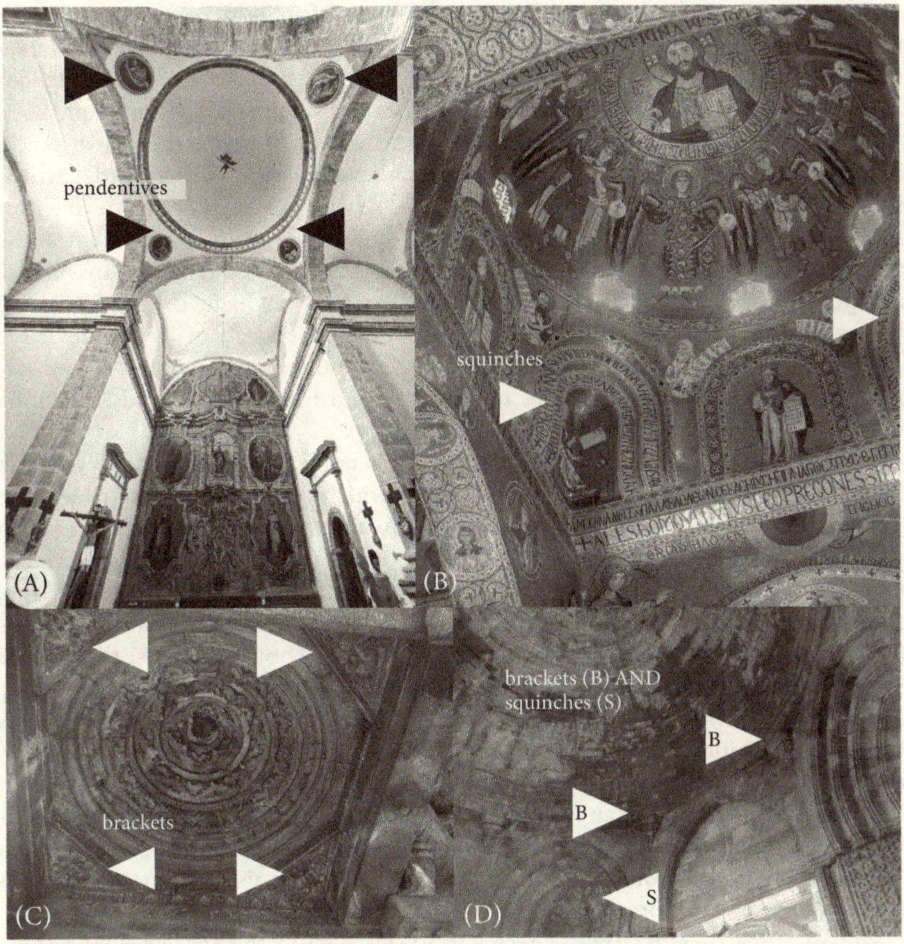

Figure 5.12 Pendentives and some "developmental" alternatives. The central metaphor of the *Spandrels* essay attempted to illustrate the way that limited developmental potential modifies standard adaptationist explanations of organismal form. (A) Gould and Lewontin suggested that there was no other way of fitting a round dome to a square building except with pendentives, the curving triangles with medallions in them around the dome, shown here in the desert oasis Misión San Ignacio Kadakaamán, Baja California Sur, Mexico. What made the central adaptationist evo-devo message in the *Spandrels* essay so hard to understand is that there are plenty of "developmental" alternatives to pendentives. For example, (B) Squinches hold up domes in the Palatine Chapel, in Palermo, Italy. (C) In the Qutb Minar complex in Delhi, India, brackets hold up small domes. (D) This Qutb Minar dome is supported by squinches *and* brackets, not to mention a supporting base that isn't square. Gould and Lewontin's argument would be that locating the medallions on the pendentives was such a harmonious design that we would easily be seduced into thinking that the pendentives were placed with the sole purpose of locating medallions. This would be a mistake, they argue, because pendentives represent an "architectural constraint" in that they are the inevitable

architecture. That awkward space beneath a staircase often used for storage (or a bedroom for an aspiring wizard) is a spandrel. The roughly triangular spaces left over between arches in an arcade are spandrels. An arcade consists of a series of contiguous arches, arranged in a linear sequence. The arches run side-by-side without intersecting or facing each other directly. This pleasing arrangement necessarily leaves spaces in between the arches (Olson, 2019b). These spaces are variously decorated, particularly fetchingly when pierced by a circular window. In adaptationist evo-devo terms, the spandrels between arches really are "developmentally" non-independent of the arches. If the arches are rounded, the spandrels are wide. If the arches are narrow and pointed, the spandrels are narrow. If the arches are trifoliate, the spandrels will be commensurately scalloped. Because the shapes of spandrels really are inevitable consequences of the shapes of the arches that define them, perhaps simply transferring the analogy from pendentives to spandrels can rescue Gould and Lewontin's intention.

Substituting "spandrels" for "pendentives" does not rescue Gould and Lewontin's intent. They wanted to illustrate a situation in which biologists mistakenly assume that developmental potential is wide and explain organismal form by appealing to an argument that the observed variant is the fittest of a wide array of possible variants. Gould and Lewontin wanted to show that biologists are in danger of constructing such adaptationist explanations even in situations where there is only a single possible variant. Pendentives do not work in their analogy, and not just because they are the variant with the highest performance out of a wide array of possible alternatives. The analogy of pendentives also does not work because, in contrast to spandrels, pendentives are not simply leftovers between other structures. They are masses of masonry, essentially a right tetrahedron with one curved face. In principle, a right tetrahedron of masonry could be built anywhere, out in St. Mark's Square if desired. A pendentive is a "part," separable to an extent from the rest of the building and with a certain "developmental" autonomy. The same can be said of organismal "parts," such as a tooth, a bone, an arm, or a leaf.

Not so with true spandrels. There is no way to build the space below a staircase independently of the staircase. Even if a builder leaves the spaces between the arches of an arcade empty and doesn't fill them with stone or masonry, the empty spaces left over in between the arches are still dictated entirely by the shapes of the arches themselves. Pendentives are thus good analogies for organismal parts,

---

Figure 5.12 *(continued)* consequence of locating a round dome on a square base. The message here is well taken—if developmental potential is restricted such that developmental alternatives are impossible, then explaining a given trait by reference to selection would need modification. But there are plenty of different ways to join a dome to a building. Out of these, because of the way they direct the mechanical loads of the dome, pendentives turn out to be the variant that maximizes "fitness," i.e. support a large dome the longest time with the least amount of material. The prevalence of the variant that performs best out of a wide array of possible variants is just selection. The fact that Gould and Lewontin's central analogy was a perfect illustration of selection made their suggestions regarding the need to explore developmental potential that much harder to extract.

and the way that natural selection shapes parts more or less independently of the rest of the organism. Spandrels are not analogous to organismal "parts." There is a risk of erroneously constructing adaptationist explanations for spandrels only if biologists erroneously think that some organismal sector is a "part" that can vary to some extent independently of the rest of the organism. So, true spandrels represent a danger only if they are mistaken for organismal parts that can be honed by natural selection to some extent independently of the rest of the building or the body.

This is a problem of trait delimitation, not lack of developmental potential (Olson, 2019b). How to delimit organismal traits or "parts," and the extent to which this can be accomplished non-arbitrarily, is an important issue in evolutionary biology (Magwene, 2001; Eble, 2005; Goswami and Polly, 2010). There is a great deal of thought regarding what species might be and how to delimit them (Templeton, 1989). As hard as the species problem is, at least species often clearly seem to represent distinct entities in nature: chimpanzees and humans certainly seem to be qualitatively different, non-imaginary entities. But organismal parts—ears, femurs, vessel elements, clamp connections—are not independent, but parts of bodies. So, while there are lots of statistical methods for identifying parts or modules, the theory regarding the biological causes of modularity and how to delimit modules non-arbitrarily lags far behind that of the species problem (Olson, 2019b). Gould and Lewontin's "spandrels" represent incorrectly delimited traits, not any sort of developmental or architectural "constraint." While the question of how to rigorously identify organismal traits and modules remains an important frontier, another, entirely different source of confusion has to do with "spandrels" versus the exotic-sounding "exaptation."

5.10.3 Spandrels can be exapted, but not all exaptations are spandrels

A final widespread confusion exists between the terms "spandrel" as used in biology and "exaptation." Exaptations are features that arose in one context and were co-opted by natural selection in a different functional context (Gould and Vrba, 1982; see also Griffiths, 1992). The term is useful for highlighting situations in which a trait would seem obviously to have evolved in its current selective context and with its current function, but actually turns out to have been co-opted from a previous, different selective context.

For example, the lung and bone structures of birds certainly appear to involve clear adaptations to an aerial lifestyle. Birds have a distinctive respiratory system in which air flows unidirectionally through their lungs, facilitated by air sacs that act as bellows, ensuring a continuous supply of oxygen-rich air during both inhalation and exhalation. This respiratory system structure supplies the intense metabolic demands of flight with a highly efficient oxygen supply. As for bones, bird bones are famously light and hollow. A hollow tube is much stiffer than a solid bar made of the same amount of material (Gere and Timoshenko, 1999). This is why bicycle frames are hollow rather than being solid bars with the same cross-sectional area of metal. Hollow spaces within bones don't automatically reduce mass, however. The hollow spaces in the bones of most animals are filled with material such as bone marrow. But in birds, many of the hollow spaces are filled by air, which dramatically reduces their weight for a given stiffness. Presumably, large, isolated, gas-filled spaces could quickly become anaerobic. Birds achieve the remarkable trick of air-filled hollow bones by innervation of the bones with extensions of the air sacs. This innervation means that even in large bones like the femur and humerus, air spaces are maintained with continually ventilated air, avoiding anaerobic conditions. If there ever were obvious adaptations to a current selective context, bird lung and bone structure in the context of flight certainly would seem to be foremost examples.

As it happens, both unidirectional ventilation and pneumatized bones evolved way before birds, in animals that couldn't have been less aerial. Pneumatized bones have been found far back in the evolution of the theropod dinosaurs, and possibly even earlier, at a time when the animals were often large and were firmly terrestrial (Figure 5.13) (Benson *et al.*, 2012). Pneumatization of bone is thought to have been favored in situations in which lower skeletal weight or metabolic demand would be favored, as in large, terrestrial animals. Such pneumatization apparently is facilitated by a certain lung structure, one involving unidirectional airflow as a precursor (compare with the evolution of high bundle sheath ratios as a prerequisite for evolution of C_4 photosynthesis in Section 2.3.9, also the evolution of drought tolerance in grasses as a precursor for invasion of cold habitats, Schat *et al.* (2025)).

As for unidirectional airflow, its origin is even more anticlimactic. It is present in contemporary crocodilians and therefore likely evolved before the divergence between crocodilians and the dinosaur–bird lineage; it is also present, apparently as an independent origin, in some lizards (Farmer, 2015). Its presence is inferred in fossils far predating birds (O'Connor and Claessens, 2005). Unidirectional airflow allows for efficient breathing, understood as maximal oxygen uptake given metabolic investment.

It is pleasing to imagine selection favoring efficient respiration with unidirectional air flow in large, mighty animals bounding across the landscape in agile pursuit of prey; such high-metabolic rate contexts would clearly favor higher oxygenation efficiency. But one possible hypothesis is that unidirectional airflow was favored as part of a suite of crypsis traits (Farmer, 2015). The scenario goes that some small, dull-colored, probably lizard-looking creature would be less likely to be noticed and eaten if, in addition to its uninspiring color pattern, it also barely moved. More efficient respiration would mean that our drab little archosauriform would not only just lie there all day, but even breathe infrequently or only shallowly. Individuals within a population that vigorously moved their bodies, panting, would be more visible to predators. So, the high-efficiency respiratory system of birds might owe its origin to an ancient extreme of sedentariness. Avian lung and bone structure exemplify features that are so clearly appropriate for their current functions that it would seem that they simply had to have arisen and been favored by selection in their current selective context. Instead, pneumatization, not to mention feathers, arose in previous selective contexts and were exapted in the evolution of birds. Recently, even the bird wrist structure, also previously thought to be a unique adaptation conferring birds' high-maneuverability flight, was taken from us; it too evolved before birds and was exapted in the evolution of flight (Napoli *et al.*, 2025). "Exaptation" helpfully picks out counterintuitive cases in which examples of seemingly clear adaptations that arose and were favored in their current roles in fact arose in a completely different selective context.

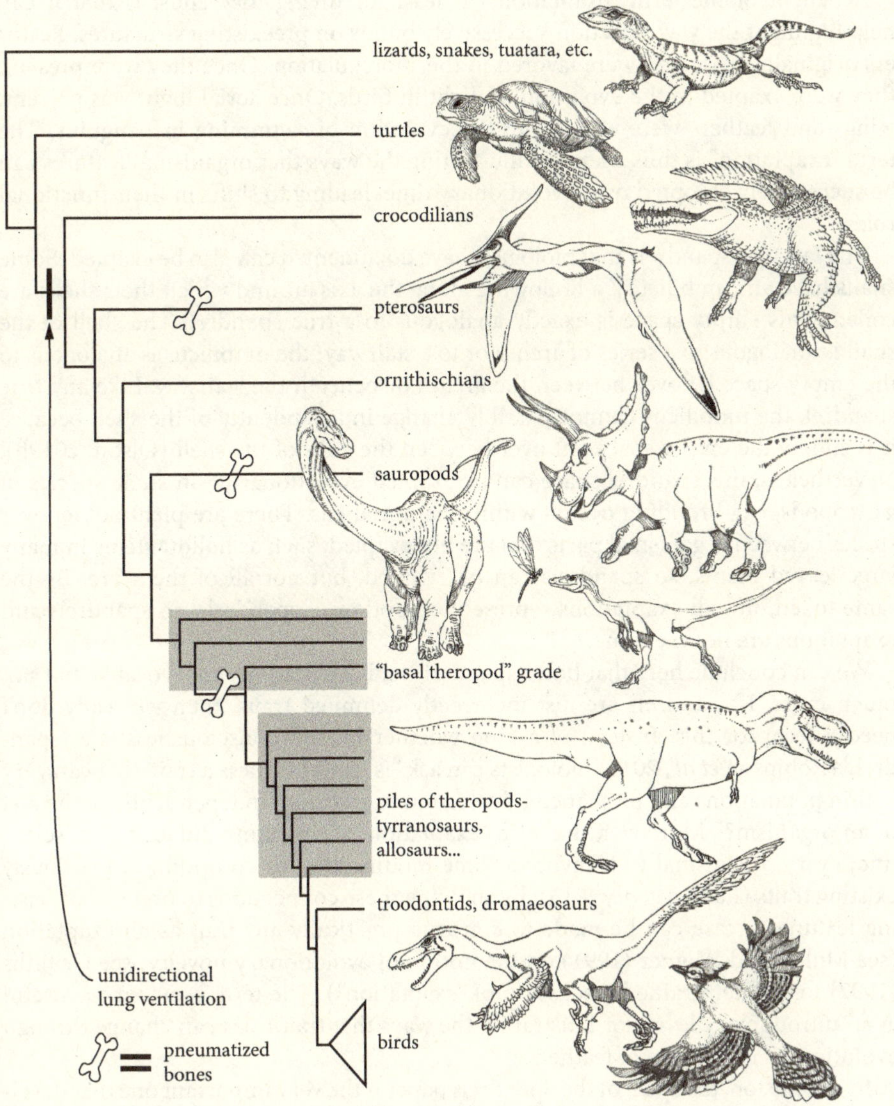

Figure 5.13 Selection co-opting existing variation through changes in selective regime. Though they would seem obvious adaptations that evolved in the context of an aerial lifestyle, unidirectional lung ventilation and hollow bones arose long before birds evolved. Bone icons show three originations of pneumatized bones: 1. in pterosaurs, where hollow bones were presumably favored in flight; 2. in sauropods, where they might have lightened the skeleton, permitting impressively long necks; and 3. early in theropod evolution, long predating flight. Without this phylogenetic context, pneumatized bones and unidirectional lung ventilation could easily be mistaken for flight-specific adaptations unique to birds; instead, it is likely that they facilitated the evolution of flight. Drawings by Darren H. Burton.

A benefit of the term "exaptation," at least for divulgatory ends, is that it can help highlight the way selection successively builds on preexisting structures. Feathers originally arose and were favored in thermoregulation. Once they were present, they were exapted in the evolution of flight in birds. Once aerial flight was present, wings and feathers were exapted in the evolution of swimming in penguins. The term "exaptation" is thus useful in illustrating the ways that organismal features can be successively co-opted over evolutionary time, leading to shifts in their functional roles.

The few true spandrels that biologists have documented can also be exapted. Some snails have an "umbilicus," a hollowing along the axis around which their shells are coiled. This empty space is exactly analogous to a true spandrel. The shell of the snail is analogous to a series of arches or to a stairway; the umbilicus is analogous to the empty space leftover between the arches or beneath the stairway. Like any true spandrel, the umbilicus cannot possibly change independently of the shell because it is simply the empty space left over between the coils of the shell (Olson, 2019b). Nevertheless, this spandrel space can be exapted evolutionarily; in some species of gastropods, egg brooding occurs within the umbilicus. There are plenty of leftover spaces between organismal parts that are not exapted, such as hollow stems in many large-leaved plants, so spandrels can be exapted, but not all of them are. By the same token, not all exaptations represent co-option of spandrels, so spandrels and exaptations are not the same.

We can conclude here that both "spandrel" and "exaptation" are evocative but not much more. If spandrels are just incorrectly delimited traits, then we really don't need a term for this. Instead of asking whether human consciousness is a "spandrel," (Robinson *et al.*, 2015) biologists can ask "is consciousness a trait, that can vary within populations and be shaped by selection to an extent independently of the rest of an organism?" Moreover, the term "exaptation" seems more didactic than scientific; every organismal trait involves some modification or co-option of previously existing traits (Laland *et al.*, 2013); by showing these connections to previously existing features, a case can be made to construe practically any trait as an exaptation (see Muller and Wagner (1991) on the notion of evolutionary novelty; see Griffiths (1992) for a finer grained discussion of "exaptation"). The term seems more useful at an introductory level, for illustrating the ways that trait roles can change through evolution, as in the case of feathers.

In conclusion, the spirit of the Spandrels paper is the very important one that developmental potential is an essential part of evolutionary explanations of organismal form. The murkiness of the Spandrels paper means that it's very hard to extract this core message.

5.11 Explanation and complementarity

The complementarity of evolutionary evidence shows that none of the subdisciplinary approaches of population biology, comparative biology, optimality modeling, and study of developmental potential can provide anything resembling a broadly supported evolutionary explanation, each on its own. Instead, each of these subdisciplines provides crucial empirical evidence, but each has glaring omissions. Population biology can examine variation, heritability, and fitness directly. However, population biology on its own cannot test the relevance of its data for evolutionary biology at large. These tests are provided by comparative biology, which, despite not being able to examine variation, heritability, and fitness directly, provides the generality that allows evolutionary biology to explain the diversity not just in a few bottles of flies but across the natural world. Recognizing this essential complementarity is an instant, forceful exhortation to overcome the tedious provincialism often found within subdisciplines. This provincialism takes the form of regarding one's own subdiscipline as providing the truly forceful theory and data regarding evolution, with the other subdisciplines providing inferior or even unnecessary data (Olson, 2024).

Instead, if different evolutionary approaches provide necessary and complementary layers of empirical evidence, then rather than viewing the weaknesses of a given approach as flaws, we biologists can direct our efforts to working together to adduce empirical evidence that is as complementary as possible, thereby reducing the relative importance of assumptions about history and unobserved individuals in our explanations. What follows are some of these considerations in light of the structure of evolutionary explanation, including a summary of some adaptationist evo-devo precepts.

6
Conclusion
Working together to build better explanations of organismal form

Adaptationist evo-devo is a completely new evolutionary biology. Just kidding—it isn't; variation has been a focus throughout evolutionary biology. The goal here isn't to claim that any narrative needs recasting but simply to give developmental variation and its close relatives useful attention as part of building well-supported explanations of organismal form. Interest in what can and can't be produced in ontogeny is not new. Darwin (1859) mentioned both what we are calling modularity or ontogenetic quasi-independence between traits, as well as "the mysterious laws of the correlation of growth," by which he meant the tendency for traits to covary in ways that often defy straightforward functional explanation (e.g. "cats with blue eyes are invariably deaf," p. 12). Ever since, a zillion different currents in evolutionary biology have reflected a focus on variation, like Bateson's call that "To collect and codify the facts of Variation is, I submit, the first duty of the naturalist" (Bateson, 1894, p. vi), to Wright's insights about how selection favors certain combinations of covarying traits over others (Wright, 1932), Pere Alberch's brilliant meditations on "internal constraints" (Alberch, 1989), and of course the foundational work of population biologists documenting heritable variation across the individuals of a population (e.g. Templeton, 2021). So, the claim is not that ontogeny has been ignored and that evolutionary biology needs an overhaul. Instead, it is that, given the history of ideas in the field, it is helpful to provide an update on advances in thinking over the past 50 years, that it remains deeply useful to think seriously about ontogeny, and that doing so with explicit attention to how different approaches complement one another (Table 5.1) is especially useful for building well-supported explanations of form.

Evolutionary biology is famously reflective about its own history, and for reasons that are not always clear, the question of what constitutes a conceptual innovation takes up a surprising amount of biologists' time. Entire books and stacks of papers are dedicated to asking whether evo-devo is a distinct discipline and whether it stands outside the Modern Synthesis or not (Raff, 2000; Laubichler, 2010; Minelli, 2010; Müller, 2011; Tamborini, 2022). The call for an "Extended Evolutionary Synthesis" invites debates about whether ideas like niche construction are truly new or just old wine in new bottles (Brodie III, 2005; Wray *et al.*, 2014). For all the course-charting value that these debates may have within specific subfields, they often seem to sow division rather than cross-pollination (Welch, 2017).

Chapter 5 makes clear why, as a field, we simply can't afford to spar across camps. We are, after all, trying to peer into the deep past to reconstruct events and processes that no one can observe directly. These reconstructions depend not on a single

Adaptationist Evo-Devo. Mark E. Olson, Oxford University Press. © Oxford University Press (2026).
DOI: 10.1093/9780197831533.003.0007

method or perspective, but on the careful assembly of thousands of small lines of evidence, from comparative biology, population biology, optimality modeling, genetics, development, ecology, paleontology, and more. No single camp, no matter how sophisticated, has anything like the full suite of tools or expertise needed to piece together such a sprawling puzzle.

That's why I've tried here to argue from the standpoint of what's useful in biological practice. No claim is being made that adaptationist evo-devo represents a revolutionary new evolutionary biology, that the Modern Synthesis was wrong, or that one group of biologists is benighted and another enlightened. The point is simply that building a good explanation requires more perspectives than any one laboratory or any one subdiscipline could ever hope to supply. Considering the enormity of the task facing us, it is a mystery why on Earth anyone would take the time to disqualify whole lines of empirical evidence and cogent reasoning because of how they are framed, given that no one is in any position to turn his or her nose up at any possible source of evidence, however small.

6.1 Adaptationist evo-devo precepts

So, why the "adaptationist evo-devo" term, then? For one, Oxford University Press wouldn't let me publish the book without a title. More importantly, some aspects of scientific terminology and scientific practice could use sharpening. Yes, we can point to a long history of evolutionary biologists talking about possible, improbable, and impossible variation. But we can point to a very large percentage of current evolutionary biology carried out without thinking about developmental potential as much as would be helpful. We can point to a fair amount of confusion around issues like "constraint," "macroevolution," "plasticity," as well as standard adaptation. Part of this neglect of developmental potential is due to a lack of a helpful and clear set of steps for thinking about form. At the same time, biologists don't spend much time talking to one another about what constitutes a solid explanation of organismal form and how to go about building one. So, it doesn't make any difference at all whether you declare adaptationist evo-devo to be a new scientific field, a useful handle, or a pointless term to describe what biologists manifestly already know (as illustrated by the case studies in Chapter 2, for example). What does make a difference is what we actually do as scientists. With this in mind, here is a summary of some of the main considerations throughout the book, now with the context of the structure of a well-supported explanation behind us.

6.1.1 Standard adaptationist accounts are non-trivial

It's often surprisingly difficult to construct standard scenarios regarding natural selection. The cases in Section 1.11 are designed to help think about what such scenarios actually involve. A particularly tricky aspect is clearly postulating the pool of variants that a given hypothesis involves, important in no small part because understanding why we observe some traits and not others requires understanding the pool of variants. Areas where difficulty in thinking about adaptation are common in the literature include the use of "macroevolution" for comparative studies that are designed to infer the population-level process of adaptation; vague uses of "constraint" for natural selection; notions of "maladaptive plasticity"; and, often, the tacit assumption that variation is unbounded and abundant, when it might not be.

6.1.2 Exploring developmental potential is a key aspect of explanations of organismal form

As part of the process of testing explanations of organismal form, an important step is the exploration of developmental potential. Without such exploration, explanations can involve a "just-so story" type assumption, either an adaptationist one or a non-adaptationist one. The non-adaptationist just-so story is "trait combination X is impossible to produce in my focal population. I know this because trait X is not observed in my focal population. It is not observed, because trait combination X is impossible to produce. . ." and so on. The adaptationist one is "this empty space corresponds to low fitness. I know this because it is empty. It is empty because it corresponds to low fitness. . ." and so on.

These are completely reasonable assumptions to make at any given point in exploring the causes of organismal trait variation. Often, information is available that makes one scenario seem more likely than another, say detailed information from a closely related species, known patterns of variation in a similar trait, or that certain traits often trade off with one another.

They are also reasonable because they are empirically testable, via the battery of approaches exemplified in Chapter 2 and summarized in Section 5.1.4. Broadly speaking, these include comparative approaches, which involve seeing whether certain "empty space" trait combinations are in fact produced under certain conditions of selection, e.g. the "empty space" around the terrestrial mammal bone-body-mass scaling line was shown to be accessible in the right circumstances by studies that examined bats and aquatic mammals (Section 2.3.8) and under artificial selection (Section 5.2.2). Explorations can use embryological strategies, which are detailed studies of variation within species, such as Galis *et al.*'s brilliant work on cervical vertebra number (Section 2.3.13) or examination of hundreds of cadavers to catalogue variation in human aortic arch configurations (Section 4.9). A wide range of approaches falls under the umbrella term "manipulation," from artificial selection in the lab, such as Jeff Conner's work on flower allometry (Section 2.3.2) to domesticated animals (Section 2.3.3–5), mutagenesis (Section 2.3.1), or surgery (Sections 2.3.7, 5.1.4.1). These techniques have been used throughout the history of evolutionary biology, so they aren't new. However, they don't really have a coherent name to refer to them; if you find calling them adaptationist evo-devo approaches is a helpful catchall, great. If you think that they already fall within standard adaptationist evolutionary biology, evo-devo, or some other domain, that's great too.

6.1.3 Developmental systems: small to large changes in research programs

The invitation in Chapter 3 was to consider, if only to see what the fuss is about, a developmental systems perspective. Entertaining such a view can lead to anything from no or only minor adjustments in research practice to full-scale shifts in a laboratory's direction. On the modest end, it relaxes the need to specify that a trait has a "genetic basis." For selection to act, it is sufficient that the trait be heritable. Shifting from "genetic" to "heritable" may be epistemically justified, but it can leave day-to-day research largely unchanged. At the other end of the spectrum, a developmental systems view can reorient an entire research program. Questions such as what constitutes the developmental system, how nongenetic elements participate in evolution, how selection shapes those elements, how to conceive of mutation beyond the genomic sequence, and how ontogeny arises through distributed causation rather than orchestrated by a central authority: all of these can become the core focus of a lab. Exploration of developmental potential, in particular, stands out as a clear mandate of the developmental systems view for any laboratory, systems-theory oriented or not.

6.1.4 Phenotypic plasticity as an adaptive phenomenon

Plasticity is a key participant in any consideration of the importance of development in evolution. For one, phenotypic plasticity is the outcome of selection shaping the breadth of developmental responses. As such, and to reiterate an important point, phenotypic plasticity is the tendency, shaped by selection, for a given developmental system (or genotype or individual or whatever the relevant unit is for you) to produce, *out of the possible configurations available to it* in a given situation, the one or ones that maximize its fitness *in that situation*. This definition excludes notions like neutral plasticity, which is just intrapopulational developmental variation that is functionally equivalent. Such variation is very important and possibly evolutionarily co-optable, but there's no reason to call all variation "plasticity." Likewise, the "out of the possible configurations available to it in a given situation, the one or ones that maximize its fitness in that situation" clause shows that maladaptive plasticity is a fallacious comparison between what a given developmental system can make in one environment as compared to a different one or to some sort of imaginary baseline. Selection shapes the variation actually present in a given population. Phenotypic accommodation is a similarly adaptive response and denotes the buffering characteristic of ontogeny. Continual detection-and-response mechanisms in ontogeny result in outcomes that are as functional as possible given the challenges faced.

The adaptive nature of plasticity is central to the plausibility of genes-as-followers evolution as a dominant mechanism of adaptive change. If nearly all members of a population show phenotypic change in a functionally relevant direction when faced with an environmental shift, and if accommodation involves individuals buffering novel perturbations in ways that yield high-functioning outcomes, then a novel environmental challenge will tend to elicit phenotypic changes across the population that move individuals in the direction of improved fit. Selection can then favor individuals whose developmental systems most reliably produce appropriate responses under the new conditions. The result is a shift in the population mean, not because a new variant arose in a single individual and swept the population, but because selection followed the lead of already-present plasticity.

6.1.5 Building explanations of organismal form: working together

The most important message of this book is that the approaches of any particular laboratory in evolutionary biology provide important insights but inevitably have their limitations. No single method in evolutionary biology is sufficient on its own, and because they complement one another's weakness, they must be combined (Table 5.1). The comparative method provides the generality we seek as evolutionary biologists attempting to account for the diversity of the living world, but little mechanistic detail. Population biology provides this mechanistic detail but its minuscule scale is only remedied by linking it to comparative data. Optimality models connect the differential performance of variants to their biophysical bases, while studies of developmental potential identify why some variants are available and others are not. To explain the diversity of life, to move from traits in populations to patterns in nature, we cannot work in isolation within subdisciplines. This isn't a generic "more data is better" posture. Instead, the very logic of evolutionary explanation demands that we draw on the strengths of each approach and confront the limitations of each in light of the others. Working together isn't just collegial; it's epistemically required.

Certainly in adaptationist evo-devo, if the case studies summarized in Chapter 2 are any indication, the spirit of cross-subdiscipline integration is a thriving one. Comparative study of centipede segment number has inspired detailed developmental work to uncover the basis of the striking only-odd pattern (Chipman *et al.*, 2004). These in turn inspired studies of centipede embryology and the search for possible teratologies across the group (Leśniewska *et al.*, 2009; Vedel *et al.*, 2010). In this case alone, we see comparative, populational, traditional embryological, and evo-devo perspectives working in concert. Jeff Conner and colleagues connected their study of radish flower proportions to variation in flower proportions across the entire plant family that radish is a part of, extrapolating the relevance of the population-level possibilities they documented in shaping floral morphological diversity across the family (Conner *et al.*, 2011; Waterman *et al.*, 2023). Frietson Galis and colleagues deeply ground their work on the causes of low cervical vertebra number in humans (Galis *et al.*, 2006, 2018, 2021, 2022) to the low variation across mammals, contrasting it with variation in vertebra number in other groups such as birds, even extending their reach to the fossil record, connecting their findings with vulnerability to extinction (Reumer *et al.*, 2014; Van Der Geer and Galis, 2017). Drake and Klingenberg's (2010) study of the remarkable amplitude of domestic dog skull shapes is compatible with the idea that not only are wolf skull proportions likely "explained by selection" in the sense of Section 5.9, but that they are across Carnivora as well. Tony Frankino, Patrícia Beldade, and their groups turn to the diversity of the butterfly genus *Bicyclus* to map which variants are experimentally producible in their focal species *B. anynana*, treating comparative diversity as a source of hypotheses (Beldade*et al.*, 2002; Brakefield *et al.*, 2009; Frankino *et al.*, 2010). Selection experiments on bone proportions in mice aim to understand these proportionalities across the mammals (Garland and Freeman, 2005; Marchini *et al.*, 2014; Sparrow *et al.*, 2017; Marchini and Rolian, 2018).

There are many other signs of an arc toward integration, with the very structure of the natural world pulling us out of our silos. Comparative study of G matrices explicitly builds bridges between comparative biology and quantitative genetic theory (Steppan *et al.*, 2002). For some comparative biologists, grounding phylogenetic comparative methods in quantitative genetics is a holy grail (Hansen and Houle, 2008). The coupling of phylogenetic, ecological, and experimental data has become a hallmark of evolutionary work that seeks maximally robust explanations for the immensity of the organismal world and the depths of time in which it formed (Larson and Losos, 1996; Garland and Rose, 2009). Reconstructing the immensity of the causes of biological variation, so much of which is lost in time, means piecing together every clue we can get. And this requires working together.

A theme throughout has been that there is more convergence and overlap in the interests and goals of biologists than disagreement over theoretical stances, even if the disagreements are often noisier. To tie these threads together to move toward a rousingly congenial conclusion, here is a reminder of the various sorts of important common ground stretching across the various interests in evolutionary biology that were touched upon above.

6.1.5.1 Common ground: complementarity of methods

The most important common ground is the complementarity of our methods, treated in Chapter 5, and that the very structure of evolutionary explanation demands mutual regard for approaches across evolutionary biology. If your faith in understanding across subdisciplines should ever flag, you could do worse than to look at Table 5.1 for inspiration.

6.1.5.2 Common ground: disputes over relative importance, not wholesale disqualification

Another thing to notice is that debates over topics like nongenetic inheritance, developmental systems, systems biology, genetic causation, genes-as-followers evolution, and niche construction that are often central to calls to "expand" evolutionary biology rarely involve outright rejection. Even in the most forceful rebuttals, biologists don't claim that these phenomena don't exist. Instead, they typically argue either that the processes are relatively unimportant in nature or that they've long been recognized and aren't especially novel. But neither of these positions amounts to saying the phenomena aren't worth studying. Meanwhile, those calling for expansion also acknowledge the crucial roles of DNA and population-level processes. Their goal is usually to shift the balance of emphasis, not to discard foundational principles such as natural selection. So, in practice, there's widespread agreement about the existence of things like nongenetic inheritance, genetic assimilation, and niche construction; the disagreements lie in how important or new these phenomena are taken to be. Setting these differences aside, there is a broad field of empirical common ground.

6.1.5.3 Common ground: systems biology

As discussed in Section 3.19.5, systems biology offers another area of shared perspective across theoretical orientations. Deborah Gordon's research on ants illustrated ways that order can emerge in colony behavior not because there is an overseer telling each ant what to do, but from simple local interactions between individual ants. Agreeing that such distributed causation can occur in one system is common ground for biologists to agree that it can occur in others, too. So, regardless of whether one begins with a focus on genes or on development, detailed empirical inquiry into how phenotypes arise eventually draws attention to the interactions among components, molecular, cellular, structural, "environmental," that give rise to form. The resulting picture does not discard DNA but reveals it as one participant in a network of relationships, rather than as a solitary director. From fields as diverse as molecular genetics to clinical work, the underlying practice of explaining development naturally leads to a systems view. In this way, systems biology seems to be a point of convergence for biologists approaching the problem of form from different angles.

6.1.5.4 Common ground: order for free

Section 3.17 introduced the expression "order for free" and argued that it, too, highlights common ground shared by biologists across theoretical divides. The phrase refers to features of behavior, metabolism, or development that are not fully specified in the genome. The metaphor suggests that encoding these features genetically would come at a "cost," and that allowing them to arise through extra-genomic processes secures them "for free." Section 3.17 pointed to examples such as protein folding and the mechanical behavior of tissues. Even under the most literal interpretation of a genetic blueprint, some aspects of development and metabolism arise through physical interactions among molecules or the biomechanics of tissues, and therefore don't need to be spelled out in the genome. Some biologists believe that relatively little is left to "order for free," while others think it accounts for a great deal. Either way, this framing shifts the focus away from camps like developmentalist versus adaptationist or internalist versus externalist, and toward the shared and fertile territory of studying how form, metabolism, and behavior emerge.

6.1.5.5 Common ground: what is possible and what is not in development

Biologists have always been concerned with what kinds of variation can arise and be exposed to selection, and with how this field of possibilities is itself shaped by selection. The language used to describe this concern varies enormously, from genetic variance-covariance matrices and studies of pleiotropy and epistasis, to talk of "constraints," "evolvability," developmental and adaptive "landscapes," attractors, tradeoffs, developmental bias, modularity, and robustness. But the underlying interest is the same: biologists recognize that some changes are possible, some are impossible, and others are possible but highly improbable, and that these distinctions matter a great deal in shaping what evolves. The vocabulary and framing may shift from one subfield to another, but if there is a biologist of any stripe who would object to the kinds of empirical studies laid out in Chapter 2, I have yet to come across one.

6.1.5.6 Common ground: universal regard for empirical data

Along these lines, Section 3.22 argues that many disagreements in evolutionary biology seem less about the data themselves and more about how those data are framed. In practice, empirical findings tend to be welcomed across theoretical lines. Most biologists, whatever their orientation, want their theories to account for as much evidence as possible, no matter where the study that produced it took its inspiration.

6.1.5.7 Common ground: the grain of research focus

Section 5.9.5 presented Wilkins and Godfrey-Smith's (2009) idea that differing views on the relative importance of "constraint" or contingency versus adaptation often stem from whether biologists focus on what these authors call the "fine grain" of population-level processes or the "coarse grain" of broad spans of time or morphospace. In their view, differences in judgments about relative importance arise naturally from the scale at which a researcher works, rather than from any deep theoretical divide. This framing points to specific areas of common ground that emerge when we recognize how different levels of granularity reveal different aspects of the same evolutionary landscape.

In a 2001 essay, Ron Amundson fretted over the "The current gulf between developmental biology and mainstream evolutionary theory" (Amundson, 2001, p. 304). Much has changed since then, and today, at least from the perch of empirical evolutionary biology, Amundson's wish for common ground seems abundantly fulfilled, in the examples in this section and beyond. From this point of view, the common ground stretches far beyond any fenced-off parcels of disagreement, perhaps even helping to reclaim some territory once undervalued as unfit for serious evolutionary thought, like telling just-so stories.

6.1.6 Just-so stories can be developmental as well as adaptationist, and they're not so bad

The term just-so story entered biology as a jab at adaptationist scenarios that relied more on narrative plausibility than on empirical support. But just-so stories aren't exclusively adaptationist, and they're not immune to testing. Consider a patchy morphospace, some regions filled with known forms, others curiously empty. One might spin an adaptationist story: the gaps represent low-fitness phenotypes, while the occupied zones are adaptive peaks. Without additional evidence, this is a thin explanation, and fair game for the "just-so" label. But the same morphospace invites a non-adaptationist counterpart: the gaps are inaccessible not because they're maladaptive, but because contingent characteristics of development impede access, and these "empty" spaces are likely of equivalent or even higher fitness than the occupied areas. That developmental story is just as speculative as the adaptationist one, and just as deserving of the "just-so" label if left untested.

The real issue with a just-so story isn't its origin point, but whether it becomes fossilized as canon without facing empirical scrutiny. Fortunately, the alternative scenarios sketched above are entirely testable using adaptationist evo-devo tools. A just-so story is, in truth, nothing more than a preliminary evolutionary explanation, a hypothesis in narrative form, offered when little evidence beyond the pattern itself is available. Explanations have to start somewhere, and as long as they are empirically accessible, they can start anywhere. So it is time to shrug off the shame around just-so stories and recognize them for what they are: honorable, and sometimes essential scouts in the landscape of evolutionary explanation.

6.1.7 From buzzwords to bedrock

Adaptive landscape, adaptive radiation, canalization, developmental "burden" and "constraint," eco-evo-devo, ecological filter, ecological niche, evolutionary novelty, evolvability, exaptation, key innovation, macroevolution-that-isn't, niche conservatism, phylogenetic constraint, phenotypic integration... ecology and evolutionary biology are full of terms that can be useful, but can also obscure more than they clarify when it comes to the biological processes actually being postulated. These terms remain understandable shorthand, but alternative vocabulary rooted in the logic of adaptationist evo-devo can provide better tools for reasoning about why traits take the forms they do. This language, in terms of what is developmentally possible or impossible, what causes that variation, and, among the possible, what is favored by selection and why, provides the perfect place to wrap up.

6.2 "Adaptation versus constraint" and alternative vocabulary

The "constraint or adaptation" dichotomy persists in part due to a lack of suitable alternative vocabulary. A major advantage of the adaptationist evo-devo perspective is that it provides the language tools to avoid the vague "explained by adaptation or constraint" dichotomy (see Pigliucci and Kaplan, 2000). By framing explanations of form explicitly in terms of developmental potential, the either/or constraint/adaptation dichotomy is obviated. It is replaced by descriptions of the range of developmental potential versus the range of the favored. In the most unequivocally "explained by adaptation" situations, the limits to developmental potential are so far outside the observed range of variation that, in most cases, it is utterly unknown where they lie. Within the space of realized developmental potential, there are no conspicuous "holes" of apparently inaccessible developmental space. Developmental alternatives adjacent to the commonly observed variants are such that functionally different configurations could readily be selected if they were favored. In standard adaptation scenarios, as one moves away from the commonly observed variance, fitness drops off. This situation, represented in Figures 5.10A and B, is just one extreme of a continuum.

Along this continuum, the field of developmental potential can become more and more circumscribed. For example, in situations of trade-offs, the space above the line that most of the data fall on is developmentally inaccessible (as in Figure 5.10C). In most cases, the space below the line is developmentally accessible but not favored by selection. Thus, a simple trade-off between two traits describes a situation in which half of the space is developmentally inaccessible. In probably most cases of organismal traits, selection favors certain trait combinations over others (e.g. in terrestrial mammals, short, skinny bones or long, thick ones, but not long, skinny ones). In these cases, selection biases ontogeny such that the favored outcome is preferentially produced. None of these situations is straightforwardly captured by either alternative in the "constraint versus adaptation" dichotomy. Adaptationist evo-devo builds explanations that document the field of developmental potential, what causes any holes, limits, or biases in it, and, of the possible, why certain variants are common whereas others are not. In such explanations, it's unnecessary to explain form by appealing to one or the other side of a "constraint-adaptation" dichotomy (see authors such as Salazar-Ciudad, 2006; Brigandt, 2020; Hordijk and Altenberg, 2020).

Tying this enormous methodological diversity together is a shared inquiry. Biologists every day are confronted with patterns of trait space occupation, in which some parts are sparsely occupied or empty and others are densely filled. In the empirical effort to understand these patterns, the logical sequence in adaptationist evo-devo studies can be summarized as:

1. Can "empty space" configurations be produced in development?
2. If so, are they of lower fitness than the commonly produced variant? This result is consistent with standard adaptation accounts.
3. If the "empty space" variants are not developmentally accessible, but can be created by surgery, do the created variants have lower fitness than the common variants? If so, this result is consistent with a standard adaptation account in which variation has been extinguished by selection or development is so biased by selection (Section 3.11) that variation is limited.
4. If empty space variants have equivalent or higher performance or fitness than the commonly observed configurations, then this suggests that some tradeoff (as in Section 2.1.4), shared developmental cascades (as in Section 2.1.5), or as in geophilomorph segment determination (Section 2.1.1), some other contingent factor of development limits the production of some variants, pointing to the study of the causes of these empty spaces.

6.3 The most useful false dichotomy

Section 5.9 sets out various reasons why the constraint-adaptation dichotomy is, at heart, untenable. First, "constraint" is a vague term, and most of its uses just denote selection in disguise. Moreover, many inherent biases in development, those that make some outcomes much more likely than others, are the very products of selection (see Section 3.11). In this sense, "constraints" are, as Schwenk and Wagner (2004, 403, 404) put it, "forged in the fires of selection." Other limits are simply the unavoidable tradeoffs imposed by the finiteness of organismal resources, where maximizing one outcome necessarily reduces another, making simultaneous maximization impossible. In addition, all species carry "contingent" developmental limits, ones that are not imposed by the nature of organismal matter itself, but that simply happened to evolve the way they did. Seven cervical vertebrae in mammals would seem to be one such example. Finally, developmental possibilities are bounded on all sides by the properties of matter. For these reasons, any explanation of form must include both selection and "constraint" in whatever way that term is interpreted. As a result, a neat delimitation between "selection" versus "constraint" as distinct causes of form is no basis for solid thinking.

In gratifying contrast to its theoretical shortcomings, a version of the constraint-adaptation dichotomy proves invaluable in early empirical research. When faced with a restricted pattern of trait variation, in which certain regions of morphospace appear empty despite seeming plausible, the working biologist has little choice but to ask whether the unoccupied forms can be developmentally produced or not, and if so, how they perform. If they apparently can't be, then the quest becomes to find out why. Experimental perturbations, artificial selection, surgical manipulations, and comparative studies can begin to fill in these blanks. Sometimes the "missing" morphologies can be produced and show lower performance or fitness. If, however, these forms cannot be readily generated, or if forced variants perform better than the wild type, the trail leads toward understanding the developmental architecture that limits variation. So, while dichotomies may not offer a coherent account of causality for building solid biological theory, the question of "accessible or not?" is an indispensable empirical first step, and the heart of adaptationist evo-devo (or whatever you want to call it).

Glossary

Abduction: A type of inference in which assumptions are accepted because they would explain the observed data well if they were true. For example, we might infer that female choice has shaped mating behavior throughout the evolutionary history of the peacock, even though we haven't observed every courtship throughout that history. This assumption is considered reasonable because it helps explain the presence of the male's conspicuous tail.

Adaptation: As a process, an evocative shorthand for how natural selection shapes the fit between form and function; there's interesting debate on this, but for all practical purposes, equivalent to selection acting on heritable variation. As an outcome, any organismal feature favored by such selection.

Adaptationist: Used here deliberately to reclaim a term that has often been used pejoratively to imply a naive and overly simplistic view. In this book, "adaptationist" refers to a perspective that takes seriously the possibility that, among the developmentally possible variants, some are favored over others. It also emphasizes that a standard adaptationist scenario is essential for detecting where a given situation resists explanation under a standard selection scenario.

Adaptive: Another shorthand for invoking the role of selection. "Adaptive evolution" is evolution by natural selection. Calling a trait "adaptive" is an economical way of saying that it is favored by selection in its current role. An "adaptive interpretation" is one that treats selection as an important cause of organismal form. An example of a "non-adaptive" feature is one that would not be favored by selection if alternative forms were developmentally available.

Additive genetic model: An additive genetic model, such as the ones used in early genome-wide association studies (GWAS), refers to a statistical framework in which the effect of each allele on the focal trait is assumed to be independent and to sum linearly with the effects of other alleles.

Additive genetic variance: A population-level parameter from quantitative genetics. It refers to the portion of the total genetic variance in a trait attributable to additive effects of alleles, the part that responds to selection and contributes to resemblance between relatives.

Ancestral: Given variation in a trait or condition, including environment, between the ingroup (the clade of interest) and an outgroup (a close relative of the ingroup), the condition shared with the outgroup is considered the "ancestral" one, that is, the state the common ancestor likely had. Sometimes called the "primitive" condition, but this is a technical term and doesn't imply anything about being somehow inferior or less functional. See *derived.*

Anthropomorphization: Describing natural processes as if they involved conscious, human-like intent, e.g., saying that trees grow tall as fast as they can "in order to get more light and outcompete their neighbors." Anthropomorphization is worth watching for, because it often masks a lack of clarity about the actual causes being proposed. See *teleology.*

Bayesian inference: A method of reasoning in which beliefs (called priors) about the likelihood of a hypothesis are updated in light of new evidence. As data come in, the prior is revised to produce a posterior, which is a new, more informed belief that can then serve as the starting point (a new prior) for further inference. This creates a back-and-forth between prior expectations and empirical data, exactly what is observed in the structure of a scientific explanation.

Biophysics: The physical and energetic principles underlying biological structures and processes, for example, how shape, size, stiffness, signaling interactions, or the energy involved in a given performance level influence survival and reproduction.

Causal parity: The principle that all developmental resources necessary for producing a trait, such as genes, cells, water, photons, sounds, temperatures, etc., should be treated as equally fundamental. While they may differ in timing or magnitude of influence at any given moment, none is privileged as the only cause.

Circularity: Circular reasoning occurs when an argument begins with a premise and eventually returns to it without introducing any genuinely new conclusion. It can sometimes hinder biological research by making patterns seem explanatory when they're just being restated. If we define an adaptive radiation as "one side of a bifurcation in a clade has 90% or more of the species," then: . . .this clade is an adaptive radiation—> I know this because of the clade imbalance—> which is observed because—> this clade is an adaptive radiation. . . Biologists often use accusations of circularity to invalidate each other's arguments. Most of the time, though, these instances of supposedly fallacious reasoning are just "loops" of reasoning in preliminary hypotheses or the legitimate acceptance of assumptions in explanations (see Sections 5.3–5.6).

Comparative approach for studying developmental potential: Searches for "empty space" trait combinations in across-species data. Finding that bat and aquatic mammal bone-body mass proportions differ from the terrestrial ones is evidence consistent with the empty space around the terrestrial scaling line being due to selection not favoring those variants, not because mammals can't produce those proportions.

Comparative method for studying adaptation: An approach that identifies evidence consistent with adaptation by examining cases of apparent convergent evolution (see *Convergent evolution*).

Configuration: A way of avoiding the term "phenotype," with all its baggage from the arbitrary genotype-phenotype dichotomy. A "configuration" refers to a particular trait value or combination of trait values. So, saying "selection tends to favor some configurations over others" focuses attention on observable outcomes without enforcing a strict division between genomic and extragenomic causes.

Conservatism, niche or phylogenetic: When closely related species resemble each other more than they resemble distant relatives, whether in a particular trait (phylogenetic conservatism, inertia, or effect) or in habitat preference (niche conservatism). Often treated as if it reflects some fancy evolutionary process, but it's better understood as a pattern that requires explanation, not a mechanism in itself.

Constraint: Probably the vaguest, least helpful term in all of biology, meaning everything from factors that oppose selection to selection itself. See Section 1.4.

Contingency: The situation of certain traits being present not because they were favored out of a broad range of possibilities, but as a result of evolutionary history unfolding in a particular way. It includes historical outcomes like seven cervical vertebrae in mammals or odd segment numbers in geophilomorph centipedes. Often contrasted with selection, but the two are intertwined: practically any condition of selection, such as the particularities of Earth's gravity or atmospheric composition, are themselves contingent outcomes of history.

Convergent evolution: The process by which similar selective contexts lead to the independent evolution of similar organismal features in different lineages, starting from functionally significantly different ancestral conditions. The ancestral states differ enough that the resulting similarities are not predicted by shared ancestry; instead, they arise under similar functional demands and in settings where developmental variation is sufficiently broad for functionally distinct alternatives to be favored by selection. Under this view, the morphological similarity between *Moloch* and *Phrynosoma* in drylands supports an adaptive explanation, as does bone-to-body mass scaling across mammals, where bone proportions correlate more strongly with body mass than with phylogenetic relatedness. What counts as "similar" and "different" are important questions (Powell, 2020). See also *parallelism*.

Cues, environmental: A specific condition detected by an organism that triggers an adaptive developmental response, with selection favoring alternative outcomes depending on the cue's state. For example, in tropical dry forest trees, high soil moisture prompts leaf retention, while low moisture triggers leaf drop: both developmental outcomes are functionally appropriate.

Deduction: A form of inference in which, given the starting conditions and the rules governing interactions, a specific conclusion necessarily follows. Science uses deduction when generating predictions from hypotheses, for example, "if X process is operating, then we must observe Y pattern." Mathematical reasoning is also deductive. While deduction plays an important role in science, full scientific explanations cannot be purely deductive; they instead have an abductive or Bayesian structure, involving inference to the best explanation based on evidence.

Derived: Given variation in a trait or condition between the ingroup (the clade of interest) and an outgroup (a close relative of the ingroup used for comparison), the condition unique to the ingroup is considered the "derived" one, that is, a trait that evolved after the ingroup split from the outgroup. Traits (including environments) are ancestral or derived, not clades. See *ancestral*.

Development: To help think about the spaces of common and uncommon variants, the possible and the impossible, "development" is used here in a broad sense to refer to all the processes by which phenotypic variation is manifested, whether in building a whole organism or just a part. This definition allows the term to point to the molecular, cellular, structural, and other mechanisms that shape the variants that organisms, whether multicellular or unicellular (or viruses), can "make." Synonymous with ontogeny.

Developmental bias: As defined by Lala *et al.* (2024), it's that "developmental systems tend to generate some characters, or trait combinations, more readily than others," a pattern with heterogeneous causes, from selection to resource allocation tradeoffs to contingent aspects of development.

Developmental potential: A term for the full space of configurations that a given individual, species, or clade can produce, from the commonly observed to the rare but still a part of everyday populations, to the teratological and variation induced experimentally. It overlaps in meaning with some uses of "evolvability" or "developmental bias," but aims to avoid the goal-directed connotation of the former and the risk of implying a single causal process in the latter.

Developmental system: The full set of resources, DNA included, required to construct a new individual that resembles its parent. As with issues such as species delimitation, how this system should be delimited represents an important empirical and conceptual matter for scientists to address.

"Empty space" variants: The variants (phenotypes or "configurations") that are not observed in a given sampling of individuals, from populations to species. Examples include variants above or below the terrestrial mammal femur length-body mass scaling line, or centipedes with even numbers of segments. Their absence prompts questions about developmental possibility and fitness consequences.

Evo-devo: Short for evolutionary developmental biology, the field that investigates how developmental processes shape the range and structure of variation, and how these processes contribute to the evolution of organismal form both within and among species. Evo-devo seeks to explain not just what variation is available to selection, but why. Used in "adaptationist evo-devo" to emphasize the focus on developmental potential.

Evolvability: Equivalent to *developmental potential* as defined here. "Evolvability" can be misleading, because it's frequently invoked in vaguely goal-directed language, e.g. "the ability of a biological system to generate adaptive phenotypic variation." It risks implying that trait diversity is inherently good or somehow favored in a forward-looking sense.

Explanation, evolutionary: The full set of considerations, empirical evidence, plus assumptions about ancestral populations and unobserved individuals, that together offer a causal account for a pattern observed in nature. Historical assumptions are accepted not because they are directly verified, but because, if true, they would plausibly account for the pattern. An explanation provides a narrative grounded in inference about what must have been the case for the world to look the way it does.

Extended inheritance: Mechanisms that contribute to reconstructing parental-type configurations beyond DNA. Jablonka and Lamb (2005) classify these as epigenetic, behavioral, and symbolic.

Externalism: The view that the range of developmentally available variation is sufficiently broad and evenly distributed that natural selection can be treated as the principal cause of organismal form. In this perspective, biases or limits imposed by development are considered minor. Externalism contrasts with *internalism*.

Fecundity: The third of the three components of fitness, generally referring to how many offspring a given variant produces. What counts as fecundity can vary depending on the system, ranging from the number of eggs or seeds produced to the number of viable offspring. It interacts with mating success to determine the actual reproductive output of a given variant.

Field of developmental possibility: A metaphor for the total range of phenotypes that could, in principle, be produced by a given developmental system, of an individual, a species, or a clade. Some outcomes are common, others are rare, and some occur only under unusual or artificial conditions. These outcomes may involve continuous or discrete traits, and their distribution helps reveal why certain trait combinations are observed and others are absent. Synonym of *developmental potential*; see also *morphospace*.

Fitness: The sum of survivorship, mating success, and fecundity that results from how the biophysical properties of a given variant interact with its selective context. Fitness emerges from these interactions in such a way that some heritable variants within populations become disproportionately represented across generations compared to others.

Form: The organismal attributes that biologists seek to explain, meant in the broadest sense, encompassing morphology, physiology, biochemistry, and other features of structure and function.

Function: For working biologists, "function" generally refers to the role of an organismal feature that was favored by natural selection and that accounts for the trait's presence. For example, the function of the heart is to pump blood. This is distinct from other real but non-functional attributes, such as incidental effects (like the sound the heart makes), dysfunctions (such as arrhythmias), malfunctions (like congenital valve defects), or pathological conditions (such as coronary artery disease or heart attacks).

"Generative" assumptions: A term used here to designate the background conditions that allow a method's findings to contribute to evolutionary explanation. They cannot be tested using additional evidence from the same method, but only by complementary approaches, e.g. the generative assumptions of population biology cannot be tested by population biology but by comparative and other methods.

"Genes-are-everything": The idea that genes, usually understood as stretches of DNA, are the only really important resource that causes development to unfold in ways that reconstruct parental configurations. It implies that genes represent the bedrock level of biological organization, and that while other factors might be necessary, they can effectively be regarded as taken-for-granted background. The term is used here simply to indicate one end of a continuum of thinking about the unit of inheritance, and isn't intended as an unfair caricature.

"Genes-aren't-everything": The view that, while genes are necessary, they are not sufficient for development, and that, because nongenetic components, from water and photons to cellular structures, epigenetic states, and environmental factors, also contribute to developmental outcomes, even across generations, they too must be considered evolutionarily significant. Used here to represent the other end of the continuum in thinking about inheritance and organismal construction from "*genes-are-everything*".

Genetic assimilation: A term that can make it sound as though some environmentally induced change is being absorbed into the genome. In fact, it describes a standard process of natural selection, in which individuals vary in how readily they produce a developmental response to a given stimulus. Selection favors those in which the response is most easily triggered, so over time the trait becomes expressed with less environmental prompting, or even without it at all.

GRN: Abbreviation for "gene regulatory network." A model describing how genes, RNAs, proteins, and other molecular components interact in gene expression. While the name might suggest a network of genes, GRN models typically represent interactions among gene products, such as transcription factors, regulatory RNAs, and other molecular agents, rather than the DNA sequences (gene loci) themselves as nodes. GRNs can also incorporate diverse and context-sensitive elements, such as chromatin state, protein modifications, cellular localization, and environmental factors.

GWAS: Genome Wide Association Studies, a method for identifying genetic variants associated with a trait by scanning the genomes of many individuals in a species. Researchers compare millions of markers, usually single-nucleotide polymorphisms, between those with a given trait and those without it. If certain variants are more common in the trait group, they may point to genomic regions associated with variation in that trait.

Heritability: A pattern, the degree to which offspring resemble their parents. Only trait variants that have some reliable degree of similarity between parents and offspring can show changes in frequency within populations between generations as a result of the interactions between the variants and their selective contexts.

Historical assumptions: Paul Griffiths's (1996b) term for claims about past evolutionary events, processes, or conditions, such as the traits, variation, and selection pressures present in unobservable ancestral populations. Historical assumptions are essential for constructing evolutionary explanations because they fill in the gaps left by empirical data. Unlike *generative assumptions*, which can be tested using present-day data from complementary methods, historical assumptions persist as necessary commitments because direct evidence is unavailable.

Inference to the best explanation: A reasoning process that involves evaluating competing explanations accounting for a pattern or phenomenon. It typically includes making and testing predictions that help distinguish among alternative explanations. Often equated with *abduction*, but perhaps better characterized as iterative applications of rounds of abductive inferences, narrowing down the pool of potential explanations. See also *underdetermination*.

Internalism: The view that organismal form is shaped not just by natural selection but by intrinsic factors like the physical properties of biological materials and the developmental processes specific to each lineage. It holds that the variation available to selection is always much more likely in some directions than others. *Structuralism* is a form of internalism; for our purposes, the two are treated as equivalent. Contrasts with *externalism*.

Macroevolution: Most helpfully, the term is synonymous with "species selection," which is a kind of selection in which species within a clade vary with respect to some attribute that causes differential tendency to survive or become extinct, or to speciate or not. "Macroevolution" sometimes includes patterns like changes in species characteristics within a lineage over geological time or just a change that strikes someone as "large." Whatever the definition, it definitely is not equivalent to the comparative method for studying adaptation, which adduces evidence consistent with adaptation, a standard population-level process, or individual-level developmental attributes, neither of which is helpfully described by simply adding "macro" to them.

Manipulation methods for studying developmental potential: Create "empty space" variants by interventions such as surgery, laboratory selection, selection under domestication, directed mutagenesis, hormonal manipulation, and other techniques. Allows directly comparing performance and fitness of "empty space" configurations with the wild-type.

Mating success: The second component of *fitness*, describing the number of opportunities for reproduction. In animals, this might mean the number of mates or matings; in plants, it can refer to the number of pollen grains exported from a flower or the number of pollen tubes initiated on a recipient stigma. Mating success, together with *fecundity*, determines overall reproductive output.

Metaphor: A metaphor uses a concept from a familiar context and projects it onto a less-known one where it doesn't literally apply. In doing so, metaphors highlight certain features of the target domain and suggest particular kinds of questions. For example, if adaptation occurs on a "landscape," we can then ask how populations move between "peaks," a question shaped by the metaphor itself. But aspects that don't project cleanly can hide alternative questions or features that fall outside the metaphor's frame. Metaphors are both tools for discovery and lenses with blind spots. They are recognizable by their evocative power and by the inevitable mismatches between source and target domains.

Missing heritability: Genome-wide association studies (*GWAS*) explain less phenotypic variation than traditional estimates of heritability suggest they should. This gap is often attributed to factors like rare variants, epistasis, or gene-environment interactions. But it is also consistent with the view that DNA is a necessary but not sufficient resource that participates in many ways within developmental systems, rather than serving as the sole "instructions," "blueprint," or "basis" of traits, as common language often implies.

Modern Synthesis: The mid-20th century effort (ca. 1930–1960) to unify evolutionary biology by linking theory, paleontology, comparative anatomy, and field and lab experiments with the emerging framework of population genetics. While often caricatured as rigidly gene-centric or simplistically reductionist (to the frustration of many biologists), the Modern Synthesis included researchers with a wide range of views, some of whom explicitly recognized the importance of development in evolutionary change.

Morphospace: A hybrid term from "morphology" and "space," referring to a coordinate system in which two or more morphological (or other phenotypic) variables are plotted against one another. Data tend to cluster in certain regions, leaving others sparsely occupied or entirely empty. Morphospaces help document the range of possible variation and prompt questions about why some trait combinations are common while others are rare or absent.

Natural selection: The population-level process in which heritable variation among individuals leads to differences in survival or reproduction (mating success and fecundity), depending on how the characteristics of variants interact with their environments. In its standard formulation, natural selection operates within populations, not between species. A "force" or "pressure" only in a metaphorical sense.

Ontogeny: Synonym of "development"

Optimality modeling: Uses basic principles such as biophysics or an engineering perspective to predict, given two or more competing considerations, what the "least bad" combination should be. Finding that nature coincides closely with optimality predictions is taken as evidence consistent with selection. Often, a lack of coincidence points to functionally relevant variables that were left out of early iterations of models. In principle, lack of coincidence could point to combinations that would be favored but are developmentally inaccessible.

***Organismal form*:** see "form"

***Parallelism*:** The evolution of similar traits in related lineages under similar selective conditions, starting from similar ancestral states. It is often seen as a sign of shared developmental potential inherited from a common ancestor. Parallelism contrasts with *convergence*, which involves different starting points. In practice, there is a continuum between the two. For example, eyes in vertebrates and insects are structurally distinct and often called convergent, but both rely on ancient, shared molecular pathways, also fitting the definition of parallelism.

***Performance*:** A measurable variable reflecting how variation in a trait relates or might relate to fitness, grounded in biomechanics, biophysics, biochemistry, or energetic tradeoffs. For example, water striders that invest more energy for the same water-striding speed could be expected to have lower fitness than those that invest less. Whether a given performance index truly reflects fitness is an issue for empirical testing.

***Phenotype*:** Used here in a broad sense to include morphological, behavioral, coloration, biochemical, and other characteristics of organisms. If organisms are understood as systems composed of interacting resources, then at any point in the life cycle, what exists is just the system, and that system is the phenotype. So, the term is used here for lack of a better widely accepted alternative and does not endorse the genotype-phenotype dichotomy as conceptually sound. Throughout, "*configurations*," "variants," and "*form*" are used as synonyms.

***Phenotypic accommodation*:** An expression of the responsiveness of ontogeny, in which developmental responses to perturbation or novel conditions help preserve or restore function, often resulting in normal or near-normal outcomes (as in common aortic arch variants), or at least the most functional outcome possible given the developmental options available (as in "brain sparing" in human development). Like *phenotypic plasticity*, an adaptive feature. While widely used in the literature, the term might be helpfully thought of as "accommodative plasticity," since it describes one aspect of the broader category of detect-and-response phenomena in development (plasticity broadly construed, encompassing both phenotypic plasticity and accommodation).

***Phenotypic plasticity*:** The tendency, favored by selection, for a developmental system to produce, out of the outcomes available to it in a given situation, the one that maximizes performance or fitness in that environment. The term emphasizes the diversity-generating side of adaptive detect-and-response mechanisms in development, in contrast to *phenotypic accommodation*, which highlights a tendency to convergence toward similar outcomes.

***Population biology (methods for studying adaptation)*:** Population biology focuses on direct studies of within-species variation in traits among individuals, the heritability of that variation, and the associated fitness consequences. It includes approaches such as laboratory selection experiments, classical population genetics, quantitative genetics, and field studies that track trait variation, reproductive output, and survivorship in natural populations. Strictly speaking, the "evo-devo" component of adaptationist evo-devo, which explores how organisms occupy trait space and why some regions remain unfilled, falls under population biology's broad concern with variation. Studies of developmental potential are treated separately here to highlight their importance and focus on the entire range of developmentally possible variation.

Resources, developmental: A component—genetic, environmental, cellular, structural, etc.—that plays a necessary role in the construction of an organismal trait or form. Developmental resources are not limited to DNA but include factors such as nutrients, physical forces, symbionts, maternal inputs, or light, whose presence enables normal development and whose absence can lead to disruption, pathology, or compensation. As with developmental systems more broadly, much remains for biologists to discuss regarding how to determine what counts as a developmental resource. See also, Cues, environmental.

Scaling: How one trait changes relative to another as size increases, like how bones get larger as body size increases, or how bone length relates to bone diameter. Allometric scaling or allometry refers to cases in which the slope of a relationship is above or below one, and isometric scaling or isometry when the slope is one.

Self-construction: The role of a developing organism in producing the conditions that contribute to the reproduction of the parental phenotype. Rather than being merely sculpted by external forces or simply executing DNA instructions, the organism helps generate key developmental resources through its own activity.

Sister group: The taxon or clade that is most closely related to a focal taxon or clade. Chimpanzees and bonobos are sister taxa, and humans are sister to the bonobo-chimpanzee clade.

Structuralism: The view that the specific developmental architectures organisms possess make evolutionary change in some directions more likely than others, and that the material properties of tissues play a large role in shaping the forms that arise. This perspective holds that such biases can so strongly influence the scope of selection, or even limit it, that they deserve at least as much attention as selection itself in explaining organismal form. Structuralism is a form of *internalism* (see also *internalism*).

Survivorship: The first component of *fitness*, followed by *mating success* and *fecundity*. Typically defined as the probability that an individual survives from birth (or another starting life stage) to a later stage, often reproductive age.

Teleology: The idea of goal-directedness, that traits exist "for" something. Two distinct uses in biology. One is legitimate: because some trait features can be explained by past selection, it can be meaningful to distinguish proper *function* from malfunction, byproduct, illness, etc. The other is misleading and pervasive: attributing purpose or conscious intention to organisms ("trees grow tall to outcompete shrubs"). Replacing both, but especially the second, with mechanistic hypotheses improves clarity.

Teratology: Rare developmental outcomes often described as "abnormalities," "malformations," or "deformities." These variants are typically uncommon because they tend to have lower fitness under ordinary selective conditions. Despite their rarity, they are just as much a part of an organism's developmental repertoire as more common outcomes.

References

Abe, G., Lee, S.-H., Chang, M., Liu, S.-C., Tsai, H.-Y., Ota, K.G. 2014. The origin of the bifurcated axial skeletal system in the twin-tail goldfish. *Nature Communications* 5: 3360.

Abrams, M. 2009. The unity of fitness. *Philosophy of Science* 76: 750–761.

Agabian, N. 1990. Trans splicing of nuclear pre-mRNAs. *Cell* 61: 1157–1160.

Akbay, G. 2022. Gene centrism, causal specificity and missing heritability. *Revista Portuguesa de Filosofia* 77: 1455–1478.

Alberch, P. 1989. The logic of monsters: Evidence for internal constraint in development and evolution. *Geobios* 22: 21–57.

Alberch, P., Gale, E. 1983. Size dependence during the development of the amphibian foot. Colchicine-induced loss and reduction. *Journal of Embryology and Experimental Morphology* 76: 177–197.

Albergante, L., Liu, D., Palmer, S., Newman, T.J. 2016. Insights into biological complexity from simple foundations. In: Leake, M. C. (ed.), *Biophysics of Infection*, Springer International Publishing, Cham, Advances in Experimental Medicine and Biology. vol. 915. pp. 295–305.

Alix, K., Gérard, P.R., Schwarzacher, T., Heslop-Harrison, J.S. 2017. Polyploidy and interspecific hybridization: Partners for adaptation, speciation and evolution in plants. *Annals of Botany* 120: 183–194.

Allan, J.D., Castillo, M.M., Capps, K.A. 2021. *Stream Ecology*, Springer International Publishing, Cham, pp. 101–139.

Allis, C.D., Jenuwein, T. 2016. The molecular hallmarks of epigenetic control. *Nature Reviews Genetics* 17: 487–500.

Alon, U. 2007. Network motifs: Theory and experimental approaches. *Nature Reviews Genetics* 8: 450–461.

Amaral, P., Carbonell-Sala, S., De La Vega, F.M., Faial, T., Frankish, A., Gingeras, T., Guigo, R., Harrow, J.L., Hatzigeorgiou, A.G., Johnson, R., Murphy, T.D., Pertea, M., Pruitt, K.D., Pujar, S., Takahashi, H., Ulitsky, I., Varabyou, A., Wells, C.A., Yandell, M., Carninci, P., Salzberg, S.L. 2023. The status of the human gene catalogue. *Nature* 622: 41–47.

Amir Afzali, Y. 2024. Asian *Mus musculus*: subspecies divergence, genetic diversity, and historical biogeography. *Journal of Mammalogy* 105: 1378–1391.

Amundson, R. 2001. Adaptation and development: On the lack of common ground. In: Orzack, S. H. and Sober, E. (eds.), *Adaptation and Optimality*, Cambridge University Press, Cambridge, UK, pp. 303–334.

Anderegg, W.R.L. 2015. Spatial and temporal variation in plant hydraulic traits and their relevance for climate change impacts on vegetation. *New Phytologist* 205: 1008–1014.

Andersen, F., Rocca, E. 2020. Underdetermination and evidence-based policy. *Studies in History and Philosophy of Science Part C: Studies in History and Philosophy of Biological and Biomedical Sciences* 84: 101335.

Anderson, B., Johnson, S.D. 2007. The geographical mosaic of coevolution in a plant-pollinator mutualism. *Evolution* 62: 220–225.

Anderson, S., Bankier, A.T., Barrell, B.G., De Bruijn, M.H.L., Coulson, A.R., Drouin, J., Eperon, I.C., Nierlich, D.P., Roe, B.A., Sanger, F., Schreier, P.H., Smith, A.J.H., Staden, R., Young, I.G. 1981. Sequence and organization of the human mitochondrial genome. *Nature* 290: 457–465.

Anfodillo, T., Petit, G., Sterck, F., Lechthaler, S., Olson, M.E. 2016. Allometric trajectories and "stress": A quantitative approach. *Frontiers in Plant Science* 7: 1681.

Angyalossy, V., Pace, M.R., Lima, A.C. 2015. Liana anatomy: A broad perspective on structural evolution of the vascular system. In: Schnitzer, S. A., Bongers, F., Burnham, R. J., and Putz, F. E. (eds.), *Ecology of Lianas*, Wiley, New York, pp. 253–287.

Archer, E., Lavie, C.J. 2023. The failure of gene-centrism. *Behavioral and Brain Sciences* 46: e209.

Archetti, E. 2015. Three kinds of constructionism: The role of metaphor in the debate over niche constructionism. *Biological Theory* 10: 103–115.

Archibald, J.D. 2014. *Aristotle's Ladder, Darwin's Tree: The Evolution of Visual Metaphors for Biological Order*. Columbia University Press, New York.

Arendt, J., Reznick, D. 2008. Convergence and parallelism reconsidered: What have we learned about the genetics of adaptation? *Trends in Ecology & Evolution* 23: 26–32.

Ariew, A., Lewontin, R.C. 2004. The confusions of fitness. *The British Journal for the Philosophy of Science* 55: 347–363.

Arnold, P.A., Kruuk, L.E.B., Nicotra, A.B. 2019. How to analyse plant phenotypic plasticity in response to a changing climate. *New Phytologist* 222: 1235–1241.

Arnold, S.J. 1983. Morphology, performance and fitness. *American Zoologist* 23: 347–361.

Arnold, S.J. 1992. Constraints on phenotypic evolution. *The American Naturalist* 140: S85–S107.

Arroyo-Santos, A., Olson, M.E., Vergara-Silva, F. 2015. Practice-oriented controversies and borrowed epistemic credibility in current evolutionary biology: Phylogeography as a case study. *Perspectives on Science* 23: 310–334.

Arthur, W. 2002. The interaction between developmental bias and natural selection: from centipede segments to a general hypothesis. *Heredity* 89: 239–246.

Arthur, W. 2004. *Biased Embryos and Evolution*. Cambridge University Press, Cambridge, UK.

Arthur, W., Farrow, M. 1999. The pattern of variation in centipede segment number as an example of developmental constraint in evolution. *Journal of Theoretical Biology* 200: 183–191.

Aspden, J.L., Wallace, E.W.J., Whiffin, N. 2023. Not all exons are protein coding: Addressing a common misconception. *Cell Genomics* 3: 100296.

Assis, L.C.S., Brigandt, I. 2009. Homology: Homeostatic property cluster kinds in systematics and evolution. *Evolutionary Biology* 36: 248–255.

Auerbach, B.M., Savell, K.R.R., Agosto, E.R. 2023. Morphology, evolution, and the whole organism imperative: Why evolutionary questions need multi-trait evolutionary quantitative genetics. *American Journal of Biological Anthropology* 181: 180–211.

Auld, J.R., Agrawal, A.A., Relyea, R.A. 2010. Re-evaluating the costs and limits of adaptive phenotypic plasticity. *Proceedings of the Royal Society B: Biological Sciences* 277: 503–511.

Ayala, F.J. 2009. Darwin and the scientific method. *Proceedings of the National Academy of Sciences* 106: 10033–10039.

Badia-i-mompel, P., Wessels, L., Müller-Dott, S., Trimbour, R., Ramirez Flores, R.O., Argelaguet, R., Saez-Rodriguez, J. 2023. Gene regulatory network inference in the era of single-cell multi-omics. *Nature Reviews Genetics* 24: 739–754.

Badyaev, A.V. 2011. Origin of the fittest: Link between emergent variation and evolutionary change as a critical question in evolutionary biology. *Proceedings of the Royal Society B: Biological Sciences* 278: 1921–1929.

Badyaev, A.V., Potticary, A.L., Morrison, E.S. 2017. Most colorful example of genetic assimilation? exploring the evolutionary destiny of recurrent phenotypic accommodation. *The American Naturalist* 190: 266–280.

Badyaev, A.V., Posner, A.B., Morrison, E.S., Higginson, D.M. 2019. Cycles of external dependency drive evolution of avian carotenoid networks. *Nature Communications* 10: 1596.

Baedke, J., Fábregas-Tejeda, A. 2023. The organism in evolutionary explanation: From early Twentieth Century to the Extended Evolutionary Synthesis. In: Dickins, T. E. and Dickins, B. J. A. (eds.), *Evolutionary Biology: Contemporary and Historical Reflections Upon Core Theory*, Springer International Publishing, Cham, Evolutionary Biology – New Perspectives on Its Development. vol. 6. pp. 121–150.

von Baer, K.E. 1827. Beiträge zur Kenntnis der niederen Thiere. *Nova Acta Academiae Caesareae Leopoldino-Carolinae Germanicae Naturae Curiosorum* 13: 523–762.

Bailer-Jones, D. 2008. Standing up against tradition: Models and theories in Nancy Cartwright's philosophy of science. In: Bovens, L., Cartwright, N., Hartmann, S., and Hoefer, C. (eds.), *Nancy Cartwright's Philosophy of Science*, Routledge, New York, Routledge studies in the philosophy of science. pp. 17–37.

Bailey, I.W., Tupper, W.W. 1918. Size variation in tracheary cells: I. A comparison between the secondary xylems of vascular cryptogams, gymnosperms and angiosperms. *Proceedings of the American Academy of Arts and Sciences* 54: 149–204.

Bak, P., Boettcher, S. 1997. Self-organized criticality and punctuated equilibria. *Physica D: Nonlinear Phenomena* 107: 143–150.

Baker, B. 2021. Natural information, factivity and nomicity. *Biology & Philosophy* 36: 26.

Baker, J., Workman, M., Bedrick, E., Frey, M.A., Hurtado, M., Pearson, O. 2010. Brains versus brawn: An empirical test of Barker's brain sparing model. *American Journal of Human Biology* 22: 206–215.

Ball, P. 2011. A metaphor too far. *Nature* Published online February 23. http://dx.doi.org/10.1038/news.2011.115.

Ball, P. 2023. *How Life Works: A User's Guide to the New Biology*. The University of Chicago Press, Chicago.

Banta, J.A., Richards, C.L. 2018. Quantitative epigenetics and evolution. *Heredity* 121: 210–224.

Barabási, A.-L. 2002. *Linked: The New Science of Networks*. Perseus Pub, Cambridge, Mass.

Barker, G. 2007. Biological levers and extended adaptationism. *Biology & Philosophy* 23: 1–25.

Bar-On, Y.M., Phillips, R., Milo, R. 2018. The biomass distribution on Earth. *Proceedings of the National Academy of Sciences* 115: 6506–6511.

Barresi, M.J.F., and Gilbert, S.F. 2023. *Developmental Biology*, International 13th edn. Oxford University Press, Oxford, UK.

Barrett, G.W., van Dyne, G.M., Odum, E.P. 1976. Stress ecology. *BioScience* 26: 192–194.

Barrio, R., Del Arco, A., Cabrera, H.L., Arribas, C. 1994. Structure and expression of the *Drosophila* ubiquitin-80-amino-acid fusion-protein gene. *Biochemical Journal* 302: 237–244.

Barseghyan, H. 2015. *The Laws of Scientific Change*. Springer, Cham.

Basolo, A.L. 1990a. Female preference predates the evolution of the sword in swordtail fish. *Science* 250: 808–810.

Basolo, A.L. 1990b. Female preference for male sword length in the green swordtail, *Xiphophorus helleri* (Pisces: Poeciliidae). *Animal Behaviour* 40: 332–338.

Bastarache, L., Delozier, S., Pandit, A., He, J., Lewis, A., Annis, A.C., LeFaive, J., Denny, J.C., Carroll, R.J., Altman, R.B., Hughey, J.J., Zawistowski, M., Peterson, J.F. 2023. The phenotype-genotype reference map: Improving biobank data science through replication. *The American Journal of Human Genetics* 110: 1522–1533.

Bateson, P., Gluckman, P. 2012. Plasticity and robustness in development and evolution. *International Journal of Epidemiology* 41: 219–223.

Bateson, W. 1894. *Materials for the Study of Variation*. Macmillan, London.

Baum, D.A., Smith, S.D. 2013. Tree thinking: an introduction to phylogenetic biology. Roberts and Co., Greenwood Village, Col.

Baverstock, K. 2021. The gene: An appraisal. *Progress in Biophysics and Molecular Biology* 164: 46–62.

Bégin, M., Roff, D.A. 2004. From micro- to macroevolution through quantitative genetic variation: positive evidence from field crickets. *Evolution* 58: 2287–2304.

Beldade, P., Brakefield, P.M. 2003. The difficulty of agreeing about constraints. *Evolution & Development* 5: 119–120.

Beldade, P., Koops, K., Brakefield, P.M. 2002. Developmental constraints versus flexibility in morphological evolution. *Nature* 416: 844–847.

Bendersky, G. 2000. Tlatilco sculptures, diprosopus, and the emergence of medical illustrations. *Perspectives in Biology and Medicine* 43: 477–501.

Benson, R.B.J., Butler, R.J., Carrano, M.T., O'Connor, P.M. 2012. Air-filled postcranial bones in theropod dinosaurs: Physiological implications and the "reptile"–bird transition. *Biological Reviews* 87: 168–193.

Berridge, M. 2023. The Zealandia Drowning Debate: Did New Zealand sink beneath the waves? *New Zealand Science Review* 70: 64.

Berry, A., Browne, J. 2022. Mendel and Darwin. *Proceedings of the National Academy of Sciences* 119: e2122144119.

Bertolaso, M., Caianiello, S., Serrelli, E., eds. 2018. *Biological Robustness: Emerging Perspectives from within the Life Sciences*. Springer, Cham.

Betts, C.R., Wootton, R.J. 1988. Wing shape and flight behaviour in butterflies (Lepidoptera: Papilionoidea and Hesperioidea): a preliminary analysis. *Journal of Experimental Biology* 138: 271–288.

Bevelhimer, M.S., Pracheil, B.M., Fortner, A.M., Saylor, R., Deck, K.L. 2019. Mortality and injury assessment for three species of fish exposed to simulated turbine blade strike. *Canadian Journal of Fisheries and Aquatic Sciences* 76: 2350–2363.

Bird, A. 2010. The historical turn in the philosophy of science. In: Psyllos, S., and Curd, M. (eds.), *The Routledge Companion to Philosophy of Science* Routledge, London, pp. 67–77.

Black, D.L. 1998. Splicing in the inner ear: A familiar tune, but what are the instruments? *Neuron* 20: 165–168.

Black, M. 1981. *Models and Metaphors: Studies in Language and Philosophy*. Cornell Univ. Press, Ithaca.

Blomberg, S.P., Garland, T., Ives, A.R. 2003. Testing for phylogenetic signal in comparative data: Behavioral traits are more labile. *Evolution* 57: 717–745.

Blumberg, M.S. 2009. *Freaks of Nature: What Anomalies Tell Us about Development and Evolution*. Oxford University Press, Oxford, UK.

Bonato, L., Drago, L., Murienne, J. 2014. Phylogeny of Geophilomorpha (Chilopoda) inferred from new morphological and molecular evidence. *Cladistics* 30: 485–507.

Bondos, S.E., Dunker, A.K., Uversky, V.N. 2022. Intrinsically disordered proteins play diverse roles in cell signaling. *Cell Communication and Signaling* 20: 20.

Bonduriansky, R., and Day, T. 2018. *Extended Heredity: A New Understanding of Inheritance and Evolution*. Princeton University Press, Princeton, NJ.

Boudry, M., Pigliucci, M. 2013. The mismeasure of machine: Synthetic biology and the trouble with engineering metaphors. *Studies in History and Philosophy of Science Part C: Studies in History and Philosophy of Biological and Biomedical Sciences* 44: 660–668.

Bourrat, P. 2015. How to read "heritability" in the recipe approach to natural selection. *The British Journal for the Philosophy of Science* 66: 883–903.

Bourrat, P., Lu, Q., Jablonka, E. 2017. Why the missing heritability might not be in the DNA. *BioEssays* 39: 1700067.

Bowler, P.J. 2020. *Evolution: The History of an Idea*. University of California Press, Berkeley.

Boyd, R. 1993. Metaphor and theory change: What is "metaphor" a metaphor for? In: Ortony, A. (ed.), *Metaphor and Thought* Cambridge University Press, Cambridge, UK, pp. 481–532.

Brakefield, P.M., Beldade, P., Zwaan, B.J. 2009. The African butterfly *Bicyclus anynana*: A model for evolutionary genetics and evolutionary developmental biology. *Cold Spring Harbor Protocols* 2009: pdb.emo122.

Breuker, C.J., Debat, V., Klingenberg, C.P. 2006. Functional evo-devo. *Trends in Ecology & Evolution* 21: 488–492.

Brigandt, I. 2020. Historical and philosophical perspectives on the study of developmental bias. *Evolution & Development* 22: 7–19.

Brodie, J.F., Mannion, P.D. 2023. The hierarchy of factors predicting the latitudinal diversity gradient. *Trends in Ecology & Evolution* 38: 15–23.

Brodie III, E.D. 2005. Caution: Niche construction ahead. *Evolution* 59: 249–251.

Broido, A.D., Clauset, A. 2019. Scale-free networks are rare. *Nature Communications* 10: 1017.

Bromberger, S. 1966. Why questions. In: Colodny, R. G. (ed.), *Mind and Cosmos: Essays in Contemporary Science and Philosophy*, University of Pittsburgh Press, Pittsburgh, PA, pp. 66–111.

Brown, R.L. 2023. Structuralism and adaptationism: Friends? Or foes? *Seminars in Cell & Developmental Biology* 145: 13–21.

Brown, S. 2008. Top billing for platypus at end of evolution tree. *Nature* 453: 138–139.

Brummer, A.B., Savage, V.M. 2021. Cancer as a model system for testing metabolic scaling theory. *Frontiers in Ecology and Evolution* 9: 691830.

Butler, M.A., King, A.A. 2004. Phylogenetic comparative analysis: A modeling approach for adaptive evolution. *The American Naturalist* 164: 683–695.

Camara, M.D., Ancell, C.A., Pigliucci, M. 2000. Induced mutations: A novel tool to study phenotypic integration and evolutionary constraints in *Arabidopsis thaliana*. *Evolutionary Ecology Research* 2: 1009–1029.

Campos-Cerda, F., Bohannan, B.J.M. 2020. The nidobiome: A framework for understanding microbiome assembly in neonates. *Trends in Ecology & Evolution* 35: 573–582.

Caporael, L.R., Griesemer, J.R., Wimsatt, W.C., eds. 2014. *Developing Scaffolds in Evolution, Culture, and Cognition*. The MIT Press, Cambridge, Massachusetts.

Carlquist, S.J. 1974. *Island Biology*. Columbia Univ. Press, New York.

Carranza Edwards, A. 2001. Grain size and sorting in modern beach sands. *Journal of Coastal Research* 17: 38–52.

Carvalho Provinciato, I.C., Araújo, M.S., Jahn, A.E. 2018. Drivers of wing shape in a widespread Neotropical bird: A dual role of sex-specific and migration-related functions. *Evolutionary Ecology* 32: 379–393.

Cavalli-Sforza, L.L., Feldman, M.W. 1973. Cultural versus biological inheritance: phenotypic transmission from parents to children. *American Journal of Human Genetics* 25: 618–637.

Chakrabortee, S., Byers, J.S., Jones, S., Garcia, D.M., Bhullar, B., Chang, A., She, R., Lee, L., Fremin, B., Lindquist, S., Jarosz, D.F. 2016. Intrinsically disordered proteins drive emergence and inheritance of biological traits. *Cell* 167: 369–381.e12.

Chen, B.-S., Wu, W.-S. 2007. Underlying principles of natural selection in network evolution: systems biology approach. *Evolutionary Bioinformatics* 3: 117693430700300010.

Chen, Z., Zhu, S., Zhang, Y., Luan, J., Li, S., Sun, P., Wan, X., Liu, S. 2020. Tradeoff between storage capacity and embolism resistance in the xylem of temperate broadleaf tree species. *Tree Physiology* 40: 1029–1042.

Cherniak, C. 1995. Neural component placement. *Trends in Neurosciences* 18: 522–527.

Chipman, A.D., Arthur, W., Akam, M. 2004. A double segment periodicity underlies segment generation in centipede development. *Current Biology* 14: 1250–1255.

Christiansen, P. 1999. Scaling of the limb long bones to body mass in terrestrial mammals. *Journal of Morphology* 239: 167–190.

Christin, P.-A., Osborne, C.P., Chatelet, D.S., Columbus, J.T., Besnard, G., Hodkinson, T.R., Garrison, L.M., Vorontsova, M.S., Edwards, E.J. 2013. Anatomical enablers and the evolution of C_4 photosynthesis in grasses. *Proceedings of the National Academy of Sciences* 110: 1381–1386.

Chrzanowska-Lightowlers, Z.M.A., Temperley, R.J., Smith, P.M., Seneca, S.H., Lightowlers, R.N. 2004. Functional polypeptides can be synthesized from human mitochondrial transcripts lacking termination codons. *Biochemical Journal* 377: 725–731.

Chung, W.-J., Oh, J.-W., Kwak, K., Lee, B.Y., Meyer, J., Wang, E., Hexemer, A., Lee, S.-W. 2011. Biomimetic self-templating supramolecular structures. *Nature* 478: 364–368.

Clarke, P.J., Lawes, M.J., Midgley, J.J., Lamont, B.B., Ojeda, F., Burrows, G.E., Enright, N.J., Knox, K.J.E. 2013. Resprouting as a key functional trait: How buds, protection and resources drive persistence after fire. *New Phytologist* 197: 19–35.

Cleland, C.E. 2001. Historical science, experimental science, and the scientific method. *Geology* 29: 987.

Cleland, C.E. 2013. Common cause explanation and the search for a smoking gun. *Rethinking the Fabric of Geology*, Geological Society of America.

Clements, R., Liew, T.-S., Vermeulen, J.J., Schilthuizen, M. 2008. Further twists in gastropod shell evolution. *Biology Letters* 4: 179–182.

Clune, J., Mouret, J.-B., Lipson, H. 2013. The evolutionary origins of modularity. *Proceedings of the Royal Society B: Biological Sciences* 280: 20122863.

Coates, M.I., Clack, J.A. 1990. Polydactyly in the earliest known tetrapod limbs. *Nature* 347: 66–69.

Cofnas, N. 2024. Natural selection requires no teleology in addition to heritable variation in fitness. *Biology and Philosophy* 39: 18.

Colombelli-Négrel, D., Hauber, M.E., Evans, C., Katsis, A.C., Brouwer, L., Adreani, N.M., Kleindorfer, S. 2021. Prenatal auditory learning in avian vocal learners and non-learners. *Philosophical Transactions of the Royal Society B: Biological Sciences* 376: 20200247.

Colosimo, P.F., Hosemann, K.E., Balabhadra, S., Villarreal, G., Dickson, M., Grimwood, J., Schmutz, J., Myers, R.M., Schluter, D., Kingsley, D.M. 2005. Widespread parallel evolution in sticklebacks by repeated fixation of ectodysplasin alleles. *Science* 307: 1928–1933.

Combes, S.A., Daniel, T.L. 2003. Flexural stiffness in insect wings I. Scaling and the influence of wing venation. *Journal of Experimental Biology* 206: 2979–2987.

Condamine, F.L., Silvestro, D., Koppelhus, E.B., Antonelli, A. 2020. The rise of angiosperms pushed conifers to decline during global cooling. *Proceedings of the National Academy of Sciences* 117: 28867–28875.

Conery, M., Grant, S.F.A. 2023. Human height: A model common complex trait. *Annals of Human Biology* 50: 258–266.

Conner, J., and Via, S. 1993. Patterns of phenotypic and genetic correlations among morphological and life-history traits in wild radish, *Raphanus raphanistrum. Evolution* 47: 704–711.

Conner, J.K. 2003. Artificial selection: A powerful tool for ecologists. *Ecology* 84: 1650–1660.

Conner, J.K., Karoly, K., Stewart, C., Koelling, V.A., Sahli, H.F., Shaw, F.H. 2011. Rapid independent trait evolution despite a strong pleiotropic genetic correlation. *The American Naturalist* 178: 429–441.

Cordeiro, I.R., Tanaka, M. 2020. Environmental oxygen is a key modulator of development and evolution: From molecules to ecology. *BioEssays* 42: 2000025.

Cordero, C. 2001. A different look at the false head of butterflies: Evolution of false heads. *Ecological Entomology* 26: 106–108.

Corning, P.A., Kauffman, S.A., Noble, D., Shapiro, J.A., Vane-Wright, R.I., Pross, A., eds. 2023. *Evolution "on Purpose": Teleonomy in Living Systems.* The MIT Press, Cambridge, Massachusetts.

Cotterell, J., Sharpe, J. 2013. Mechanistic explanations for restricted evolutionary paths that emerge from gene regulatory networks. *PLoS ONE* 8: e61178.

Coyne, J.A. 2009. Evolution's challenge to genetics. *Nature* 457: 382–383.

Crisp, M., Cook, L. 2005. Do early branching lineages signify ancestral traits? *Trends in Ecology & Evolution* 20: 122–128.

Crisp, M.D., Cook, L.G. 2012. Phylogenetic niche conservatism: What are the underlying evolutionary and ecological causes? *New Phytologist* 196: 681–694.

Cullinane, D.M. 2000. Axial versus appendicular: constraint versus selection. *American Zoologist* 40: 136–145.

Currie, A., Levy, A. 2019. Why experiments matter. *Inquiry* 62: 1066–1090.

Cuxart-Erruz, R., Van Dooren, T.J.M., Van Der Geer, A.A.E., Galis, F. 2024. Increased incidences of cervical ribs in deer indicate extinction risk. *Proceedings of the National Academy of Sciences* 121: e2406670121.

Dahms, H., Fornshell, J., Fornshell, B. 2006. Key for the identification of crustacean nauplii. *Organisms Diversity & Evolution* 6: 47–56.

Darwin, C. 1859. *On the Origin of Species.* John Murray, London.

Dautant, A., Meier, T., Hahn, A., Tribouillard-Tanvier, D., Di Rago, J.-P., Kucharczyk, R. 2018. ATP synthase diseases of mitochondrial genetic origin. *Frontiers in Physiology* 9: 329.

Dawkins, R. 1976. *The Selfish Gene.* Oxford University Press, Oxford, UK.

Dawkins, R. 1990. *The Extended Phenotype: The Long Reach of the Gene*, Reprinted. Oxford University Press, Oxford, UK.

Dawkins, R. 2024. *The Genetic Book of the Dead: A Darwinian Reverie.* Head of Zeus, London.

De Jong, G. 2005. Evolution of phenotypic plasticity: Patterns of plasticity and the emergence of ecotypes. *New Phytologist* 166: 101–118.

De Jong, G. 2007. Much ado... *Ethology* 113: 730–732.

De Jong, G., Crozier, R.H. 2003. A flexible theory of evolution. *Nature* 424: 16–17.

De Queiroz, K., and Poe, S. 2003. Failed refutations: Further comments on parsimony and likelihood methods and their relationship to Popper's degree of corroboration. *Systematic Biology* 52: 352–367.

De Santis, M.D. 2021a. Scientific explanation and systematics. *Systematics and Biodiversity* 19: 312–321.

De Santis, M.D. 2021b. Misconceptions about historical sciences in evolutionary biology. *Evolutionary Biology* 48: 94–99.

De Santis, M.D. 2024. On the nature of evolutionary explanations: A critical appraisal of Walter Bock's approach with a new revised proposal. *History and Philosophy of the Life Sciences* 46: 3.

Deffner, D. 2023. Construction "on purpose": How niche construction affects natural selection. In: Corning, P. A., Kauffman, S. A., Noble, D., Shapiro, J. A., Vane-Wright, R. I., and Pross, A. (eds.), *Evolution "On Purpose": Teleonomy in Living Systems*, MIT Press, Cambridge, Mass, Vienna Series In Theoretical Biology. pp. 59–77.

Delhey, K., Valcu, M., Dale, J., Kempenaers, B. 2023. The evolution of carotenoid-based plumage colours in passerine birds. *Journal of Animal Ecology* 92: 66–77.

Depew, D.J., Weber, B.H. 1997. *Darwinism Evolving: Systems Dynamics and the Genealogy of Natural Selection*. MIT Press, Cambridge, Mass.

Detrain, C., Deneubourg, J. 2006. Self-organized structures in a superorganism: Do ants "behave" like molecules? *Physics of Life Reviews* 3: 162–187.

Dewey, K.G., Begum, K. 2011. Long-term consequences of stunting in early life. *Maternal & Child Nutrition* 7: 5–18.

Dibus, M., Joshi, O., Ivaska, J. 2024. Novel tools to study cell-ECM interactions, cell adhesion dynamics and migration. *Current Opinion in Cell Biology* 88: 102355.

Dickins, T.E., Dickins, B.J.A. 2018. The extent of the Modern Synthesis: the foundational framework for evolutionary biology. In: Burggren, W. and Dubansky, B. (eds.), *Development and Environment*, Springer International Publishing, Cham, pp. 155–176.

Dietrich, M.R., Skipper, R.A. 2007. Manipulating underdetermination in scientific controversy: the case of the molecular clock. *Perspectives on Science* 15: 295–326.

Dietrich, M.R., Skipper, R.A. 2012. A shifting terrain: A brief history of the adaptive landscape. In: Svensson, E. I. and Calsbeek, R. (eds.), *The Adaptive Landscape In Evolutionary Biology*, Oxford University Press, Oxford, UK, pp. 283–295.

Diggle, P.K., Miller, J.S. 2013. Developmental plasticity, genetic assimilation, and the evolutionary diversification of sexual expression in *Solanum*. *American Journal of Botany* 100: 1050–1060.

Dill, K.A., MacCallum, J.L. 2012. The protein-folding problem, 50 years on. *Science* 338: 1042–1046.

Discher, D.E., Mooney, D.J., Zandstra, P.W. 2009. Growth factors, matrices, and forces combine and control stem cells. *Science* 324: 1673–1677.

Dobzhansky, T. 1970. *Genetics of the Evolutionary Process*. Columbia University Press, New York.

Domyan, E.T., Shapiro, M.D. 2017. Pigeonetics takes flight: Evolution, development, and genetics of intraspecific variation. *Developmental Biology* 427: 241–250.

Downes, D.J., Davis, M.A., Wong, K.H., Kreutzberger, S.D., Hynes, M.J., Todd, R.B. 2014. Dual DNA binding and coactivator functions of *Aspergillus nidulans* TamA, a Zn(II)2Cys6 transcription factor. *Molecular Microbiology* 92: 1198–1211.

Doyle, J.J., Coate, J.E. 2019. Polyploidy, the nucleotype, and novelty: The impact of genome doubling on the biology of the cell. *International Journal of Plant Sciences* 180: 1–52.

Drake, A.G., Klingenberg, C.P. 2010. Large-scale diversification of skull shape in domestic dogs: disparity and modularity. *The American Naturalist* 175: 289–301.

Drenckhahn, J. 2009. Growth plasticity of the embryonic and fetal heart. *BioEssays* 31: 1288–1298.

Dudley, R. 2018. *The Biomechanics of Insect Flight: Form, Function, Evolution*. Princeton University Press, Princeton, NJ.

Dunker, A.K., Cortese, M.S., Romero, P., Iakoucheva, L.M., Uversky, V.N. 2005. Flexible nets: The roles of intrinsic disorder in protein interaction networks. *The FEBS Journal* 272: 5129–5148.

Durand, M. 2006. Architecture of optimal transport networks. *Physical Review E* 73: 016116.

Durrant, R., Haig, B.D. 2001. How to pursue the adaptationist program in psychology. *Philosophical Psychology* 14: 357–380.

Dwyer, P.D. 1984. Functionalism and structuralism: Two programs for evolutionary biologists. *The American Naturalist* 124: 745–750.

Eble, G.J. 2005. Morphological modularity and macroevolution: Conceptual and empirical aspects. In: Callebaut, W., and Rasskin-Gutman, D. (eds.), *Modularity: Understanding the Development and Evolution of Natural Complex Systems*, MIT Press, Cambridge, Massachusetts, pp. 221–238.

Echeverría, A., Petrone-Mendoza, E., Anfodillo, T., Brodribb, T., Torres-San Miguel, C.R., Rueda Arreguín, J.L., Olson, M.E. 2023. Testing Carlquistian hypotheses on the functional significance of vessel element length. *IAWA Journal* 44: 399–428.

Eco, U. 2008. *Trattato di Semiotica Generale*, 19th edn., Bompiani, Milano.

Edelaar, P., Otsuka, J., Luque, V.J. 2023. A generalised approach to the study and understanding of adaptive evolution. *Biological Reviews* 98: 352–375.

Edward, D.P., Kaufman, L.M. 2003. Anatomy, development, and physiology of the visual system. *Pediatric Clinics of North America* 50: 1–23.

Edwards, E.J. 2019. Evolutionary trajectories, accessibility and other metaphors: the case of C_4 and CAM photosynthesis. *New Phytologist* 223: 1742–1755.

Edwards, M., Shiota, K., Smith, M.S.R., Walsh, D.A. 1995. Hyperthermia and birth defects. *Reproductive Toxicology* 9: 411–425.

Egli, D., Birkhoff, G., Eggan, K. 2008. Mediators of reprogramming: Transcription factors and transitions through mitosis. *Nature Reviews Molecular Cell Biology* 9: 505–516.

Ellis, A.G., Midgley, J.J. 1996. A new plant-animal mutualism involving a plant with sticky leaves and a resident hemipteran insect. *Oecologia* 106: 478–481.

Eloy, C. 2013. On the best design for undulatory swimming. *Journal of Fluid Mechanics* 717: 48–89.

Elsaafien, K., Harden, S.W., Johnson, D.N., Kimball, A.K., Sheng, W., Smith, J.A., Scott, K.A., Frazier, C.J., De Kloet, A.D., Krause, E.G. 2022. A novel organ-specific approach to selectively target sensory afferents innervating the aortic arch. *Frontiers in Physiology* 13: 841078.

Emlen, D.J. 2001. Costs and the diversification of exaggerated animal structures. *Science* 291: 1534–1536.

Evans, S.R., Postma, E. 2025. Counting chicks before they hatch: Extending the observed lifetime to better characterize evolutionary processes in the wild. *Evolution* 79: 155–163.

Fábregas-Tejeda, A., Vergara-Silva, F. 2018. Hierarchy theory of evolution and the Extended Evolutionary Synthesis: Some epistemic bridges, some conceptual rifts. *Evolutionary Biology* 45: 127–139.

Facoetti, M. 2021. United in diversity: An organic overview of non-adaptationist evolutionary epistemology. *Journal for General Philosophy of Science* 52: 211–225.

Falconer, D.S., and Mackay, T. 2009. *Introduction to Quantitative Genetics*, 4th . Pearson, Prentice Hall, Harlow.

Falk, R. 2010. What is a gene?—Revisited. *Studies in History and Philosophy of Science Part C: Studies in History and Philosophy of Biological and Biomedical Sciences* 41: 396–406.

Falkner, D. 2016. Metaphors of life: Reflections on metaphors in the debate on synthetic biology. In: Hagen, K., Engelhard, M., and Toepfer, G. (eds.), *Ambivalences of Creating Life*, Springer, Heidelberg, vol. 45. pp. 251–265.

Farmer, C.G. 2015. The evolution of unidirectional pulmonary airflow. *Physiology* 30: 260–272.

Feil, R., Fraga, M.F. 2012. Epigenetics and the environment: Emerging patterns and implications. *Nature Reviews Genetics* 13: 97–109.

Félix, M.-A., Barkoulas, M. 2015. Pervasive robustness in biological systems. *Nature Reviews Genetics* 16: 483–496.

Felsenstein, J. 1985. Phylogenies and the comparative method. *The American Naturalist* 125: 1–15.

Fish, F.E., Lauder, G.V. 2006. Passive and active flow control by swimming fishes and mammals. *Annual Review of Fluid Mechanics* 38: 193–224.

Fisher, D.C. 1985. Evolutionary morphology: Beyond the analogous, the anecdotal, and the ad hoc. *Paleobiology* 11: 120–138.

Fisher, R.A. 1930. *The Genetical Theory of Natural Selection*. Oxford University Press, Oxford, UK.

Flanagan, J.M. 2015. Epigenome-wide association studies (EWAS): Past, present, and future. In: Verma, M. (ed.), *Cancer Epigenetics*, Springer New York, New York, NY, Methods in Molecular Biology. vol. 1238. pp. 51–63.

Flatt, T. 2020. Life-history evolution and the genetics of fitness components in *Drosophila melanogaster*. *Genetics* 214: 3–48.

Flores-Renteria, L., Vazquez-Lobo, A., Whipple, A.V., Pinero, D., Marquez-Guzman, J., Dominguez, C.A. 2011. Functional bisporangiate cones in *Pinus johannis* (Pinaceae): Implications for the evolution of bisexuality in seed plants. *American Journal of Botany* 98: 130–139.

Folkestad, L., Hald, J.D., Canudas-Romo, V., Gram, J., Hermann, A.P., Langdahl, B., Abrahamsen, B., Brixen, K. 2016. Mortality and causes of death in patients with osteogenesis imperfecta: A register-based nationwide cohort study. *Journal of Bone and Mineral Research* 31: 2159–2166.

Frank, S.A. 2012. Natural selection. IV. The Price equation. *Journal of Evolutionary Biology* 25: 1002–1019.

Frankino, W.A., Emlen, D.J., Shingleton, A.W. 2010. Experimental approaches to studying the evolution of animal form: The shape of things to come. In: Garland, T. and Rose, M. R. (eds.), *Experimental Evolution: Concepts, Methods, and Applications of Selection Experiments*, University of California Press, Berkeley, pp. 419–478.

Frankino, W.A., Zwaan, B.J., Stern, D.L., Brakefield, P.M. 2007. Internal and external constraints in the evolution of morphological allometries in a butterfly. *Evolution* 61: 2958–2970.

Friedman, M. 2022. The macroevolutionary history of bony fishes: A paleontological view. *Annual Review of Ecology, Evolution, and Systematics* 53: 353–377.

Fuller, R.C. 2005. How and when selection experiments might actually be useful. *Integrative and Comparative Biology* 45: 391–404.

Fusco, G. 2001. How many processes are responsible for phenotypic evolution? *Evolution & Development* 3: 279–286.

Fusco, G. 2022. Serial homology. *Biological Theory* 17: 114–119.

Fusco, G., Carrer, R., Serrelli, E. 2014. The landscape metaphor in development. In: Minelli, A. and Pradeu, T. (eds.), *Towards a Theory of Development*, Oxford University Press, Oxford, UK, pp. 114–128.

Futuyma, D.J. 2017. Evolutionary biology today and the call for an extended synthesis. *Interface Focus* 7: 20160145.

Gage, S.H., Davey Smith, G., Ware, J.J., Flint, J., Munafò, M.R. 2016. G=E: What GWAS can tell us about the environment. *PLOS Genetics* 12: e1005765.

Galilei, G. 1638. *Discorsi e Dimostrazioni Matematiche Intorno a Due Nuove Scienze*. Appresso gli Elsevirii, Leiden, archive.org/details/bub_gb_E9BhikF658wC/page/n141/mode/2up.

Galileo, G. 1954. *Dialogues Concerning Two New Sciences*, Unabridged and unaltered republ. of the transl. by Henry Crew and Alfonso de Salvio, orig. publ. in New York, Macmillan, 1914. Dover, New York, NY.

Galis, F., Metz, J.A.J., van Alphen, J.J.M. 2018. Development and evolutionary constraints in animals. *Annual Review of Ecology, Evolution, and Systematics* 49: 499–522.

Galis, F., Van Dooren, T.J.M., Van Der Geer, A.A.E. 2022. Breaking the constraint on the number of cervical vertebrae in mammals: On homeotic transformations in lorises and pottos. *Evolution & Development* 24: 196–210.

Galis, F., Schut, P.C., Cohen-Overbeek, T.E., and Ten Broek, C.M.A. 2021. Evolutionary and Developmental issues of cervical ribs. In: Illig, K. A., Thompson, R. W., Freischlag, J. A., Donahue, D. M., Jordan, S. E., Lum, Y. W., and Gelabert, H. A. (eds.), *Thoracic Outlet Syndrome*, Springer International Publishing, Cham, pp. 23–35.

Galis, F., Van Dooren, T.J.M., Feuth, J.D., Metz, J.A.J., Witkam, A., Ruinard, S., Steigenga, M.J., and Wijnaendts, L.C.D. 2006. Extreme selection in humans against homeotic transformations of cervical vertebrae. *Evolution* 60: 2643–2654.

Gamermann, D., Triana-Dopico, J., Jaime, R. 2019. A comprehensive statistical study of metabolic and protein–protein interaction network properties. *Physica A: Statistical Mechanics and Its Applications* 534: 122204.

García-de Blas, E., Mateo, R., Alonso-Alvarez, C. 2016. Specific carotenoid pigments in the diet and a bit of oxidative stress in the recipe for producing red carotenoid-based signals. *PeerJ* 4: e2237.

Gardiner, H.M. 2005. Response of the fetal heart to changes in load: From hyperplasia to heart failure. *Heart* 91: 871–873.

Garland, T., and Losos, J.B. 1994. Ecological morphology of locomotor performance in squamate reptiles. In: Wainwright, P. C., and Reilly, S. M. (eds.), *Ecological Morphology: Integrative Organismal Biology*, University of Chicago Press, Chicago, pp. 240–302.

Garland, T., Freeman, P.W. 2005. Selective breeding for high endurance running increases hindlimb symmetry. *Evolution* 59: 1851–1854.

Garland, T., Rose, M.R., eds. 2009. *Experimental Evolution: Concepts, Methods, and Applications of Selection Experiments*. University of California Press, Berkeley.

Garland, T., Downs, C.J., Ives, A.R. 2022. Trade-offs (and constraints) in organismal biology. *Physiological and Biochemical Zoology* 95: 82–112.

Garson, J. 2016. *A Critical Overview of Biological Functions*. Springer, Berlin.
Gavrilets, S. 1997. Evolution and speciation on holey adaptive landscapes. *Trends in Ecology & Evolution* 12: 307–312.
Gawne, R., McKenna, K.Z., Nijhout, H.F. 2018. Unmodern synthesis: Developmental hierarchies and the origin of phenotypes. *BioEssays* 40: 1600265.
Geoffroy de Saint-Hilaire, I.G. 1832. *Traité de Tératologie*. J.-B. Baillière, Paris.
Gere, J.M., Timoshenko, S.P. 1999. *Mechanics of Materials 4th SI Edition*. Tanley Thornes, Cheltenham, UK.
Gerstein, M.B., Bruce, C., Rozowsky, J.S., Zheng, D., Du, J., Korbel, J.O., Emanuelsson, O., Zhang, Z.D., Weissman, S., Snyder, M. 2007. What is a gene, post-ENCODE? History and updated definition. *Genome Research* 17: 669–681.
Ghalambor, C.K., McKAY, J.K., Carroll, S.P., Reznick, D.N. 2007. Adaptive versus non-adaptive phenotypic plasticity and the potential for contemporary adaptation in new environments. *Functional Ecology* 21: 394–407.
Ghiselin, M.T. 1966. On psychologism in the logic of taxonomic controversies. *Systematic Biology* 15: 207–215.
Ghiselin, M.T. 1974. A radical solution to the species problem. *Systematic Biology* 23: 536–544.
Ghiselin, M.T. 1984. *The Triumph of the Darwinian Method: With a New Preface*. University of Chicago Press, Chicago.
Giannini, N.P. 2003. Canonical phylogenetic ordination. *Systematic Biology* 52: 684–695.
Gianoli, E., Valladares, F. 2012. Studying phenotypic plasticity: The advantages of a broad approach. *Biological Journal of the Linnean Society* 105: 1–7.
Gilbert, S.F. 2000. *Developmental Biology*, 6th edn. Sinauer Associates, Sunderland, Mass.
Gilbert, S.F., and Bard, J. 2014. Formalizing theories of development: A fugue on the orderliness of change. In: Minelli, A., and Pradeu, T. (eds.), *Towards a Theory of Development*, Oxford University Press, Oxford, UK, pp. 129–143.
Gilbert, S.F., Epel, D. 2015. *Ecological Developmental Biology*, 2nd edn. Sinauer Associates, Inc. Publishers, Sunderland, Massachusetts, U.S.A.
Givnish, T.J. 2015. New evidence on the origin of carnivorous plants. *Proceedings of the National Academy of Sciences* 112: 10–11.
Gjuvsland, A.B., Vik, J.O., Beard, D.A., Hunter, P.J., Omholt, S.W. 2013. Bridging the genotype-phenotype gap: what does it take? *The Journal of Physiology* 591: 2055–2066.
Glazebrook, T., and Goldsby, M. 2018. Climate science, indeterminacy, and food security in sub-Saharan Africa. In: Scott, R. H. (ed.), *The Significance of Interdeterminacy: Perspectives from Asian and Continental philosophy*, Taylor & Francis, New York, Routledge studies in contemporary philosophy. pp. 148–167.
Gluckman, P.D., and Hanson, M. 2005. *The Fetal Matrix: Evolution, Development, and Disease*. Cambridge University Press, Cambridge, UK.
Godfrey-Smith, P. 2009. *Darwinian Populations and Natural Selection*. Oxford University Press, Oxford, UK.
Goldwasser, S. 2023. Standard aberration: Cancer biology and the modeling account of normal function. *Biology & Philosophy* 38: 4.
Goodwin, B. 1990. Structuralism in biology. *Science Progress* 74: 227–243.
Goodwin, B.C. 2001. *How the Leopard Changed Its Spots: The Evolution of Complextiy*, 1st Princeton ed., with new preface. Princeton University Press, Princeton, NJ.
Gordon, D. 2010. *Ant Encounters: Interaction Networks and Colony Behavior*. Princeton University Press, Princeton, NJ.
Gordon, D.M. 2002. The regulation of foraging activity in red harvester ant colonies. *The American Naturalist* 159: 509–518.
Gordon, D.M. 2023. *The Ecology of Collective Behavior*, 1st edn. Princeton University Press, Princeton, NJ.
Gordon, J.E. 2003. *Structures: Or Why Things Don't Fall Down*, 2nd edn. Da Capo Press, New York.
Gosline, W.A. 1997. Functional morphology of the caudal skeleton in teleostean fishes. *Ichthyological Research* 44: 137–141.

Goswami, A., Polly, P.D. 2010. Methods for studying morphological integration and modularity. *The Paleontological Society Papers* 16: 213–243.

Gottlieb, G. 1971. *Development of Species Identification in Birds: An Inquiry into the Prenatal Determinants of Perception.* University of Chicago Press, Chicago.

Gould, G.M., Pyle, W.L. 1897. *Anomalies and Curiosities of Medicine.* The Rebman Publishing Co., London, UK.

Gould, S.J. 1977. The return of hopeful monsters. *Natural History* 86: 22–30.

Gould, S.J. 1978. Sociobiology: The art of storytelling. *New Scientist* 16 (November): 530–533.

Gould, S.J. 1980a. Is a new and general theory of evolution emerging? *Paleobiology* 6: 119–130.

Gould, S.J. 1980b. The promise of paleobiology as a nomothetic, evolutionary discipline. *Paleobiology* 6: 96–118.

Gould, S.J. 1985. The paradox of the first tier: An agenda for paleobiology. *Paleobiology* 11: 2–12.

Gould, S.J. 1986. The hardening of the Modern Synthesis. In: Grene, M. T. (ed.), *Dimensions of Darwinism,* Cambridge University Press, Cambridge, pp. 71–93.

Gould, S.J. 1993. Male nipples and clitoral ripples. *Columbia: A Journal of Literature and Art* 20: 80–96.

Gould, S.J. 1997. The exaptive excellence of spandrels as a term and prototype. *Proceedings of the National Academy of Sciences* 94: 10750–10755.

Gould, S.J. 2002. *The Structure of Evolutionary Theory.* Belknap Press of Harvard University Press, Cambridge, Mass.

Gould, S.J., and Lewontin, R.C. 1979. The spandrels of San Marco and the Panglossian paradigm: a critique of the adaptationist programme. *Proceedings of the Royal Society B: Biological Sciences* 205: 581–598.

Gould, S.J., Vrba, E.S. 1982. Exaptation—a missing term in the science of form. *Paleobiology* 8: 4–15.

Gould, S.J., Lloyd, E.A. 1999. Individuality and adaptation across levels of selection: How shall we name and generalize the unit of Darwinism? *Proceedings of the National Academy of Sciences* 96: 11904–11909.

Grandcolas, P., D'Haese, C. 2003. Testing adaptation with phylogeny: how to account for phylogenetic pattern and selective value together. *Zoologica Scripta* 32: 483–490.

Grantham, T. 2007. Is macroevolution more than successive rounds of microevolution? *Palaeontology* 50: 75–85.

Grantham, T.A. 1995. Hierarchical approaches to macroevolution: Recent work on species selection and the "effect hypothesis." *Annual Review of Ecology and Systematics* 26: 301–321.

Greene, M.J., Gordon, D.M. 2003. Cuticular hydrocarbons inform task decisions. *Nature* 423: 32–32.

Greene, M.J., Gordon, D.M. 2007. Interaction rate informs harvester ant task decisions. *Behavioral Ecology* 18: 451–455.

Gregory, T.R. 2009. Understanding natural selection: Essential concepts and common misconceptions. *Evolution: Education and Outreach* 2: 156–175.

Greilhuber, J., Borsch, T., Müller, K., Worberg, A., Porembski, S., Barthlott, W. 2006. Smallest angiosperm genomes found in Lentibulariaceae, with chromosomes of bacterial size. *Plant Biology* 8: 770–777.

Griffiths, P., Stotz, K. 2013. *Genetics and Philosophy: An Introduction.* Cambridge University Press, Cambridge, UK; New York.

Griffiths, P., Stotz, K. 2018. *Developmental Systems Theory as a Process Theory.* Oxford University Press, Oxford, UK.

Griffiths, P.E. 1992. Adaptive explanation and the concept of a vestige. In: Griffiths, P. E. (ed.), *Trees of Life,* Springer Netherlands, Dordrecht, pp. 111–131.

Griffiths, P.E. 1993. Functional analysis and proper functions. *The British Journal for the Philosophy of Science* 44: 409–422.

Griffiths, P.E. 1996a. Darwinism, process structuralism, and natural kinds. *Philosophy of Science* 63: S1–S9.

Griffiths, P.E. 1996b. The historical turn in the study of adaptation. *The British Journal for the Philosophy of Science* 47: 511–532.

Griffiths, P.E. 2017. Genetic, epigenetic and exogenetic information in development and evolution. *Interface Focus* 7: 20160152.

Griffiths, P.E., Gray, R.D. 1994. Developmental systems and evolutionary explanation. *Journal of Philosophy* 91: 277–304.

Griffiths, P.E., Knight, R.D. 1998. What is the developmentalist challenge? *Philosophy of Science* 65: 253–258.

Griffiths, P.E., Gray, R.D. 2005. Three ways to misunderstand Developmental Systems Theory. *Biology & Philosophy* 20: 417–425.

Griffiths, P.E., Stotz, K. 2006. Genes in the postgenomic era. *Theoretical Medicine and Bioethics* 27: 499–521.

Griffiths, P.E., Stotz, K. 2008. Experimental philosophy of science. *Philosophy Compass* 3: 507–521.

Gross, P.R. 2012. Review: Freaks of Nature: What Anomalies Tell Us about Development and Evolution. *Reports of the National Center for Science Education* 32: 12–14.

Guan, D., Shao, J., Zhao, Z., Wang, P., Qin, J., Deng, Y., Boheler, K.R., Wang, J., Yan, B. 2014. PTH-GRN: Unraveling post-translational hierarchical gene regulatory networks using PPI, ChIP-seq and gene expression data. *Nucleic Acids Research* 42: W130–W136.

Guyer, C., Slowinski, J.B. 1993. Adaptive radiation and the topology of large phylogenies. *Evolution* 47: 253–263.

Haig, B., Durrant, R. 2002. Adaptationism and inference to the best explanation. *Behavioral and Brain Sciences* 25: 520–521.

Hallgrímsson, B., Hall, B.K., eds. 2005. *Variation*. Elsevier Academic Press, Amsterdam; Boston.

Hansen, T.F. 2014. Use and misuse of comparative methods in the study of adaptation. In: Garamszegi, L. Z. (ed.), *Modern Phylogenetic Comparative Methods and Their Application in Evolutionary Biology*, Springer Berlin, pp. 351–379.

Hansen, T.F., Houle, D. 2008. Measuring and comparing evolvability and constraint in multivariate characters. *Journal of Evolutionary Biology* 21: 1201–1219.

Harland, M., Taylor, C.F., Bass, S., Churchman, M., Randerson-Moor, J.A., Holland, E.A., Mann, G.J., Bishop, D.T., Newton Bishop, J.A. 2005. Intronic sequence variants of the *CDKN2A* gene in melanoma pedigrees. *Genes, Chromosomes and Cancer* 43: 128–136.

Haskell, E.F. 1940. Mathematical systematization of "environment," "organism" and "habitat." *Ecology* 21: 1–16.

Hautier, L., Weisbecker, V., Sánchez-Villagra, M.R., Goswami, A., and Asher, R.J. 2010. Skeletal development in sloths and the evolution of mammalian vertebral patterning. *Proceedings of the National Academy of Sciences* 107: 18903–18908.

Hautmann, M. 2020. What is macroevolution? *Palaeontology* 63: 1–11.

Held, Jr., L.I. 2021. *Animal Anomalies: What Abnormal Anatomies Reveal about Normal Development*, 1st edn. Cambridge University Press: UK.

Held, Jr., L.I. 2009. *Quirks of Human Anatomy: An Evo-Devo Look at the Human Body*. Cambridge University Press, Cambridge [England]; New York.

Hendrikse, J.L., Parsons, T.E., Hallgrímsson, B. 2007. Evolvability as the proper focus of evolutionary developmental biology: Evolvability as the central question of evo-devo. *Evolution & Development* 9: 393–401.

Henfrey, A., Huxley, T.H., eds. 1853. *Scientific Memoirs: Selected from the Transactions of Foreign Academies of Science and Learned Societies, and from Foreign Journals*. Taylor & Francis, London.

Herai, R.H., Yamagishi, M.E.B. 2010. Detection of human interchromosomal trans-splicing in sequence databanks. *Briefings in Bioinformatics* 11: 198–209.

Herhaus, L., Dikic, I. 2015. Expanding the ubiquitin code through post-translational modification. *EMBO Reports* 16: 1071–1083.

Hernández, U., Posadas-Vidales, L., Espinosa-Soto, C. 2022. On the effects of the modularity of gene regulatory networks on phenotypic variability and its association with robustness. *Biosystems* 212: 104586.

Hintze, A., Adami, C. 2008. Evolution of complex modular biological networks. *PLoS Computational Biology* 4: e23.

Hodges, S.A. 1995. Spurring plant diversification: Are floral nectar spurs a key innovation? *Proceedings of the Royal Society of London. Series B: Biological Sciences* 262: 343–348.

Hodges, S.A. 1997. Floral nectar spurs and diversification. *International Journal of Plant Sciences* 158: S81–S88.

Hoekstra, H.E., Hirschmann, R.J., Bundey, R.A., Insel, P.A., Crossland, J.P. 2006. A single amino acid mutation contributes to adaptive beach mouse color pattern. *Science* 313: 101–104.

Holcomb, H.R. 1996. Just so stories and inference to the best explanation in evolutionary psychology. *Minds and Machines* 6: 525–540.

Holehouse, A.S., Kragelund, B.B. 2024. The molecular basis for cellular function of intrinsically disordered protein regions. *Nature Reviews Molecular Cell Biology* 25: 187–211.

Hopkin, K. 2009. The evolving definition of a gene. *BioScience* 59: 928–931.

Hordijk, W., Altenberg, L. 2020. Developmental structuring of phenotypic variation: A case study with a cellular automata model of ontogeny. *Evolution & Development* 22: 20–34.

Hotzy, C., Polak, M., Rönn, J.L., Arnqvist, G. 2012. Phenotypic engineering unveils the function of genital morphology. *Current Biology* 22: 2258–2261.

Houle, D. 2024. Niches for neglected ideas? *Evolution* 78: 1883–1887.

Houston, A.I. 1997. Are the spandrels of San Marco really Panglossian pendentives? *Trends in Ecology & Evolution* 12: 125.

Houston, A.I. 2009. San Marco and evolutionary biology. *Biology & Philosophy* 24: 215–230.

Hu, Y., Albertson, R.C. 2017. Baby fish working out: An epigenetic source of adaptive variation in the cichlid jaw. *Proceedings of the Royal Society B: Biological Sciences* 284: 20171018.

Hutchinson, G.E. 1948. Circular causal systems in ecology. *Annals of the New York Academy of Sciences* 50: 221–246.

Hutter, D. 2018. The human embryo and its environment: In utero challenges and physiological outcomes. In: Burggren, W. and Dubansky, B. (eds.), *Development and Environment*, Springer International Publishing, Cham, pp. 403–420.

Ingram, P.J., Stumpf, M.P., Stark, J. 2006. Network motifs: Structure does not determine function. *BMC Genomics* 7: 108.

Ioannou, C.C., Krause, J. 2009. Interactions between background matching and motion during visual detection can explain why cryptic animals keep still. *Biology Letters* 5: 191–193.

Islas, P., Platnich, C.M., Gidi, Y., Karimi, R., Ginot, L., Saliba, D., Luo, X., Cosa, G., Sleiman, H.F. 2024. Automated synthesis of DNA nanostructures. *Advanced Materials* 36: 2403477.

Jablonka, E., Lamb, M.J. 2005. *Evolution in Four Dimensions: Genetic, Epigenetic, Behavioral, and Symbolic Variation in the History of Life*. MIT Press, Cambridge, Mass.

Jablonski, D. 2007. Scale and hierarchy in macroevolution. *Palaeontology* 50: 87–109.

Jablonski, D. 2022. Evolvability and macroevolution: Overview and synthesis. *Evolutionary Biology* 49: 265–291.

Jacob, F., Monod, J. 1961. Genetic regulatory mechanisms in the synthesis of proteins. *Journal of Molecular Biology* 3: 318–356.

Jaenisch, R., Bird, A. 2003. Epigenetic regulation of gene expression: How the genome integrates intrinsic and environmental signals. *Nature Genetics* 33: 245–254.

Jahn, M., Seebacher, F. 2022. Variations in cost of transport and their ecological consequences: A review. *Journal of Experimental Biology* 225: jeb243646.

Jaksić, F.M. 1981. Recognition of morphological adaptations in animals: The hypothetico-deductive method. *BioScience* 31: 667–670.

Johnson, B. 2020. Abiotic ecosystems?: A critical examination of Arthur Tansley's ecosystem definition. *Environmental Ethics* 42: 39–53.

Johnson, D., Eckart, P., Alsamadisi, N., Noble, H., Martin, C., Spicer, R. 2018. Polar auxin transport is implicated in vessel differentiation and spatial patterning during secondary growth in *Populus*. *American Journal of Botany* 105: 186–196.

Johnson, Z.I., Chisholm, S.W. 2004. Properties of overlapping genes are conserved across microbial genomes. *Genome Research* 14: 2268–2272.

Johnston, R.F. 1992. Evolution in the rock dove: Skeletal morphology. *The Auk* 109: 530–542.

Jordan, C.Y., Harder, L.D. 2006. Manipulation of bee behavior by inflorescence architecture and its consequences for plant mating. *The American Naturalist* 167: 496–509.

Juravel, K., Porras, L., Höhna, S., Pisani, D., Wörheide, G. 2023. Exploring genome gene content and morphological analysis to test recalcitrant nodes in the animal phylogeny. *PLOS ONE* 18: e0282444.

Kaplan, J. 2008. The end of the adaptive landscape metaphor? *Biology & Philosophy* 23: 625–638.

Karacan, A., Türkvatan, A., Karacan, K. 2014. Anatomical variations of aortic arch branching: evaluation with computed tomographic angiography. *Cardiology in the Young* 24: 485–493.

Karlebach, G., Shamir, R. 2008. Modelling and analysis of gene regulatory networks. *Nature Reviews Molecular Cell Biology* 9: 770–780.

Karp, P.D., Midford, P.E., Billington, R., Kothari, A., Krummenacker, M., Latendresse, M., Ong, W.K., Subhraveti, P., Caspi, R., Fulcher, C., Keseler, I.M., Paley, S.M. 2021. Pathway Tools version 23.0 update: Software for pathway/genome informatics and systems biology. *Briefings in Bioinformatics* 22: 109–126.

Kauffman, S.A. 1996. *At Home in the Universe: The Search for Laws of Self-Organization and Complexity*. Oxford University Press, Oxford, UK.

Kawecki, T.J. 1994. Accumulation of deleterious mutations and the evolutionary cost of being a generalist. *The American Naturalist* 144: 833–838.

Keddy, P.A. 2001. *Competition*, 2nd edn. Springer, Dordrecht.

Keet, K., Gunston, G., Alexander, R. 2019. Variations in the branching pattern of the aortic arch: an African perspective. *European Journal of Anatomy* 23: 91–102.

Keller, E.F. 2005. Revisiting "scale-free" networks. *BioEssays* 27: 1060–1068.

Keller, L. 2003. Changing the world. *Nature* 425: 769–770.

Ketterson, E.D., Nolan, Jr., V. 1999. Adaptation, exaptation, and constraint: A hormonal perspective. *The American Naturalist* 154: S4–S25.

Khanin, R., Wit, E. 2006. How scale-free are biological networks. *Journal of Computational Biology* 13: 810–818.

King, R.B. 1989. Sexual dimorphism in snake tail length: Sexual selection, natural selection, or morphological constraint? *Biological Journal of the Linnean Society* 38: 133–154.

Kipling, R. 1902. *Just So Stories*. Doubleday Page and Company, Garden City, NY.

Kirschner, M., Gerhart, J. 1998. Evolvability. *Proceedings of the National Academy of Sciences* 95: 8420–8427.

Kirschner, M., Gerhart, J., Mitchison, T. 2000. Molecular "vitalism." *Cell* 100: 79–88.

Klein, D.E. 2004. The structure of colleters in several species of *Simira* (Rubiaceae). *Annals of Botany* 94: 733–740.

Klein, R.J., Zeiss, C., Chew, E.Y., Tsai, J.-Y., Sackler, R.S., Haynes, C., Henning, A.K., SanGiovanni, J.P., Mane, S.M., Mayne, S.T., Bracken, M.B., Ferris, F.L., Ott, J., Barnstable, C., Hoh, J. 2005. Complement factor H polymorphism in age-related macular degeneration. *Science* 308: 385–389.

Klingenberg, C.P. 2014. Studying morphological integration and modularity at multiple levels: Concepts and analysis. *Philosophical Transactions of the Royal Society B: Biological Sciences* 369: 20130249.

Kluge, A.G. 2005. Testing lineage and comparative methods for inferring adaptation. *Zoologica Scripta* 34: 653–663.

Koçillari, L., Olson, M.E., Suweis, S., Rocha, R.P., Lovison, A., Cardin, F., Dawson, T.E., Echeverría, A., Fajardo, A., Lechthaler, S., Martínez-Pérez, C., Marcati, C.R., Chung, K.-F., Rosell, J.A., Segovia-Rivas, A., Williams, C.B., Petrone-Mendoza, E., Rinaldo, A., Anfodillo, T., Banavar, J.R., Maritan, A. 2021. The Widened Pipe Model of plant hydraulic evolution. *Proceedings of the National Academy of Sciences* 118: 1–8.

Kodituwakku, P., Kodituwakku, E.L. 2022. Fetal alcohol syndrome. In: Pfaff, D. W., Volkow, N. D., and Rubenstein, J. L. (eds.), *Neuroscience in the 21st Century*, Springer International Publishing, Cham, pp. 3627–3647.

Kong, H., Sun, B., Yu, F., Wang, Q., Xia, K., Jiang, D. 2023. Exploring the potential of three-dimensional DNA crystals in nanotechnology: Design, optimization, and applications. *Advanced Science* 10: 2302021.

Kooter, J.M., De Lange, T., Borst, P. 1984. Discontinuous synthesis of mRNA in trypanosomes. *The EMBO Journal* 3: 2387–2392.

Kovaka, K. 2019. Underdetermination and evidence in the developmental plasticity debate. *The British Journal for the Philosophy of Science* 70: 127–152.

Kovaka, K. 2021. Fighting about frequency. *Synthese* 199: 7777–7797.

Kovaka, K., Novick, R. 2024. Relative frequency controversies and the growth of biological knowledge. *Philosophy, Theory, and Practice in Biology* 16: 1–19.

Kozłowski, J., Konarzewski, M. 2005. West, Brown and Enquist's model of allometric scaling again: The same questions remain. *Functional Ecology* 19: 739–743.

Kreidberg, L., Bean, J.L., Désert, J.-M., Benneke, B., Deming, D., Stevenson, K.B., Seager, S., Berta-Thompson, Z., Seifahrt, A., Homeier, D. 2014. Clouds in the atmosphere of the super-Earth exoplanet GJ 1214b. *Nature* 505: 69–72.

Krell, F.-T., Cranston, P.S. 2004. Which side of the tree is more basal? *Systematic Entomology* 29: 279–281.

Kühnen, U., Kitayama, S. 2021. Agency-detection. In: Shackelford, T. K. and Weekes-Shackelford, V. A. (eds), *Encyclopedia of Evolutionary Psychological Science*, Springer, Cham, Springer Nature Living Reference Behavioral Science and Psychology. pp. 128–135.

Kupiec, J.-J. 2014. Cell differentiation is a stochastic process subjected to natural selection. In: Minelli, A. and Pradeu, T. (eds.), *Towards a Theory of Development*, Oxford University Press, Oxford, UK, pp. 155–173.

Lachowiec, J., Queitsch, C., Kliebenstein, D.J. 2016. Molecular mechanisms governing differential robustness of development and environmental responses in plants. *Annals of Botany* 117: 795–809.

Ladyman, J. 2002. *Understanding Philosophy of Science*. Routledge, London.

Lafuente, E., Duneau, D., and Beldade, P. 2018. Genetic basis of thermal plasticity variation in *Drosophila melanogaster* body size. *PLOS Genetics* 14: e1007686.

Lai, X., Wolkenhauer, O., Vera, J. 2016. Understanding microRNA-mediated gene regulatory networks through mathematical modelling. *Nucleic Acids Research* 44: 6019–6035.

Laird, P.W. 1989. Trans splicing in trypanosomes — archaism or adaptation? *Trends in Genetics* 5: 204–208.

Lala, K.N., Uller, T., Feiner, N., Feldman, M.W., and Gilbert, S.F. 2024. *Evolution Evolving: The Developmental Origins of Adaptation and Biodiversity*. Princeton University Press, Princeton, NJ.

Laland, K.N., Odling-Smee, J., Hoppitt, W., Uller, T. 2013. More on how and why: cause and effect in biology revisited. *Biology & Philosophy* 28: 719–745.

Laland, K.N., Uller, T., Feldman, M.W., Sterelny, K., Müller, G.B., Moczek, A., Jablonka, E., Odling-Smee, J. 2015. The extended evolutionary synthesis: Its structure, assumptions and predictions. *Proceedings of the Royal Society B: Biological Sciences* 282: 20151019.

Lancaster, J.A.T. 2011. The semantic structure of evolutionary biology as an argument against intelligent design. *Zygon* 46: 26–46.

Landeen, E.A., Badyaev, A.V. 2012. Developmental integration of feather growth and pigmentation and its implications for the evolution of diet-derived coloration. *Journal of Experimental Zoology Part B: Molecular and Developmental Evolution* 318B: 59–70.

Landrum, J.V. 2006. Wide-band tracheids in genera of Portulacaceae: Novel, non-xylary tracheids possibly evolved as an adaptation to water stress. *Journal of Plant Research* 119: 497–504.

Lardner, B. 1998. Plasticity or fixed adaptive traits? Strategies for predation avoidance in *Rana arvalis* tadpoles. *Oecologia* 117: 119–126.

Larson, A., Losos, J. 1996. Phylogenetic systematics of adaptation. In: Rose, M. R. and Lauder, G. V. (eds.), *Adaptation*, Academic Press, San Diego, pp. 187–221.

Laubichler, M.D. 2010. Evolutionary developmental biology offers a significant challenge to the Neo-Darwinian paradigm. In: Ayala, F. J. and Arp, R. (eds), *Contemporary Debates in Philosophy of Biology*, Wiley-Blackwell, Chichester, UK; Malden, MA, pp. 199–212.

Lebedev, O.A., and Coates, M.I. 1995. The postcranial skeleton of the Devonian tetrapod *Tulerpeton curtum* Lebedev. *Zoological Journal of the Linnean Society* 114: 307–348.

Lennox, J.G. 1991. Darwinian thought experiments: A function for just-so stories. In: Horowitz, T. and Massey, G. J. (eds.), *Thought Experiments in Science and Philosophy*, Rowman and Littlefield, Savage, Maryland, USA, pp. 223–245.

Leroi, A.M., Rose, M.R., Lauder, G.V. 1994. What does the comparative method reveal about adaptation? *The American Naturalist* 143: 381–402.

Leśniewska, M., Bonato, L., Minelli, A., Fusco, G. 2009. Trunk anomalies in the centipede *Stigmatogaster subterranea* provide insight into late-embryonic segmentation. *Arthropod Structure & Development* 38: 417–426.

Levine, M., Davidson, E.H. 2005. Gene regulatory networks for development. *Proceedings of the National Academy of Sciences* 102: 4936–4942.

Levis, N.A., Isdaner, A.J., Pfennig, D.W. 2018. Morphological novelty emerges from pre-existing phenotypic plasticity. *Nature Ecology & Evolution* 2: 1289–1297.

Levy, A. 2011. Information in biology: A fictionalist account. *Noûs* 45: 640–657.

Lewontin, R.C. 2000a. *The Triple Helix: Gene, Organism, and Environment.* Harvard University Press, Cambridge, Mass.

Lewontin, R.C. 2000b. *It Ain't Necessarily so: The Dream of the Human Genome and Other Illusions.* New York Review Books, New York.

Li, G., Liu, H., Müller, U.K., Voesenek, C.J., Van Leeuwen, J.L. 2021. Fishes regulate tail-beat kinematics to minimize speed-specific cost of transport. *Proceedings of the Royal Society B: Biological Sciences* 288: 20211601.

Liang, X., Haritan, A., Uliel, S., Michaeli, S. 2003. *trans* and *cis* splicing in trypanosomatids: mechanism, factors, and regulation. *Eukaryotic Cell* 2: 830–840.

Lickliter, R., Harshaw, C. 2010. Canalization and malleability reconsidered: The developmental basis of phenotypic stability and variability. In: Hood, K. E., Halpern, C. T., Greenberg, G., and Lerner, R. M. (eds.), *Handbook of Developmental Science, Behavior, and Genetics*, Wiley-Blackwell, Oxford, UK, pp. 491–525.

Lindeman, R.L. 1942. The trophic-dynamic aspect of ecology. *Ecology* 23: 399–417.

Lindhe Norberg, U.M. 2002. Structure, form, and function of flight in engineering and the living world. *Journal of Morphology* 252: 52–81.

Linh, N.M., Scarpella, E. 2022. Leaf vein patterning is regulated by the aperture of plasmodesmata intercellular channels. *PLOS Biology* 20: e3001781.

Lipton, P. 2001. What good is an explanation? In: Hon, G. and Rakover, S. S. (eds.), *Explanation*, Springer Netherlands, Dordrecht, Synthese Library. pp. 43–59.

Lipton, P. 2008. Inference to the best explanation. In: Psillos, S. and Curd, M. (eds.), *The Routledge Companion to the Philosophy of Science*, Routledge, London, UK, pp. 193–202.

Lloyd, E.A. 2005. *The Case of the Female Orgasm: Bias in the Science of Evolution.* Harvard University Press, Cambridge, Mass.

Lloyd, E.A., Gould, S.J. 1993. Species selection on variability. *Proceedings of the National Academy of Sciences* 90: 595–599.

Longman, D., Stock, J.T., Wells, J.C.K. 2017. A trade-off between cognitive and physical performance, with relative preservation of brain function. *Scientific Reports* 7: 13709.

López Beltrán, C., ed. 2011. *Genes (&) Mestizos: Genómica y Raza En La Biomedicina Mexicana*, Primera edición. Ficticia: Universidad Nacional Autónoma de México, México, D.F.

de Lorenzo, V. 2011. Beware of metaphors: Chasses and orthogonality in synthetic biology. *Bioengineered Bugs* 2: 3–7.

Losos, J.B. 2008. Phylogenetic niche conservatism, phylogenetic signal and the relationship between phylogenetic relatedness and ecological similarity among species. *Ecology Letters* 11: 995–1003.

Losos, J.B. 2017. *Improbable Destinies: Fate, Chance, and the Future of Evolution.* Riverhead Books, New York.

Losos, J.B. 2023. *The Cat's Meow: How Cats Evolved from the Savanna to Your Sofa.* Viking, New York.

Love, A.C. 2013. Theory is as theory does: Scientific practice and theory structure in biology. *Biological Theory* 7: 325–337.

Lovell, J.H. 1918. *The Flower and the Bee.* Charles Scribner's Sons, New York.

Lu, Q. 2023. Epigenetic inheritance and the modern evolutionary synthesis. *Journal of Human Cognition* 7: 29–41.

Lubbock, J. 1891. On stipules, their form and function. *Botanical Journal of the Linnean Society* 28: 217–243.

MacColl, A.D.C. 2011. The ecological causes of evolution. *Trends in Ecology & Evolution* 26: 514–522.

MacNutt, M.J., Laursen, P.B., Kedia, S., Neupane, M., Parajuli, P., Pokharel, J., Sheel, A.W. 2012. Acclimatisation in trekkers with and without recent exposure to high altitude. *European Journal of Applied Physiology* 112: 3287–3294.

Maddison, W.P., FitzJohn, R.G. 2015. The unsolved challenge to phylogenetic correlation tests for categorical characters. *Systematic Biology* 64: 127–136.

Magwene, P. 2001. New tools for studying integration and modularity. *Evolution* 55: 1734–1745.

Mahner, M., Bunge, M. 1997. *Foundations of Biophilosophy*. Springer, Berlin.

Maki, K.C., Dicklin, M.R., Kirkpatrick, C.F. 2021. Saturated fats and cardiovascular health: Current evidence and controversies. *Journal of Clinical Lipidology* 15: 765–772.

Malagón, J.N., Ahuja, A., Sivapatham, G., Hung, J., Lee, J., Muñoz-Gómez, S.A., Atallah, J., Singh, R.S., Larsen, E. 2014. Evolution of *Drosophila* sex comb length illustrates the inextricable interplay between selection and variation. *Proceedings of the National Academy of Sciences* 111: E4103–E4109.

Mangan, S., Zaslaver, A., Alon, U. 2003. The coherent feedforward loop serves as a sign-sensitive delay element in transcription networks. *Journal of Molecular Biology* 334: 197–204.

Mangelsdorf, P.C. 1974. *Corn: Its Origin, Evolution and Improvement*. Harvard University Press, Cambridge, Massachusetts.

Manolio, T.A., Collins, F.S., Cox, N.J., Goldstein, D.B., Hindorff, L.A., Hunter, D.J., McCarthy, M.I., Ramos, E.M., Cardon, L.R., Chakravarti, A., Cho, J.H., Guttmacher, A.E., Kong, A., Kruglyak, L., Mardis, E., Rotimi, C.N., Slatkin, M., Valle, D., Whittemore, A.S., Boehnke, M., Clark, A.G., Eichler, E.E., Gibson, G., Haines, J.L., Mackay, T.F.C., McCarroll, S.A., Visscher, P.M. 2009. Finding the missing heritability of complex diseases. *Nature* 461: 747–753.

Marchini, M., Rolian, C. 2018. Artificial selection sheds light on developmental mechanisms of limb elongation. *Evolution* 72: 825–837.

Marchini, M., Sparrow, L.M., Cosman, M.N., Dowhanik, A., Krueger, C.B., Hallgrimsson, B., Rolian, C. 2014. Impacts of genetic correlation on the independent evolution of body mass and skeletal size in mammals. *BMC Evolutionary Biology* 14: 258.

Martins, E.P. 2000. Adaptation and the comparative method. *Trends in Ecology & Evolution* 15: 296–299.

Martins, E.P., Hansen, T.F. 1997. Phylogenies and the comparative method: A general approach to incorporating phylogenetic information into the analysis of interspecific data. *The American Naturalist* 149: 646–667.

Matthaei, J.H., Nirenberg, M.W. 1961. Characteristics and stabilization of DNAase-sensitive protein synthesis in *E. coli* extracts. *Proceedings of the National Academy of Sciences* 47: 1580–1588.

Matthen, M., Ariew, A. 2002. Two ways of thinking about fitness and natural selection. *Journal of Philosophy* 99: 55–83.

Maynard Smith, J., Burian, R., S. Kauffman, Alberch, P., Campbell, J., Goodwin, B., Lande, R., Raup, D., Wolpert, L. 1985. Developmental constraints and evolution: a perspective from the Mountain Lake Conference on Development and Evolution. *The Quarterly Review of Biology* 60: 265–287.

Maynard Smith, J.M. 2000. The concept of information in biology. *Philosophy of Science* 67: 177–194.

Mayr, E. 1982. *The Growth of Biological Thought: Diversity, Evolution, and Inheritance*. Belknap Press, Cambridge, Mass.

McDaniel, S.F. 2021. Bryophytes are not early diverging land plants. *New Phytologist* 230: 1300–1304.

McGhee, G.R. 2007. *The Geometry of Evolution: Adaptive Landscapes and Theoretical Morphospaces*. Cambridge University Press, Cambridge, UK; New York.

McGraw, K.J., Hill, G.E., Parker, R.S. 2005. The physiological costs of being colourful: Nutritional control of carotenoid utilization in the American goldfinch, *Carduelis tristis*. *Animal Behaviour* 69: 653–660.

McKenna, K.Z., Gawne, R., Nijhout, H.F. 2022. The genetic control paradigm in biology: What we say, and what we are entitled to mean. *Progress in Biophysics and Molecular Biology* 169–170: 89–93.

McKinney, B.A., Pajewski, N.M. 2012. Six degrees of epistasis: Statistical network models for GWAS. *Frontiers in Genetics* 2: 109.

McLuhan, M. 2017. The medium is the message. In: Mortensen, C. D. (ed.), *Communication Theory* Routledge: New York, pp. 390–402.

Medawar, P.B. 1967. *The Art of the Soluble*. Methuen & Co., Madison, WI.

Meier, J.I., Marques, D.A., Mwaiko, S., Wagner, C.E., Excoffier, L., Seehausen, O. 2017. Ancient hybridization fuels rapid cichlid fish adaptive radiations. *Nature Communications* 8: 14363.

Menezes, B.F., Vigoder, F.M., Peixoto, A.A., Varaldi, J., Bitner-Mathé, B.C. 2013. The influence of male wing shape on mating success in *Drosophila melanogaster*. *Animal Behaviour* 85: 1217–1223.
Milo, R., Shen-Orr, S., Itzkovitz, S., Kashtan, N., Chklovskii, D., Alon, U. 2002. Network motifs: simple building blocks of complex networks. *Science* 298: 824–827.
Mincer, S.T., Russo, G.A. 2020. Substrate use drives the macroevolution of mammalian tail length diversity. *Proceedings of the Royal Society B: Biological Sciences* 287: 20192885.
Minelli, A. 2009. *Forms of Becoming: The Evolutionary Biology of Development*. Princeton University Press, Princeton.
Minelli, A. 2010. Evolutionary developmental biology does not offer a significant challenge to the Neo-Darwinian paradigm. In: Ayala, F. J., and Arp, R. (eds.), *Contemporary Debates in Philosophy of Biology*, Wiley-Blackwell, Chichester, UK, pp. 213–226.
Minelli, A. 2019. An evo-devo perspective on analogy in biology. *Philosophies* 4: 5.
Mitchison, G. 1991. Neuronal branching patterns and the economy of cortical wiring. *Proceedings of the Royal Society of London. Series B: Biological Sciences* 245: 151–158.
Moczek, A.P. 2020. Biases in the study of developmental bias. *Evolution & Development* 22: 3–6.
Moger-Reischer, R.Z., Glass, J.I., Wise, K.S., Sun, L., Bittencourt, D.M.C., Lehmkuhl, B.K., Schoolmaster, D.R., Lynch, M., Lennon, J.T. 2023. Evolution of a minimal cell. *Nature* 620: 122–127.
Monaghan, P., Ivimey-Cook, E.R. 2023. No time to die: Evolution of a post-reproductive life stage. *Journal of Zoology* 321: 1–21.
Montes-Cartas, C.G., Padilla, P., Rosell, J.A., Domínguez, C.A., Fornoni, J., Olson, M.E. 2017. Testing the hypothesis that biological modularity is shaped by adaptation: Xylem in the *Bursera simaruba* clade of tropical trees. *Evolution & Development* 19: 111–123.
Moore, D.S., Lickliter, R. 2023. Development as explanation: Understanding phenotypic stability and variability after the failure of genetic determinism. *Progress in Biophysics and Molecular Biology* 178: 72–77.
Moore, W.S., Overton, L.C., Miglia, K.J. 2011. Mitochondrial DNA based phylogeny of the woodpecker genera *Colaptes* and *Piculus*, and implications for the history of woodpecker diversification in South America. *Molecular Phylogenetics and Evolution* 58: 76–84.
Morange, M. 2009. Articulating different modes of explanation: The present boundary in biological research. In: Barberousse, A., Morange, M., and Pradeu, T. (eds.), *Mapping the Future of Biology*, Springer Netherlands, Dordrecht, vol. 266. pp. 15–26.
Morgan, T.H. 1917. The theory of the gene. *The American Naturalist* 51: 513–544.
Morris, M.R.J., Rogers, S.M. 2013. Overcoming maladaptive plasticity through plastic compensation. *Current Zoology* 59: 526–536.
Morrison, E.S., Badyaev, A.V. 2016. Structuring evolution: biochemical networks and metabolic diversification in birds. *BMC Evolutionary Biology* 16: 168.
Morrison, E.S., Badyaev, A.V. 2017. Beyond topology: Coevolution of structure and flux in metabolic networks. *Journal of Evolutionary Biology* 30: 1796–1809.
Morse, A. 1998. Searching for the Holy Grail: The Human Genome Project and its implications. *Journal of Law and Health* 13: 219–256.
Moses, D., Ginell, G.M., Holehouse, A.S., Sukenik, S. 2023. Intrinsically disordered regions are poised to act as sensors of cellular chemistry. *Trends in Biochemical Sciences* 48: 1019–1034.
Mottus, R.C., Whitehead, I.P., O'Grady, M., Sobel, R.E., Burr, R.H.L., Spiegelman, G.B., Grigliatti, T.A. 1997. Unique gene organization: Alternative splicing in *Drosophila* produces two structurally unrelated proteins. *Gene* 198: 229–236.
Mueller, P., Chien, T.F., Rudy, B. 1983. Formation and properties of cell-size lipid bilayer vesicles. *Biophysical Journal* 44: 375–381.
Mukherjee, S. 2022. *The Song of the Cell*. Scribner, New York, NY.
Müller, G., Newman, S., eds. 2003. *Origination of Organismal Form: Beyond the Gene in Developmental and Evolutionary Biology*. MIT Press, Cambridge, Mass.
Müller, G.B. 2011. Evo-devo as a discipline. In: Minelli, A. and Fusco, G. M. (eds.), *Evolving Pathways: Key Themes in Evolutionary Developmental Biology*, Cambridge Univ. Press, Cambridge, pp. 5–30.
Muller, G.B., Wagner, G.P. 1991. Novelty in evolution: Restructuring the concept. *Annual Review of Ecology and Systematics* 22: 229–256.

Müller, M., Schmitz, B.L., Pauls, S., Schick, M., Röhrer, S., Kapapa, T., Schlötzer, W. 2011. Variations of the aortic arch – a study on the most common branching patterns. *Acta Radiologica* 52: 738–742.

Munar-Delgado, G., Araya-Ajoy, Y.G., Edelaar, P. 2023. Estimation of additive genetic variance when there are gene–environment correlations: Pitfalls, solutions and unexplored questions. *Methods in Ecology and Evolution* 14: 1245–1258.

Muravnik, L.E., Kostina, O.V., Shavarda, A.L. 2014. Development, structure and secretion compounds of stipule colleters in *Pentas lanceolata* (Rubiaceae). *South African Journal of Botany* 93: 27–36.

Murren, C.J., Auld, J.R., Callahan, H., Ghalambor, C.K., Handelsman, C.A., Heskel, M.A., Kingsolver, J.G., Maclean, H.J., Masel, J., Maughan, H., Pfennig, D.W., Relyea, R.A., Seiter, S., Snell-Rood, E., Steiner, U.K., Schlichting, C.D. 2015. Constraints on the evolution of phenotypic plasticity: Limits and costs of phenotype and plasticity. *Heredity* 115: 293–301.

Nachman, M.W., Hoekstra, H.E., D'Agostino, S.L. 2003. The genetic basis of adaptive melanism in pocket mice. *Proceedings of the National Academy of Sciences* 100: 5268–5273.

Napoli, J.G., Fabbri, M., Ruebenstahl, A.A., O'Connor, J.K., Bhullar, B.-A.S., Norell, M.A. 2025. Reorganization of the theropod wrist preceded the origin of avian flight. *Nature* 644: 699–705.

Narita, Y., Kuratani, S. 2005. Evolution of the vertebral formulae in mammals: A perspective on developmental constraints. *Journal of Experimental Zoology Part B: Molecular and Developmental Evolution* 304B: 91–106.

Natsis, K.I., Tsitouridis, I.A., Didagelos, M.V., Fillipidis, A.A., Vlasis, K.G., Tsikaras, P.D. 2009. Anatomical variations in the branches of the human aortic arch in 633 angiographies: clinical significance and literature review. *Surgical and Radiologic Anatomy* 31: 319–323.

Nesteruk, I., Passoni, G., Redaelli, A. 2014. Shape of aquatic animals and their swimming efficiency. *Journal of Marine Biology* 2014: 1–9.

Neto, C. 2020. When imprecision is a good thing, or how imprecise concepts facilitate integration in biology. *Biology & Philosophy* 35: 58.

Neves, G., Zucker, J., Daly, M., Chess, A. 2004. Stochastic yet biased expression of multiple *Dscam* splice variants by individual cells. *Nature Genetics* 36: 240–246.

Newman, S.A. 2003. From physics to development: The evolution of morphogenetic mechanisms. In: Müller, G. B. and Newman, S. A. (eds.), *Origination of Organismal Form*, The MIT Press, Cambridge, Mass., pp. 221–240.

Newman, S.A. 2014. Physico-genetics of morphogenesis: The hybrid nature of developmental mechanisms. In: Minelli, A. and Pradeu, T. (eds.), *Towards a Theory of Development*, Oxford University Press, Oxford, UK, pp. 95–113.

Newman, S.A. 2015. Development and evolution: The physics connection. In: Love, A. C. (ed.), *Conceptual Change in Biology*, Springer, New York, Boston, Studies in the Philosophy and History of Science. pp. 421–440.

Newsome, T.M., Wolf, C., Nimmo, D.G., Kopf, R.K., Ritchie, E.G., Smith, F.A., Ripple, W.J. 2020. Constraints on vertebrate range size predict extinction risk. *Global Ecology and Biogeography* 29: 76–86.

Ng, C.S., Kopp, A. 2008. Sex combs are important for male mating success in *Drosophila melanogaster*. *Behavior Genetics* 38: 195–201.

Ng, S.F., Frankel, J. 1977. 180 degrees rotation of ciliary rows and its morphogenetic implications in *Tetrahymena pyriformis*. *Proceedings of the National Academy of Sciences* 74: 1115–1119.

Nguyen, M.-T.T., Vemaraju, S., Nayak, G., Odaka, Y., Buhr, E.D., Alonzo, N., Tran, U., Batie, M., Upton, B.A., Darvas, M., Kozmik, Z., Rao, S., Hegde, R.S., Iuvone, P.M., Van Gelder, R.N., Lang, R.A. 2019. An opsin 5–dopamine pathway mediates light-dependent vascular development in the eye. *Nature Cell Biology* 21: 420–429.

Niklas, K.J. 1992. *Plant Biomechanics: An Engineering Approach to Plant Form and Function*. University of Chicago Press, Chicago.

Niklas, K.J. 2009. Deducing plant function from organic form: challenges and pitfalls. In: Laubichler, M. D. and Maienschein, J. (eds), *Form and Function in Developmental Evolution*, Cambridge University Press, Leiden, pp. 47–82.

Niklas, K.J., Bondos, S.E., Dunker, A.K., Newman, S.A. 2015. Rethinking gene regulatory networks in light of alternative splicing, intrinsically disordered protein domains, and post-translational modifications. *Frontiers in Cell and Developmental Biology* 3: 8.

Nirenberg, M.W., Matthaei, J.H. 1961. The dependence of cell-free protein synthesis in *E. coli* upon naturally occurring or synthetic polyribonucleotides. *Proceedings of the National Academy of Sciences* 47: 1588–1602.

Noble, D. 2008. *The Music of Life: Biology beyond Genes*. Oxford University Press, Oxford, UK.

Noble, D. 2012. A theory of biological relativity: no privileged level of causation. *Interface Focus* 2: 55–64.

Noble, D. 2013. A biological relativity view of the relationships between genomes and phenotypes. *Progress in Biophysics and Molecular Biology* 111: 59–65.

Noble, D. 2015. Evolution beyond neo-Darwinism: A new conceptual framework. *Journal of Experimental Biology* 218: 7–13.

Norberg, U.M. 1989. *Vertebrate Flight: Mechanics, Physiology, Morphology, Ecology and Evolution*. Springer, Berlin.

Ochoterena, H., Vrijdaghs, A., Smets, E., Claßen-Bockhoff, R. 2019. The search for common origin: homology revisited. *Systematic Biology* 68: 767–780.

O'Connor, P.M., Claessens, L.P.A.M. 2005. Basic avian pulmonary design and flow-through ventilation in non-avian theropod dinosaurs. *Nature* 436: 253–256.

Odling-Smee, F.J., Laland, K.N., Feldman, M.W. 2003. *Niche Construction: The Neglected Process in Evolution*. Princeton University Press, Princeton.

Odum, E.P. 1953. *Fundamentals of Ecology*. W. B. Saunders Co., Philadelphia.

Okasha, S. 2008. *Evolution and the Levels of Selection*. Clarendon Press, Oxford, UK.

Oldham, S., Böhni, R., Stocker, H., Brogiolo, W., Hafen, E. 2000. Genetic control of size in *Drosophila*. *Philosophical Transactions of the Royal Society of London. Series B: Biological Sciences* 355: 945–952.

Olson, M.E. 2012a. The developmental renaissance in adaptationism. *Trends in Ecology & Evolution* 27: 278–287.

Olson, M.E. 2012b. Linear trends in botanical systematics and the Major Trends of Xylem Evolution. *The Botanical Review* 78: 154–183.

Olson, M.E. 2019a. Overcoming the constraint-adaptation dichotomy: Long live the constraint-adaptation dichotomy. In: Fusco, G. (ed.), *Perspectives on Evolutionary and Developmental Biology*, University of Padova Press, Padova, Italy, pp. 78–94.

Olson, M.E. 2019b. Spandrels and trait delimitation: no such thing as "architectural constraint." *Evolution & Development* 21: 59–71.

Olson, M.E. 2021. The comparative method is not macroevolution: Across-species evidence for within-species process. *Systematic Biology* 70: 1272–1281.

Olson, M.E. 2023. A skeptic's guide to Sherwin Carlquist's inferences of xylem function. *IAWA Journal* 44: 1–17.

Olson, M.E. 2024. Is population genetics really relevant to evolutionary biology? *Evolutionary Biology* 51: 235–243.

Olson, M.E., Arroyo-Santos, A. 2015. How to study adaptation (and why to do it that way). *The Quarterly Review of Biology* 90: 167–191.

Olson, M.E., Arroyo-Santos, A., Vergara-Silva, F. 2019. A user's guide to metaphors in ecology and evolution. *Trends in Ecology & Evolution* 34: 605–615.

Olson, M.E., Rosell, J.A., Zamora Muñoz, S., Castorena, M. 2018. Carbon limitation, stem growth rate and the biomechanical cause of Corner's rules. *Annals of Botany* 122: 583–592.

O'Malley, M.A., Dupré, J. 2007. Size doesn't matter: Towards a more inclusive philosophy of biology. *Biology & Philosophy* 22: 155–191.

Omland, K.E., Cook, L.G., Crisp, M.D. 2008. Tree thinking for all biology: The problem with reading phylogenies as ladders of progress. *BioEssays* 30: 854–867.

de Onís, M., Monteiro, C., Akré, J., Glugston, G. 1993. The worldwide magnitude of protein-energy malnutrition: An overview from the WHO Global Database on Child Growth. *Bulletin of the World Health Organization* 71: 703–712.

Orzack, S.H. 1981. The Modern Synthesis is partly Wright. *Paleobiology* 7: 128–131.

Orzack, S.H., Sober, E. 1994. Optimality models and the test of adaptationism. *The American Naturalist* 143: 361–380.

Orzack, S.H., Sober, E. 2001. Adaptation, phylogenetic inertia, and the method of controlled comparisons. In: Orzack, S. H. and Sober, E. (eds.), *Adaptationism and Optimality*, Cambridge University Press, Cambridge, UK, pp. 45–63.

Oyama, S., Griffiths, P.E., Gray, R.D. 2003. *Cycles of Contingency: Developmental Systems and Evolution*. MIT Press, Cambridge, Mass.

Palacios, M.G., Sparkman, A.M., Bronikowski, A.M. 2011. Developmental plasticity of immune defence in two life-history ecotypes of the garter snake, *Thamnophis elegans* - a common-environment experiment: *Journal of Animal Ecology* 80: 431–437.

Palmer, D.S., Zhou, W., Abbott, L., Wigdor, E.M., Baya, N., Churchhouse, C., Seed, C., Poterba, T., King, D., Kanai, M., Bloemendal, A., Neale, B.M. 2023. Analysis of genetic dominance in the UK Biobank. *Science* 379: 1341–1348.

Panchen, A.L. 1992. *Classification, Evolution, and the Nature of Biology*. Cambridge University Press, Cambridge, UK.

Parke, E.C. 2014. Experiments, simulations, and epistemic privilege. *Philosophy of Science* 81: 516–536.

Parker, G.A., Maynard Smith, J. 1990. Optimality theory in evolutionary biology. *Nature* 348: 27–33.

Patterson, C. 1982. Morphological characters and homology. In: Joysey, K. A. and Friday, A. E. (eds.), *Problems of phylogenetic reconstruction*, Published for the Systematics Association by Academic Press, London, Systematics Association special vol. pp. 21–74.

Patnaik, R., Auffray, J.C., Jaeger, J.J., Sahni, A. 1996. House mouse ancestor from Late Pliocene Siwalik sediments of India. *Comptes Rendus de l'Academie Des Sciences-Serie III-Sciences de La Vie* 319: 431–434.

Pausas, J.G. 2015. Bark thickness and fire regime. *Functional Ecology* 29: 315–327.

Pavličev, M., Wagner, G.P. 2016. The evolutionary origin of female orgasm. *Journal of Experimental Zoology Part B: Molecular and Developmental Evolution* 326: 326–337.

Payne, J.L., Wagner, A. 2019. The causes of evolvability and their evolution. *Nature Reviews Genetics* 20: 24–38.

Petit, G., Anfodillo, T. 2009. Plant physiology in theory and practice: An analysis of the WBE model for vascular plants. *Journal of Theoretical Biology* 259: 1–4.

Petit, W.A., Adamec, C.A. 2005. *The Encyclopedia of Endocrine Diseases and Disorders*. Facts On File, New York, NY.

Pfennig, D.W., ed. 2021a. *Phenotypic Plasticity & Evolution: Causes, Consequences, Controversies*. CRC Press, Boca Raton, FL.

Pfennig, D.W. 2021b. Key Questions about Phenotypic Plasticity. In: Pfennig, D. W. (ed.), *Phenotypic Plasticity & Evolution* CRC Press, Boca Raton, pp. 55–88.

Pham, K.K., Hipp, A.L., Manos, P.S., Cronn, R.C. 2017. A time and a place for everything: phylogenetic history and geography as joint predictors of oak plastome phylogeny. *Genome* 60: 720–732.

Pigliucci, M. 2001. *Phenotypic Plasticity: Beyond Nature and Nurture*. Johns Hopkins University Press, Baltimore, MD.

Pigliucci, M. 2005. Evolution of phenotypic plasticity: Where are we going now? *Trends in Ecology & Evolution* 20: 481–486.

Pigliucci, M. 2006. Genetic variance–covariance matrices: A critique of the evolutionary quantitative genetics research program. *Biology & Philosophy* 21: 1–23.

Pigliucci, M. 2008. The proper role of population genetics in modern evolutionary theory. *Biological Theory* 3: 316–324.

Pigliucci, M. 2013. Landscapes, surfaces, and morphospaces: What are they good for? In: Svensson, E., and Calsbeek, R. (eds.), *The Adaptive Landscape in Evolutionary Biology*, Oxford University Press, Oxford, UK, pp. 26–38.

Pigliucci, M., Kaplan, J. 2000. The fall and rise of Dr Pangloss: Adaptationism and the Spandrels paper 20 years later. *Trends in Ecology & Evolution* 15: 66–70.

Pigliucci, M., Murren, C.J. 2003. Genetic assimilation and a possible evolutionary paradox: Can macroevolution sometimes be so fast as to pass us by? *Evolution* 57: 1455.

Pigliucci, M., and Kaplan, J. 2006. *Making Sense of Evolution: The Conceptual Foundations of Evolutionary Biology*. Univ. of Chicago Press, Chicago, IL.

Pigliucci, M., Müller, G., eds. 2010. *Evolution, the Extended Synthesis*. MIT Press, Cambridge, Mass.

Pigliucci, M., Murren, C.J., Schlichting, C.D. 2006. Phenotypic plasticity and evolution by genetic assimilation. *Journal of Experimental Biology* 209: 2362–2367.

Piovesan, G., Biondi, F. 2021. On tree longevity. *New Phytologist* 231: 1318–1337.

Platnick, N.I., Gaffney, E.S. 1978. Evolutionary biology: A Popperian perspective. *Systematic Zoology* 27: 138.

Poe, S., Donald, L.A.H., Anderson, C. 2021. What constrains adaptive radiation? Documentation and explanation of under-evolved morphologies in *Anolis* lizards. *Proceedings of the Royal Society B: Biological Sciences* 288: 20210340.

Poelwijk, F.J., Kiviet, D.J., Weinreich, D.M., Tans, S.J. 2007. Empirical fitness landscapes reveal accessible evolutionary paths. *Nature* 445: 383–386.

Polak, M., Rashed, A. 2010. Microscale laser surgery reveals adaptive function of male intromittent genitalia. *Proceedings of the Royal Society B: Biological Sciences* 277: 1371–1376.

Polak, M., McEvey, S.F. 2022. Refutation of traumatic insemination in the *Drosophila bipectinata* species complex. *Biology Letters* 18: 20210625.

Pontarotti, G., Mossio, M., Pocheville, A. 2022. The genotype–phenotype distinction: from Mendelian genetics to 21st century biology. *Genetica* 150: 223–234.

Popieluszko, P., Henry, B.M., Sanna, B., Hsieh, W.C., Saganiak, K., Pękala, P.A., Walocha, J.A., Tomaszewski, K.A. 2018. A systematic review and meta-analysis of variations in branching patterns of the adult aortic arch. *Journal of Vascular Surgery* 68: 298–306.e10.

Portin, P., Wilkins, A. 2017. The evolving definition of the term "gene." *Genetics* 205: 1353–1364.

Potochnik, A. 2009. Optimality modeling in a suboptimal world. *Biology & Philosophy* 24: 183–197.

Potochnik, A. 2017. *Idealization and the Aims of Science*. The University of Chicago Press, Chicago.

Potticary, A.L., Morrison, E.S., Badyaev, A.V. 2020. Turning induced plasticity into refined adaptations during range expansion. *Nature Communications* 11: 3254.

Powell, R. 2020. *Contingency and Convergence: Toward a Cosmic Biology of Body and Mind*. The MIT Press, Cambridge, Massachusetts.

Pratt, R.B., Jacobsen, A.L. 2017. Conflicting demands on angiosperm xylem: Tradeoffs among storage, transport and biomechanics. *Plant, Cell & Environment* 40: 897–913.

Priya, S., Thomas, R., Nagpal, P., Sharma, A., Steigner, M. 2018. Congenital anomalies of the aortic arch. *Cardiovascular Diagnosis and Therapy* 8: S26–S44.

Qi, Y., Ge, H. 2006. Modularity and dynamics of cellular networks. *PLoS Computational Biology* 2: e174.

Quelle, D.E., Zindy, F., Ashmun, R.A., Sherr, C.J. 1995. Alternative reading frames of the INK4a tumor suppressor gene encode two unrelated proteins capable of inducing cell cycle arrest. *Cell* 83: 993–1000.

Rabuffo, C., Schmidt, M.R., Yadav, P., Tong, P., Carloni, R., Barcons-Simon, A., Cosentino, R.O., Krebs, S., Matthews, K.R., Allshire, R.C., Siegel, T.N. 2024. Inter-chromosomal transcription hubs shape the 3D genome architecture of African trypanosomes. *Nature Communications* 15: 10716.

Raff, R.A. 2000. Evo-devo: The evolution of a new discipline. *Nature Reviews Genetics* 1: 74–79.

Ramirez, A., Vyzas, C.A., Zhao, H., Eng, K., Degenhardt, K., Astrof, S. 2024. Buffering mechanism in aortic arch artery formation and congenital heart disease. *Circulation Research* 134: e112–e132.

Ranade, S.C., Sarfaraz Nawaz, Md., Kumar Rambtla, P., Rose, A.J., Gressens, P., Mani, S. 2012. Early protein malnutrition disrupts cerebellar development and impairs motor coordination. *British Journal of Nutrition* 107: 1167–1175.

Rao, S., Chun, C., Fan, J., Kofron, J.M., Yang, M.B., Hegde, R.S., Ferrara, N., Copenhagen, D.R., Lang, R.A. 2013. A direct and melanopsin-dependent fetal light response regulates mouse eye development. *Nature* 494: 243–246.

Raup, D.M., Michelson, A. 1965. Theoretical morphology of the coiled shell. *Science* 147: 1294–1295.

Raup, D.M. 1961. The geometry of coiling in gastropods. *Proceedings of the National Academy of Sciences* 47: 602–609.

Raup, D.M. 1966. Geometric analysis of shell coiling: general problems. *Journal of Paleontology* 40: 1178–1190.

Raup, D.M. 1967. Geometric analysis of shell coiling: coiling in ammonoids. *Journal of Paleontology* 41: 43–65.

Ray, R.P., Nakata, T., Henningsson, P., Bomphrey, R.J. 2016. Enhanced flight performance by genetic manipulation of wing shape in *Drosophila. Nature Communications* 7: 10851.

Reed, R.D., Papa, R., Martin, A., Hines, H.M., Counterman, B.A., Pardo-Diaz, C., Jiggins, C.D., Chamberlain, N.L., Kronforst, M.R., Chen, R., Halder, G., Nijhout, H.F., McMillan, W.O. 2011. *optix* drives the repeated convergent evolution of butterfly wing pattern mimicry. *Science* 333: 1137–1141.

Relyea, R.A. 2002. Costs of phenotypic plasticity. *The American Naturalist* 159: 272–282.

Reumer, J.W.F., Ten Broek, C.M.A., Galis, F. 2014. Extraordinary incidence of cervical ribs indicates vulnerable condition in Late Pleistocene mammoths. *PeerJ* 2: e318.

Rice, G., David, J.R., Kamimura, Y., Masly, J.P., Mcgregor, A.P., Nagy, O., Noselli, S., Nunes, M.D.S., O'Grady, P., Sánchez-Herrero, E., Siegal, M.L., Toda, M.J., Rebeiz, M., Courtier-Orgogozo, V., Yassin, A. 2019. A standardized nomenclature and atlas of the male terminalia of *Drosophila melanogaster. Fly* 13: 51–64.

Rieppel, O. 2003. Popper and systematics. *Systematic Biology* 52: 259–271.

Rincón-Rubio, V.A., Székely, T., Liker, A., Gonzalez-Voyer, A. 2023. Carotenoid-dependent plumage coloration is associated with reduced male care in passerine birds. *Behavioral Ecology* 34: 872–880.

Robert, J.S. 2004. *Embryology, Epigenesis, and Evolution: Taking Development Seriously.* Cambridge University Press, Cambridge, UK.

Roberts, H.R., Warren, J.M., Provan, J. 2018. Evidence for facultative protocarnivory in *Capsella bursa-pastoris* seeds. *Scientific Reports* 8: 10120.

Robinson, Z., Maley, C.J., Piccinini, G. 2015. Is consciousness a spandrel? *Journal of the American Philosophical Association* 1: 365–383.

Rodriguez-Zaccaro, F.D., Henry, I.M., Groover, A. 2021. Genetic regulation of vessel morphology in *Populus. Frontiers in Plant Science* 12: 705596.

Rollo, C.D. 1995. *Phenotypes: Their Epigenetics, Ecology and Evolution.* Chapman & Hall, London.

Rose, M.R., Lauder, G.V. 1996. Post-spandrel adaptationism. In: Rose, M. R. and Lauder, G. V. (eds.), *Adaptation*, Academic Press, San Diego, pp. 1–8.

Rosell, J.A. 2016. Bark thickness across the angiosperms: More than just fire. *New Phytologist* 211: 90–102.

Rosell, J.A. 2019. Bark in woody plants: Understanding the diversity of a multifunctional structure. *Integrative and Comparative Biology* 59: 535–547.

Rosell, J.A., Olson, M.E. 2014. The evolution of bark mechanics and storage across habitats in a clade of tropical trees. *American Journal of Botany* 101: 764–777.

Rosell, J.A., Olson, M.E., Aguirre-Hernández, R., Sánchez-Sesma, F.J. 2012. Ontogenetic modulation of branch size, shape, and biomechanics produces diversity across habitats in the *Bursera simaruba* clade of tropical trees. *Evolution & Development* 14: 437–449.

Rosenberg, A., Williams, M. 1986. Fitness as primitive and propensity. *Philosophy of Science* 53: 412–418.

Rosenblatt, K.P., Sun, Z.-P., Heller, S., Hudspeth, A.J. 1997. Distribution of Ca2+-activated K+ channel isoforms along the tonotopic gradient of the chicken's cochlea. *Neuron* 19: 1061–1075.

Rueffler, C., Van Dooren, T.J.M., Leimar, O., Abrams, P.A. 2006. Disruptive selection and then what? *Trends in Ecology & Evolution* 21: 238–245.

Ruse, M. 1979. *Sociobiology, Sense or Nonsense?* D. Reidel Pub. Co, Dordrecht.

Ruse, M. 2005. *The Evolution-Creation Struggle.* Harvard University Press, Cambridge, Mass.

Russell, E.S. 1916. *Form and Function.* John Murray, London.

Ryerson, W.G., Deban, S.M. 2010. Buccal pumping mechanics of *Xenopus laevis* tadpoles: Effects of biotic and abiotic factors. *Journal of Experimental Biology* 213: 2444–2452.

Saadat, M., Fish, F.E., Domel, A.G., Di Santo, V., Lauder, G.V., Haj-Hariri, H. 2017. On the rules for aquatic locomotion. *Physical Review Fluids* 2: 083102.

Safwan-Zaiter, H., Wagner, N., Wagner, K.-D. 2022. P16INK4A—more than a senescence marker. *Life* 12: 1332.

Salazar-Ciudad, I. 2006. Developmental constraints vs. variational properties: How pattern formation can help to understand evolution and development. *Journal of Experimental Zoology Part B: Molecular and Developmental Evolution* 306B: 107–125.

Salmon, W.C. 1989. Four decades of scientific explanation. In: Kitcher, P. and Salmon, W. C. (eds.), *Scientific Explanation*, University of Minnesota Press, Minneapolis, pp. 3–219.

Samhita, L. 2021. The Boggarts of biology: How non-genetic changes influence the genotype. *Current Genetics* 67: 65–77.

Sansom, R. 2009. The nature of developmental constraints and the difference-maker argument for externalism. *Biology & Philosophy* 24: 441–459.

Santiago, E., Moreno, D.F., Acar, M. 2022. Phenotypic plasticity as a facilitator of microbial evolution. *Environmental Epigenetics* 8: dvac020.

Santos-Durán, G.N., Cooper, R.L., Jahanbakhsh, E., Timin, G., Milinkovitch, M.C. 2024. Self-organized patterning of crocodile head scales by compressive folding. *Nature* 637: 375–383.

Sappington, A., Mohanty, V. 2025. Probabilistic genotype-phenotype maps reveal mutational robustness of RNA folding, spin glasses, and quantum circuits. *Physical Review Research* 7: 013118.

Sarkar, S. 1996. Biological information: A skeptical look at some central dogmas of molecular biology. In: Sarkar, S. (ed.), *The Biology and History of Molecular Biology: New Perspectives*, Kluwer, Dordrecht, pp. 187–232.

Sather, B.A., Hageman, D., Wagenseil, J.E. 2012. Murray's Law in elastin haploinsufficient (Eln+/−) and wild-type (WT) mice. *Journal of Biomechanical Engineering* 134: 124504.

Sawatari, E., Hashimoto, H., Matsumura, T., Iwata, Y., Yamamoto, N., Yokoyama, Y., Wakamatsu, Y. 2009. Cell growth-promoting activity of fluid from eye sacs of the bubble-eye goldfish (*Carassius auraitus*). *Zoological Science* 26: 254.

Schachtzabel, E. 1906. *Illustriertes Prachtwerk Sämtlicher Tauben-Rassen*. Königl. Universitätsdruckerei H. Stürtz a.g, Würzburg, biodiversitylibrary.org/bibliography/50313.

Schat, L., Schubert, M., Fjellheim, S., Humphreys, A.M. 2025. Drought tolerance as an evolutionary precursor to frost and winter tolerance in grasses. *Evolution* 79: 541–556.

Scheiner, S.M., and Levis, N.A. 2021. The loss of phenotypic plasticity via natural selection: Genetic assimilation. In: Pfennig, D. W. (ed.), *Phenotypic Plasticity & Evolution: Causes, Consequences, Controversies*, CRC Press, New York, CRC Press. pp. 161–181.

Schlichting, C., Pigliucci, M. 1998. *Phenotypic Evolution: A Reaction Norm Perspective.* Sinauer, Sunderland, Mass.

Schlosser, G., Wagner, G.P., eds. 2004. *Modularity in Development and Evolution.* University of Chicago Press, Chicago.

Schluter, D. 1996. Adaptive radiation along genetic lines of least resistance. *Evolution* 50: 1766–1774.

Schmalhausen, I.I. 1949. *Factors of Evolution: The Theory of Stabilizing Selection.* Blakiston, Philadelphia.

Schraiber, J.G., Edge, M.D., Pennell, M. 2024. Unifying approaches from statistical genetics and phylogenetics for mapping phenotypes in structured populations. *PLOS Biology* 22: e3002847.

Schubert, S.A., Morreau, H., De Miranda, N.F.C.C., Van Wezel, T. 2020. The missing heritability of familial colorectal cancer. *Mutagenesis* 35: 221–231.

Schultz, D.T., Haddock, S.H.D., Bredeson, J.V., Green, R.E., Simakov, O., Rokhsar, D.S. 2023. Ancient gene linkages support ctenophores as sister to other animals. *Nature* 618: 110–117.

Schut, P.C., Eggink, A.J., Cohen-Overbeek, T.E., Van Dooren, T.J.M., De Borst, G.J., Galis, F. 2020. Miscarriage is associated with cervical ribs in thoracic outlet syndrome patients. *Early Human Development* 144: 105027.

Schwander, T., Leimar, O. 2011. Genes as leaders and followers in evolution. *Trends in Ecology & Evolution* 26: 143–151.

Schwarz, Z., Jolly, S.O., Steinmetz, A.A., Bogorad, L. 1981. Overlapping divergent genes in the maize chloroplast chromosome and *in vitro* transcription of the gene for tRNA. *Proceedings of the National Academy of Sciences* 78: 3423–3427.

Schwenk, K., Wagner, G.P. 2004. The relativism of constraints on phenotypic evolution. In: *Phenotypic Integration. Studying the Ecology and Evolution of Complex Phenotypes*, Oxford University Press, Oxford, UK, pp. 390–408.

Scotland, R.W. 2010. Deep homology: A view from systematics. *BioEssays* 32: 438–449.

Scotland, R.W. 2011. What is parallelism? *Evolution & Development* 13: 214–227.

Scriven, M. 1959. Explanation and prediction in evolutionary theory. *Science* 130: 477–482.

Selzer, J., ed. 1993. *Understanding Scientific Prose.* University of Wisconsin Press, Madison.

Semba, R.D. 2016. The rise and fall of protein malnutrition in global health. *Annals of Nutrition and Metabolism* 69: 79–88.

Serban, M. 2020. Exploring modularity in biological networks. *Philosophical Transactions of the Royal Society B: Biological Sciences* 375: 20190316.

Serrelli, E. 2015. Visualizing macroevolution: from adaptive landscapes to compositions of multiple spaces. In: Serrelli, E. and Gontier, N. (eds.), *Macroevolution*, Springer International Publishing, Cham, vol. 2. pp. 113–162.

Serrelli, E., Gontier, N., eds. 2015. *Macroevolution: Explanation, Interpretation and Evidence.* Springer, Cham.

Shanahan, T. 2008. Why don't zebras have machine guns? Adaptation, selection, and constraints in evolutionary theory. *Studies in History and Philosophy of Science Part C: Studies in History and Philosophy of Biological and Biomedical Sciences* 39: 135–146.

Sharpless, N.E. 2005. *INK4a/ARF*: A multifunctional tumor suppressor locus. *Mutation Research* 576: 22–38.

Sherr, C.J. 2001. The *INK4a/ARF* network in tumour suppression. *Nature Reviews Molecular Cell Biology* 2: 731–737.

Shimizu, T., Cerpa, M., Lenke, L.G. 2021. Understanding sagittal compensation in adult spinal deformity patients: Relationship between pelvic tilt and lower-extremity position. *Journal of Neurosurgery: Spine* 35: 616–623.

Shin, C., Manley, J.L. 2004. Cell signalling and the control of pre-mRNA splicing. *Nature Reviews Molecular Cell Biology* 5: 727–738.

Shipley, B. 2000. *Cause and Correlation in Biology: A User's Guide to Path Analysis, Structural Equations, and Causal Inference.* Cambridge University Press, Cambridge, UK.

Silva-Rocha, R., De Lorenzo, V. 2010. Noise and robustness in prokaryotic regulatory networks. *Annual Review of Microbiology* 64: 257–275.

Silvertown, J., Gordon, D.M. 1989. A framework for plant behavior. *Annual Review of Ecology and Systematics* 20: 349–366.

Simmons, R.E., Scheepers, L. 1996. Winning by a neck: Sexual selection in the evolution of giraffe. *The American Naturalist* 148: 771–786.

Simpson, C., Halling, A., Leventhal, S. 2024. Levels of selection and macroevolution in organisms, colonies, and species. *Paleobiology* 51: 62–70.

Simpson, G.G. 1953. *The Major Features of Evolution.* Columbia University Press, New York.

Sinervo, B., Licht, P. 1991. Proximate constraints on the evolution of egg size, number, and total clutch mass in lizards. *Science* 252: 1300–1302.

Singh, P., Ahi, E.P. 2022. The importance of alternative splicing in adaptive evolution. *Molecular Ecology* 31: 1928–1938.

Skinner, M.K., Nilsson, E.E. 2021. Role of environmentally induced epigenetic transgenerational inheritance in evolutionary biology: Unified Evolution Theory. *Environmental Epigenetics* 7: dvab012.

Slik, J.W.F., Arroyo-Rodríguez, V., Aiba, S.-I., Alvarez-Loayza, P., Alves, L.F., Ashton, P., Balvanera, P., Bastian, M.L., Bellingham, P.J., van den Berg, E., Bernacci, L., da Conceição Bispo, P., Blanc, L., Böhning-Gaese, K., Boeckx, P., Bongers, F., Boyle, B., Bradford, M., Brearley, F.Q., Breuer-Ndoundou Hockemba, M., Bunyavejchewin, S., Calderado Leal Matos, D., Castillo-Santiago, M., Catharino, E.L.M., Chai, S.-L., Chen, Y., Colwell, R.K., Chazdon, R.L., Clark, C., Clark, D.B., Clark, D.A., Culmsee, H., Damas, K., Dattaraja, H.S., Dauby, G., Davidar, P., DeWalt, S.J., Doucet, J.-L., Duque, A., Durigan, G., Eichhorn, K.A.O., Eisenlohr, P.V., Eler, E., Ewango, C., Farwig, N., Feeley, K.J., Ferreira, L., Field, R., de Oliveira Filho, A.T., Fletcher, C., Forshed, O., Franco, G., Fredriksson, G., Gillespie, T., Gillet, J.-F., Amarnath, G., Griffith, D.M., Grogan, J., Gunatilleke, N., Harris, D., Harrison, R., Hector, A., Homeier, J., Imai, N., Itoh, A., Jansen, P.A., Joly, C.A., de Jong, B.H.J., Kartawinata, K., Kearsley, E., Kelly, D.L., Kenfack, D., Kessler, M., Kitayama, K., Kooyman, R., Larney, E., Laumonier, Y., Laurance, S., Laurance, W.F., Lawes, M.J., Amaral, I.L. do, Letcher, S.G.,

Lindsell, J., Lu, X., Mansor, A., Marjokorpi, A., Martin, E.H., Meilby, H., Melo, F.P.L., Metcalfe, D.J., Medjibe, V.P., Metzger, J.P., Millet, J., Mohandass, D., Montero, J.C., de Morisson Valeriano, M., Mugerwa, B., Nagamasu, H., Nilus, R., Ochoa-Gaona, S., Onrizal, Page, N., Parolin, P., Parren, M., Parthasarathy, N., Paudel, E., Permana, A., Piedade, M.T.F., Pitman, N.C.A., Poorter, L., Poulsen, A.D., Poulsen, J., Powers, J., Prasad, R.C., Puyravaud, J.-P., Razafimahaimodison, J.-C., Reitsma, J., Dos Santos, J.R., Roberto Spironello, W., Romero-Saltos, H., Rovero, F., Rozak, A.H., Ruokolainen, K., Rutishauser, E., Saiter, F., Saner, P., Santos, B.A., Santos, F., Sarker, S.K., Satdichanh, M., Schmitt, C.B., Schöngart, J., Schulze, M., Suganuma, M.S., Sheil, D., da Silva Pinheiro, E., Sist, P., Stevart, T., Sukumar, R., Sun, I.-F., Sunderland, T., Suresh, H.S., Suzuki, E., Tabarelli, M., Tang, J., Targhetta, N., Theilade, I., Thomas, D.W., Tchouto, P., Hurtado, J., Valencia, R., van Valkenburg, J.L.C.H., Van Do, T., Vasquez, R., Verbeeck, H., Adekunle, V., Vieira, S.A., Webb, C.O., Whitfeld, T., Wich, S.A., Williams, J., Wittmann, F., Wöll, H., Yang, X., Adou Yao, C.Y., Yap, S.L., Yoneda, T., Zahawi, R.A., Zakaria, R., Zang, R., de Assis, R.L., Garcia Luize, B., Venticinque, E.M. 2015. An estimate of the number of tropical tree species. *Proceedings of the National Academy of Sciences* 112: 7472–7477.

Smartt, J. 2001. *Goldfish Varieties and Genetics: A Handbook for Breeders.* Fishing News Books, Oxford, UK.

Smith, R.J. 2016. Explanations for adaptations, just-so stories, and limitations on evidence in evolutionary biology. *Evolutionary Anthropology: Issues, News, and Reviews* 25: 276–287.

Smocovitis, V.B. 2023. Every evolutionist their own historian: The importance of history, context, and the Extended Evolutionary Synthesis. In: Dickins, T. E. and Dickins, B. J. A. (eds.), *Evolutionary Biology: Contemporary and Historical Reflections Upon Core Theory*, Springer, Cham, vol. 6. pp. 25–54.

Snell-Rood, E.C., Ehlman, S.M. 2023. Developing the genotype-to-phenotype relationship in evolutionary theory: A primer of developmental features. *Evolution & Development* 25: 393–409.

Sol, D. 2008. Artificial selection, naturalization, and fitness: Darwin's pigeons revisited. *Biological Journal of the Linnean Society* 93: 657–665.

Solé, R.V., Goodwin, B., Goodwin, B.C. 2000. *Signs of Life: How Complexity Pervades Biology.* Basic Books, New York, NY.

Song, J., Chen, C., Zhu, S., Zhu, M., Dai, J., Ray, U., Li, Y., Kuang, Y., Li, Y., Quispe, N., Yao, Y., Gong, A., Leiste, U.H., Bruck, H.A., Zhu, J.Y., Vellore, A., Li, H., Minus, M.L., Jia, Z., Martini, A., Li, T., Hu, L. 2018. Processing bulk natural wood into a high-performance structural material. *Nature* 554: 224–228.

Song, R., Llaca, V., Linton, E., and Messing, J. 2001. Sequence, regulation, and evolution of the maize 22-kD α zein gene family. *Genome Research* 11: 1817–1825.

Sparrow, L.M., Pellatt, E., Yu, S.S., Raichlen, D.A., Pontzer, H., Rolian, C. 2017. Gait changes in a line of mice artificially selected for longer limbs. *PeerJ* 5: e3008.

Stadler, P.F., Stadler, B.M.R. 2006. Genotype-phenotype maps. *Biological Theory* 1: 268–279.

Stanley, S.M. 1975. A theory of evolution above the species level. *Proceedings of the National Academy of Sciences* 72: 646–650.

Stearns, S.C. 1989. The evolutionary significance of phenotypic plasticity. *BioScience* 39: 436–445.

Steinberg, C. 2011. *Stress Ecology: Environmental Stress as Ecological Driving Force and Key Player in Evolution.* Springer, Dordrecht.

Steinberg, M. 2003. Cell adhesive interactions and tissue self-organization. In: Müller, G. B. and Newman, S. A. (eds.), *Origination of Organismal Form*, The MIT Press, Cambridge, Mass., pp. 137–164.

Stenseth, N.C., Andersson, L., Hoekstra, H.E. 2022. Gregor Johann Mendel and the development of modern evolutionary biology. *Proceedings of the National Academy of Sciences* 119: e2201327119.

Steppan, S.J., Phillips, P.C., Houle, D. 2002. Comparative quantitative genetics: Evolution of the G matrix. *Trends in Ecology & Evolution* 17: 320–327.

Sterelny, K., Griffiths, P. 1999. *Sex and Death: An Introduction to Philosophy of Biology.* University of Chicago Press, Chicago, IL.

Stevens, W.C. 1910. *Plant Anatomy from the Standpoint of the Development and Functions of the Tissues*, 2nd edn. Blakiston's Son & Co., Philadelphia, Pennsylvania.

Stone, L., Simberloff, D., Artzy-Randrup, Y. 2019. Network motifs and their origins. *PLOS Computational Biology* 15: e1006749.

Stotz, K. 2009. Experimental philosophy of biology: Notes from the field. *Studies in History and Philosophy of Science Part A* 40: 233–237.

Stotz, K. 2017. Why developmental niche construction is not selective niche construction: And why it matters. *Interface Focus* 7: 20160157.

Stotz, K., Griffiths, P. 2004. Genes: Philosophical analyses put to the test. *History & Philosophy of the Life Sciences* 26: 5–28.

Strate, L. 2017. Understanding the message of *Understanding Media*. *Atlantic Journal of Communication* 25: 244–254.

Striedter, G.F. 1998. Stepping into the same river twice: Homologues as recurring attractors in epigenetic landscapes. *Brain, Behavior and Evolution* 52: 218–231.

Stringer, S., Wray, N.R., Kahn, R.S., Derks, E.M. 2011. Underestimated effect sizes in GWAS: fundamental limitations of single SNP analysis for dichotomous phenotypes. *PLoS ONE* 6: e27964.

Sugrue, K.F., Zohn, I.E. 2019. Reduced maternal vitamin A status increases the incidence of normal aortic arch variants. *Genesis* 57: e23326.

Sutera, S.P., Skalak, R. 1993. The history of Poiseuille's Law. *Annual Review of Fluid Mechanics* 25: 1–20.

Svensson, E.I. 2023. The structure of evolutionary theory: Beyond neo-Darwinism, neo-Lamarckism and biased historical narratives about the Modern Synthesis. In: Dickins, T. E., and Dickins, B. J. A. (eds.), *Evolutionary Biology: Contemporary and Historical Reflections Upon Core Theory*,. vol. 6. pp. 173–217. Cham, Springer.

Swartz, S.M., Middleton, K.M. 2008. Biomechanics of the bat limb skeleton: Scaling, materials properties and mechanics. *Cells, Tissues, Organs* 187: 59–84.

Tabery, J.G. 2004. The "evolutionary synthesis" of George Udny Yule. *Journal of the History of Biology* 37: 73–101.

Taipale, M., Jarosz, D.F., Lindquist, S. 2010. HSP90 at the hub of protein homeostasis: Emerging mechanistic insights. *Nature Reviews Molecular Cell Biology* 11: 515–528.

Tamborini, M. 2022. *The Architecture of Evolution: The Science of Form in Twentieth-Century Evolutionary Biology*. University of Pittsburgh Press, Pittsburgh, PA.

Tando, Y., Matsui, Y. 2023. Inheritance of environment-induced phenotypic changes through epigenetic mechanisms. *Environmental Epigenetics* 9: dvad008.

Tawfik, A.M., Sobh, D.M., Ashamallah, G.A., Batouty, N.M. 2019. Prevalence and types of aortic arch variants and anomalies in congenital heart diseases. *Academic Radiology* 26: 930–936.

Taylor, C.H., Watson, D.J.G., Skelhorn, J., Bell, D., Burdett, S., Codyre, A., Cooley, K., Davies, J.R., Dawson, J.J., D'Cruz, T., Gandhi, S.R., Jackson, H.J., Lowe, R., Ogilvie, E., Pond, A.L., Rees, H., Richardson, J., Sains, J., Short, F., Brignell, C., Davidson, G.L., Rowland, H.M., East, M., Goodridge, R., Gilbert, F., Reader, T. 2025. Mapping the adaptive landscape of Batesian mimicry using 3D-printed stimuli. *Nature* 644: 706–713.

Tello-Ramos, M.C., Harper, L., Tortora-Brayda, I., Guillette, L.M., Capilla-Lasheras, P., Harrison, X.A., Young, A.J., Healy, S.D. 2024. Architectural traditions in the structures built by cooperative weaver birds. *Science* 385: 1004–1009.

Templeton, A.R. 1989. The meaning of species and speciation: A genetic perspective. In: Otte, D., and Endler, J. A. (eds.), *Speciation and its Consequences*, Sinauer, Sunderland, Mass, pp. 3–27.

Templeton, A.R. 2021. *Population Genetics and Microevolutionary Theory*, 2nd edn. Wiley-Blackwell, Hoboken, NJ.

Ten Broek, C.M.A., Bakker, A.J., Varela-Lasheras, I., Bugiani, M., Van Dongen, S., Galis, F. 2012. Evo-devo of the human vertebral column: On homeotic transformations, pathologies and prenatal selection. *Evolutionary Biology* 39: 456–471.

Thomas, V., Dave, Y. 1990. Structure and necrosis of stipular colleters in *Mitragyna parvifolia*. *Belgian Journal of Botany* 123: 67–72.

Thomson, K.S. 1997. Natural selection and evolution's smoking gun. *American Scientist* 85: 516–518.

Thorvaldsen, S., Øhrstrøm, P., Hössjer, O. 2024. The representation, quantification, and nature of genetic information. *Synthese* 204: 15.

Torkamani, A., Wineinger, N.E., Topol, E.J. 2018. The personal and clinical utility of polygenic risk scores. *Nature Reviews Genetics* 19: 581–590.

Trejo, L., Rosell, J.A., Olson, M.E. 2018. Nearly 200 years of sustained selection have not overcome the leaf area-stem size relationship in the poinsettia. *Evolutionary Applications* 11: 1401–1411.

True, J.R. 2008. Combing evolution. *Evolution & Development* 10: 400–402.

Tsongalis, G.J. 2018. Integrative systems biology. In: Coleman, W. B. and Tsongalis, G. J. (eds), *Molecular Pathology*, Elsevier: London, pp. 205–215.

Turnbull, M.G. 2018. Underdetermination in science: What it is and why we should care. *Philosophy Compass* 13: e12475.

Turner, T.L., Stewart, A.D., Fields, A.T., Rice, W.R., Tarone, A.M. 2011. Population-based resequencing of experimentally evolved populations reveals the genetic basis of body size variation in *Drosophila melanogaster*. *PLoS Genetics* 7: e1001336.

Tuttle, L.M., Pacheco, D., Warfield, L., Luo, J., Ranish, J., Hahn, S., Klevit, R.E. 2018. Gcn4-mediator specificity Is mediated by a large and dynamic fuzzy protein-protein complex. *Cell Reports* 22: 3251–3264.

Uebel, T. 2024. Vienna Circle. *The Stanford Encyclopedia of Philosophy*. Available at <https://plato.stanford.edu/archives/sum2024/entries/vienna-circle/>.

Uller, T., Moczek, A.P., Watson, R.A., Brakefield, P.M., Laland, K.N. 2018. Developmental bias and evolution: A regulatory network perspective. *Genetics* 209: 949–966.

UNICEF 2019. *The State of the World's Children 2019. Children, Food and Nutrition: Growing Well in a Changing World*. UNICEF, New York, NY.

Urry, L.A., Cain, M.L., Wasserman, S.A., Minorsky, P.V., Orr, R.B., Campbell, N.A. 2021. *Campbell Biology*, 12th edn. Pearson, New York, NY.

Uyeda, J.C., Bone, N., McHugh, S., Rolland, J., Pennell, M.W. 2021. How should functional relationships be evaluated using phylogenetic comparative methods? A case study using metabolic rate and body temperature. *Evolution* 75: 1097–1105.

Van Der Geer, A.A.E., Galis, F. 2017. High incidence of cervical ribs indicates vulnerable condition in Late Pleistocene woolly rhinoceroses. *PeerJ* 5: e3684.

Van Kleunen, M., Fischer, M. 2005. Constraints on the evolution of adaptive phenotypic plasticity in plants. *New Phytologist* 166: 49–60.

Varela-Lasheras, I., Bakker, A.J., Van Der Mije, S.D., Metz, J.A., Van Alphen, J., Galis, F. 2011. Breaking evolutionary and pleiotropic constraints in mammals: On sloths, manatees and homeotic mutations. *EvoDevo* 2: 11.

Varella, M.A.C. 2018. The biology and evolution of the three psychological tendencies to anthropomorphize biology and evolution. *Frontiers in Psychology* 9: 1839.

Vedel, V., Apostolou, Z., Arthur, W., Akam, M., Brena, C. 2010. An early temperature-sensitive period for the plasticity of segment number in the centipede *Strigamia maritima*. *Evolution & Development* 12: 347–352.

Veeramachaneni, V., Makalowski, W., Galdzicki, M., Sood, R., Makalowska, I. 2004. Mammalian overlapping genes: The comparative perspective. *Genome Research* 14: 280–286.

Vemaraju, S., Rao, S., Lang, R.A. 2025. Formation and regression of the primary vitreous and hyaloid vascular system. In: D'Amore, P. A. (ed.), *Reference Module in Neuroscience and Biobehavioral Psychology*, Elsevier: Amsterdam, 4: pp. 183–193.

Verna, C., Ravichandran, S.J., Sawchuk, M.G., Linh, N.M., Scarpella, E. 2019. Coordination of tissue cell polarity by auxin transport and signaling. *eLife* 8: e51061.

Vidair, C.A., Doxsey, S.J., Dewey, W.C. 1993. Heat shock alters centrosome organization leading to mitotic dysfunction and cell death. *Journal of Cellular Physiology* 154: 443–455.

Vincent, T.L., Brown, J.S. 2005. *Evolutionary Game Theory, Natural Selection, and Darwinian Dynamics*. Cambridge University Press, Cambridge, UK.

Vinogradov, A.E. 2004. Genome size and extinction risk in vertebrates. *Proceedings of the Royal Society of London. Series B: Biological Sciences* 271: 1701–1705.

Visscher, P.M., Hill, W.G., Wray, N.R. 2008. Heritability in the genomics era — concepts and misconceptions. *Nature Reviews Genetics* 9: 255–266.

Visser, M.E. 2008. Keeping up with a warming world; assessing the rate of adaptation to climate change. *Proceedings of the Royal Society B: Biological Sciences* 275: 649–659.

Vlcek, C., Marande, W., Teijeiro, S., Lukeš, J., Burger, G. 2011. Systematically fragmented genes in a multipartite mitochondrial genome. *Nucleic Acids Research* 39: 979–988.

Vogel, S. 1994. *Life in Moving Fluids: The Physical Biology of Flow*, 2nd edn., rev. expanded. Princeton University Press, Princeton, NJ.

Vogt, G. 2017. Facilitation of environmental adaptation and evolution by epigenetic phenotype variation: Insights from clonal, invasive, polyploid, and domesticated animals. *Environmental Epigenetics* 3: dvx002.

Vrba, E.S., Gould, S.J. 1986. The hierarchical expansion of sorting and selection: sorting and selection cannot be equated. *Paleobiology* 12: 217–228.

Waddington, C.H. 1942. Canalization of development and the evolution of acquired characteristics. *Nature* 150: 563–565.

Waddington, C.H. 1953. Genetic assimilation of an acquired character. *Evolution* 7: 118–126.

Waddington, C.H. 1956. Genetic assimilation of the bithorax phenotype. *Evolution* 10: 1–13.

Waddington, C.H. 1957. *The Strategy of the Genes*. Allen & Unwin, London, UK.

Waddington, C.H. 1961. *The Nature of Life*. Allen & Unwin, London, UK.

Wade, M.J., Sultan, S.E. 2023. Niche construction and the environmental term of the price equation: How natural selection changes when organisms alter their environments. *Evolution & Development* 25: 451–469.

Wagner, A. 2008. Robustness and evolvability: A paradox resolved. *Proceedings of the Royal Society B: Biological Sciences* 275: 91–100.

Wagner, A., Fell, D.A. 2001. The small world inside large metabolic networks. *Proceedings of the Royal Society of London. Series B: Biological Sciences* 268: 1803–1810.

Wagner, G.P. 1989. The origin of morphological characters and the biological basis of homology. *Evolution* 43: 1157–1171.

Wagner, G.P. 1996. Homologues, natural kinds and the evolution of modularity. *American Zoologist* 36: 36–43.

Wagner, G.P. 2014. *Homology, Genes, and Evolutionary Innovation*. Princeton University Press, Princeton, NJ.

Wagner, G.P., and Chiu, C. 2003. Genetic and epigenetic factors in the origin of the tetrapod limb. In: Müller, G. B., and Newman, S. (eds.), *Origination of Organismal Form*, MIT Press, Cambridge, Mass., pp. 265–285.

Wagner, G.P., Pavličev, M. 2017. Origin, function, and effects of female orgasm: All three are different. *Journal of Experimental Zoology Part B: Molecular and Developmental Evolution* 328: 299–303.

Wagner, G.P., Pavlicev, M., Cheverud, J.M. 2007. The road to modularity. *Nature Reviews Genetics* 8: 921–931.

Wagner, G.P., Erkenbrack, E.M., Love, A.C. 2019. Stress-induced evolutionary innovation: A mechanism for the origin of cell rypes. *BioEssays* 41: 1800188.

Walsh, D.M. 2015. *Organisms, Agency, and Evolution*. Cambridge University Press, Cambridge.

Wang, Z., Wang, Z., Wang, J., Sui, Y., Zhang, J., Liao, D., Wu, R. 2012. A quantitative genetic and epigenetic model of complex traits. *BMC Bioinformatics* 13: 274.

Waterman, R., Sahli, H., Koelling, V.A., Karoly, K., Conner, J.K. 2023. Strong evidence for positive and negative correlational selection revealed by recreating ancestral variation. *Evolution* 77: 264–275.

Waters, J.M., Craw, D. 2006. Goodbye Gondwana? New Zealand biogeography, geology, and the problem of circularity. *Systematic Biology* 55: 351–356.

Watt, W.B. 1986. Power and efficiency as indexes of fitness in metabolic organization. *The American Naturalist* 127: 629–653.

Webb, P.W. 1984. Body form, locomotion and foraging in aquatic vertebrates. *American Zoologist* 24: 107–120.

Weismann, A. 1891. *Essays upon Heredity and Kindred Biological Problems*. Clarendon Press, Oxford, UK.

Weiss, K.M., Buchanan, A.V. 2011. Is life law-like? *Genetics* 188: 761–771.

Welch, J.J. 2017. What's wrong with evolutionary biology? *Biology & Philosophy* 32: 263–279.

West, G.B., Brown, J.H., Enquist, B.J. 1997. A general model for the origin of allometric scaling laws in biology. *Science* 276: 122–126.

West, G.B., Brown, J.H., Enquist, B.J. 1999. A general model for the structure and allometry of plant vascular systems. *Nature* 400: 664–667.

West, G.B., Savage, V.M., Gillooly, J., Enquist, B.J., Woodruff, W.H., Brown, J.H. 2003. Why does metabolic rate scale with body size? *Nature* 421: 713–713.

West-Eberhard, M.J. 2003. *Developmental Plasticity and Evolution*. Oxford University Press, Oxford, UK.

Wiens, J.J., Graham, C.H. 2005. Niche conservatism: Integrating evolution, ecology, and conservation biology. *Annual Review of Ecology, Evolution, and Systematics* 36: 519–539.

Wiens, J.J., Ackerly, D.D., Allen, A.P., Anacker, B.L., Buckley, L.B., Cornell, H.V., Damschen, E.I., Jonathan Davies, T., Grytnes, J., Harrison, S.P., Hawkins, B.A., Holt, R.D., McCain, C.M., Stephens, P.R. 2010. Niche conservatism as an emerging principle in ecology and conservation biology. *Ecology Letters* 13: 1310–1324.

Wikström, N., Kainulainen, K., Razafimandimbison, S.G., Smedmark, J.E.E., Bremer, B. 2015. A revised time tree of the asterids: Establishing a temporal framework for evolutionary studies of the coffee family (Rubiaceae). *PLOS ONE* 10: e0126690.

Wilder, H.H. 1908. The morphology of cosmobia: Speculations concerning the significance of certain types of monsters. *American Journal of Anatomy* 8: 355–440.

Wiley, E.O. 1975. Karl R. Popper, systematics, and classification: A reply to Walter Bock and other evolutionary taxonomists. *Systematic Zoology* 24: 233.

Wilkins, J.F., Godfrey-Smith, P. 2009. Adaptationism and the adaptive landscape. *Biology & Philosophy* 24: 199–214.

Williams, C.B., Anfodillo, T., Crivellaro, A., Lazzarin, M., Dawson, T.E., Koch, G.W. 2019. Axial variation of xylem conduits in the Earth's tallest trees. *Trees* 33: 1299–1311.

Wilmut, I., Schnieke, A.E., McWhir, J., Kind, A.J., Campbell, K.H.S. 1997. Viable offspring derived from fetal and adult mammalian cells. *Nature* 385: 810–813.

Wimsatt, W.C. 1997. Aggregativity: Reductive heuristics for finding emergence. *Philosophy of Science* 64: S372–S384.

Wolf, J., Allen, C.E., Frankino, W.A. 2004. Multivariate phenotypic evolution in developmental hyperspace. In: Pigliucci, M. and Preston, K. (eds.), *Phenotypic Integration: Studying the Ecology and Evolution of Complex Phenotypes*, Oxford University Press, Oxford, UK, pp. 366–389.

Woltereck, R. 1913. Weitere experimentelle untersuchungen über Artänderung, speziell über das Wesen quantitativer Artunterschiede bei Daphniden. *Zeitschrift für Induktive Abstammungs- und Vererbungslehre* 9: 146–146.

Won, Y.-J., Sivasundar, A., Wang, Y., Hey, J. 2005. On the origin of Lake Malawi cichlid species: A population genetic analysis of divergence. *Proceedings of the National Academy of Sciences* 102: 6581–6586.

Woodger, J.H. 1930. The "concept of organism" and the relation between embryology and genetics. Part I. *The Quarterly Review of Biology* 5: 1–22.

Wray, G.A., Hoekstra, H.E., Futuyma, D.J., Lenski, R.E., Mackay, T., Schluter, D., Strassman, J.E. 2014. Does evolutionary theory need a rethink? No, all is well. *Nature* 524: 161–164.

Wray, N.R., Purcell, S.M., Visscher, P.M. 2011. Synthetic associations created by rare variants do not explain most GWAS results. *PLoS Biology* 9: e1000579.

Wright, B.W., Molloy, M.P., Jaschke, P.R. 2022. Overlapping genes in natural and engineered genomes. *Nature Reviews Genetics* 23: 154–168.

Wright, S. 1931. Evolution in Mendelian populations. *Genetics* 16: 97–159.

Wright, S. 1932. The roles of mutation, inbreeding, crossbreeding and selection in evolution. *Proceedings of the Sixth International Congress of Genetics* 1: 356–366.

Wright, S. 1934. Physiological and evolutionary theories of dominance. *The American Naturalist* 68: 24–53.

Xi, Z., Liu, L., Rest, J.S., Davis, C.C. 2014. Coalescent versus concatenation methods and the placement of *Amborella* as sister to water lilies. *Systematic Biology* 63: 919–932.

Xie, P., Kocur, O.M., Cheung, S., Ng, L., Albertini, D.F., Rosenwaks, Z., Palermo, G.D. 2023. Sperm centriolar factors and genetic defects that can predict pregnancy. *Fertility and Sterility* 120: 720–728.

Yablo, S. 2000. A paradox of existence. In: Everett, A. J. and Hofweber, T. (eds.), *Empty Names, Fiction, and the Puzzles of Non-Existence*, CSLI Publications, Stanford, Calif, pp. 275–312.

Ye, T., Ronse De Craene, L. 2024. Stipules in angiosperms. *Botanical Journal of the Linnean Society* 205: 191–242.

Yengo, L., Sidorenko, J., Kemper, K.E., Zheng, Z., Wood, A.R., Weedon, M.N., Frayling, T.M., Hirschhorn, J., Yang, J., Visscher, P.M., the GIANT Consortium 2018. Meta-analysis of genome-wide association studies for height and body mass index in ~700000 individuals of European ancestry. *Human Molecular Genetics* 27: 3641–3649.

Young, N.M., Hallgrímsson, B. 2005. Serial homology and the evolution of mammalian limb covariation structure. *Evolution* 59: 2691–2704.

Yule, G.U. 1902. Mendel's laws and their probable relations to intra-racial heredity. *New Phytologist* 1: 222–238.

Zedek, F., Šmerda, J., Halasová, A., Adamec, L., Veleba, A., Plačková, K., Bureš, P. 2024. The smallest angiosperm genomes may be the price for effective traps of bladderworts. *Annals of Botany* 134: 1131–1138.

Zelditch, M., ed. 2004. *Geometric Morphometrics for Biologists: A Primer*. Elsevier Academic Press, Amsterdam, Netherlands.

Zhang, K., Ji, Y., Yao, L., Liu, H., Zhang, Y., Baskin, J.M., Baskin, C.C., Zhang, L., Xu, C., Tao, J., Prinzing, A. 2024. Paternal intergenerational plasticity in the plant species *Paeonia ostii*: Implications for parental fitness and offspring performance. *Functional Ecology* 38: 832–847.

Zhang, Z., Zhang, J. 2009. A big world inside small-world networks. *PLoS ONE* 4: e5686.

Zhao, M., Kurtis, S.M., White, N.D., Moncrieff, A.E., Leite, R.N., Brumfield, R.T., Braun, E.L., Kimball, R.T. 2023. Exploring conflicts in whole genome phylogenetics: A case study within manakins (Aves: Pipridae). *Systematic Biology* 72: 161–178.

Zobel, B.J., Sprague, J.R. 1998. *Juvenile Wood in Forest Trees*. Springer, Berlin, Germany.

Index

For the benefit of digital users, indexed terms that span two pages (e.g., 52–53) may, on occasion, appear on only one of those pages.

Tables, figures, and boxes are indicated by an italic *t*, *f*, or *b*.